ACS SYMPOSIUM SERIES **437**

Novel Materials in Heterogeneous Catalysis

R. Terry K. Baker, EDITOR
Auburn University

Larry L. Murrell, EDITOR
Engelhard Corporation

Developed from a symposium sponsored
by the Divisions of Colloid and Surface Chemistry;
Fuel Industry; Industrial and Engineering
Chemistry, Inc.; and Petroleum Chemistry, Inc.,
at the 198th National Meeting
of the American Chemical Society,
Miami Beach, Florida,
September 10–15, 1989

American Chemical Society, Washington, DC 1990

Library of Congress Cataloging-in-Publication Data

Novel materials in heterogeneous catalysis / R. Terry K. Baker,
    editor, Larry L. Murrell, editor.

    p.   cm.—(ACS symposium series; 437)

    "Developed from a symposium sponsored by the Divisions of Colloid
and Surface Chemistry . . . [et al.] at the 198th National Meeting of the
American Chemical Society, Miami Beach, Florida, September 10–15,
1989."

    Papers presented at the Symposium on New Catalytic Materials and
Techniques, held at the ACS meeting in Miami Beach, September 1989.

    Includes bibliographical references and index.

    ISBN 0–8412–1863–3

    1. Heterogeneous catalysis—Congresses.

    I. Baker, R. T. K., 1938–   . II. Murrell, Larry L., 1942– .
III. Symposium on New Catalytic Materials and Techniques (1989:
Miami Beach, Fla.) IV. American Chemical Society. Meeting (198th:
Miami Beach, Fla.) V. American Chemical Society. Division of Colloid
and Surface Chemistry. VI. Series.

QD505.N69   1990
541.3'95—dc20   90–1209
                                            CIP

The paper used in this publication meets the minimum requirements of American National
Standard for Information Sciences—Permanence of Paper for Printed Library Materials, ANSI
Z39.48–1984.   ∞

ACS Symposium Series

M. Joan Comstock, *Series Editor*

1990 ACS Books Advisory Board

Foreword

THE ACS SYMPOSIUM SERIES was founded in 1974 to provide a medium for publishing symposia quickly in book form. The format of the Series parallels that of the continuing ADVANCES IN CHEMISTRY SERIES except that, in order to save time, the papers are not typeset, but are reproduced as they are submitted by the authors in camera-ready form. Papers are reviewed under the supervision of the editors with the assistance of the Advisory Board and are selected to maintain the integrity of the symposia. Both reviews and reports of research are acceptable, because symposia may embrace both types of presentation. However, verbatim reproductions of previously published papers are not accepted.

Contents

vi

Preface

CATALYSTS PREPARED IN UNCONVENTIONAL FORMS can yield many benefits. Researchers are no longer content merely to investigate the behavior of a supported metal dispersed on either a powdered or pelleted carrier. This stereotypical approach has been replaced by a desire to exploit the opportunity afforded either by producing metal particles via novel routes or by supporting them in unusual locations on a carrier material. Using sophisticated molecular beam techniques, researchers now produce clusters consisting of a few atoms of a metal, which can possess quite different chemical and electronic properties from that of the bulk metal. As an example, the degree to which hydrogen and alkanes are chemisorbed on metal clusters is extremely sensitive to the number of atoms constituting the cluster and, as a consequence, this aspect can induce massive changes in the selectivity of such a catalyst system.

The supports used in heterogeneous catalysis are also at a stage where major innovations are on the horizon. Control of pore size and volume, in combination with acidity and high-temperature stability of surface-modified aluminas, promise years of fertile research. New support materials do not necessarily require a unique bulk phase, because designed surface structures on conventional materials can produce a wide spectrum of novel carriers. By taking advantage of the aforementioned features, major advances are expected in this area.

Improvements in the resolving power of transmission electron microscopes, coupled with the advent of the exciting new technique of scanning tunneling microscopy, are now providing an abundance of information on the growth and structural characteristics of individual metal clusters supported on a variety of substrate materials. These and other techniques have shown that the melting temperature of small metal particles decreases in a systematic fashion with decreasing particle size. This phenomenon is one of the key factors in determining the sintering behavior, and hence the activity maintenance, of a supported metal catalyst system.

Ceramic membranes, a relatively new class of material, offer numerous advantages over conventional catalysts as support media. They have a controlled, stable, pore-size distribution. This allows for selective removal of certain gas-phase products from the reaction zone and makes it possible to enhance the yield of products from thermodynamically limited or product-inhibited reactions. Additionally, the ability to conduct two sequential catalytic reactions at the same time is possible simply by depositing different metals on either side of the membrane.

Molecular sieves continue to be a fruitful field for study, and the search for new routes for the synthesis of both aluminophosphate and aluminosilicate (zeolates) structures is an area of considerable activity. Recent crystallization studies performed in space raise the question of whether a higher quality molecular sieve catalyst could be produced under these conditions. The laminated sandwich structures displayed by pillared clay materials provide certain advantages over the caged structures found in molecular sieves. The former contain relatively large pore openings and allow access to larger reactant molecules. Unfortunately, their thermal stability, especially in a wet environment, is a limiting factor, and development of more stable structures is currently being pursued.

In the areas of fuel production and utilization the focus is in two directions: the search for an unsupported catalyst for use in direct coal liquefaction processing, and the development of catalysts capable of generating specialty byproduct chemicals from coal and petroleum feedstocks.

Acknowledgments

We want to thank all who made the symposium possible, and in particular E. G. Kugler (West Virginia University), M. E. Davis (Virginia Polytechnic Institute and State University), T. J. Pinnavaia (Michigan State University), J. S. Bradley (Exxon), W. R. Moser (Worcester Polytechnic Institute), C. W. Curtis (Auburn University), and H. P. Stephens (Sandia National Laboratories), who were responsible for organizing the various sessions.

R. Terry K. BAKER
Auburn University
Auburn University, AL 36849

LARRY L. MURRELL
Engelhard Corporation
Edison, NJ 08818

June 23, 1990

ZEOLITIC   MATERIALS

Chapter 1

Crystallization in Space

Implications for Molecular Sieve Synthesis

Charles W. J. Scaife[1], S. Richard Cavoli[1,3], and Steven L. Suib[2]

[1]Department of Chemistry, Union College, Schenectady, NY 12308
[2]Departments of Chemistry and Chemical Engineering and Institute
of Materials Science, University of Connecticut, Storrs, CT 06269-3060

Crystals of lead iodide were grown on the September 1988
NASA Discovery mission in order to ascertain some of the
important conditions for growth of high purity materials in
space. Solutions of lead and of iodide ions were allowed
to diffuse from opposite sides of a cellulose membrane at
zero gravity aboard the spacecraft. Control experiments
done under earth's gravity were also carried out. A comparison of the composition, surface, structural and electronic properties of the crystals grown on earth and in
space has been made. In addition, videotaping of the
crystallization processes has shown that crystals grown on
earth only form on the lower half of the membrane while
space grown crystals form over all of the membrane.
Space grown crystals also grow in isolated regions of solution away from the membrane in contrast to earth grown
crystals. Results of characterization studies suggest that
the space grown crystals are purer and better insulators
than the earth grown crystals. The implications of these
results in relation to the growth in space of other materials
such as molecular sieves are discussed.

Research in space is important in several areas including preparation
of new materials, new physical phenomena, biological science, and
crystal growth. Several recent efforts in the area of crystal growth in

[3]Current address: Medical School, State University of New York, Buffalo, NY 14150

0097–6156/90/0437–0002$06.00/0

space and in simulations of zero gravity on earth have been made. Manufacturing and research efforts in space have recently been reviewed (1). The role of space travel and experiments in the chemistry curriculum has also been addressed (2).

Convection effects (3 – 4) due to gravitational fields are often thought to be important in crystal growth processes. One specific example is the floating zone process for the production of doped semiconductors and alloys (4). Particle contaminants and oxygenated species aboard space vehicles are also important factors in the crystal growing process (5 – 6). Several clever characterization experiments such as the use of holograms during crystal growth in space (7) have been developed to determine the mechanisms of nucleation and growth.

Lead iodide (8) and related materials (9) like $PbBr_2$ are typically grown from melt techniques in order to acquire large single crystals. In fact, special furnaces (10) were used aboard Spacelab III for preparation of related compounds like HgI_2 . Such materials are often used for films in dental and astronomical applications (11) and larger single crystals may provide better cathodoluminescent materials that are used to develop film that is in contact with these materials. The most common materials typically grown in space are protein crystals and considerable progress has been made in the area of vapor diffusion, and dialysis growth of proteins in space (12 – 13).

The research reported here involves the use of a cellulose membrane to prepare crystals of lead iodide. The membrane orients crystal growth and minimizes random crystal formation (18). The effects of temperature and structure of cellulose with respect to transport properties are well known (14). Our goal was to prepare large pure single crystals of lead iodide on cellulose membranes by doing experiments in space. In turn, these results may be relevant to the preparation of other inorganic materials such as molecular sieves. In fact, there are known to be gravitational effects in the growth of zeolite A (15).

The crystal growth of lead iodide in space and on earth is the subject of this paper. All materials were characterized on earth although the growth mechanism was filmed at both locations in order to better understand differences between earth and space grown crystals.

EXPERIMENTAL

APPARATUS DESIGN. Parts to construct five four-chambered, acrylic Plexiglas containers shown in Figure 1 were designed and machined at Union College. An 8.3 cm outside diameter by 6.4 cm inside diameter acrylic Plexiglas tube as well as solid acrylic pieces for endplates, interior bulkheads, and valves were used. Bulkheads and endplates were sealed with ethylene propylene Parker O-rings and Delrin screws. Valves had ethylene propylene rubber gaskets. Valve handles and valve stem collars were aluminum. The total length of the apparatus was about 36 cm. The membrane was of natural cellulose with 6,000-8,000 molecular weight cutoff.

SYNTHESIS OF LEAD IODIDE CRYSTALS. The four chambers were filled, respectively, through the fill plugs with 0.0624 M lead (II) acetate, $Pb(C_2H_3O_2)_2$; separated by a valve from deionized water; separated by the membrane from deionized water; and separated by a valve from either 0.360 M, 0.180 M, or 0.090 M potassium iodide, KI. A nitrogen bubble about 2 cm in diameter was left in each chamber during filling to allow for easy compression or expansion when the interior valves were opened. The bubbles were at the top of each chamber during earth experiments, but were more nearly centered in each chamber in the shuttle experiments.

Three different KI concentrations as listed above were used in the shuttle experiments. Similar control experiments were done on earth. Videotaping of the crystal growth procedure was done with a television camera and still shots were also taken at various stages throughout the crystal growth experiments in both the shuttle and the laboratory.

Crystal growth was initiated by diffusion of the salt solutions toward the membrane by opening the two valves using exterior valve handles. Crystals typically appear within 30 sec to 120 sec after opening the valves depending on the iodide concentrations. Iodide concentrations were chosen to cause crystallization only on the lead(II) side of the membrane so that photographs would be facilitated. Crystal growth was allowed to proceed for at least 40 hours.

AUGER ELECTRON SPECTROSCOPY AND SCANNING AUGER MICROSCOPY. Auger electron spectroscopy and scanning Auger microscopy experiments were done on both the shuttle and earth grown crystals. Samples were mounted by pressing crystals into indium foil. A PHI model 610 scanning Auger microscope was used for all experiments. This instrument is equipped with a secondary electron detector for imaging purposes and a cylindrical mirror analyzer for Auger electron microscopy experiments. A beam voltage of 2 keV and a target current of 10 nanoamperes were used in all experiments. Auger electron spectra were collected in the absorption mode and then differentiated after data collection. A sample tilt of 30° was used during analysis and the pressure in the chamber was less than 1×10^{-8} torr. A 100 microampere emission current was used. Secondary electron detection experiments were run in an analog mode whereas AES experiments were done by control of a PDP-1123 computer. Further details of Auger experimental conditions can be found elsewhere (16).

RESULTS

SYNTHESIS OF LEAD IODIDE CRYSTALS. In the three shuttle experiments, crystals grew symmetrically over the entire membrane, and a few crystals also grew out in the chamber away from the membrane as shown in Figure 2. Some crystals were dislodged from the membrane during reentry or during removal of the solutions, but many remained attached. On the other hand, with the apparatus lying horizontally in

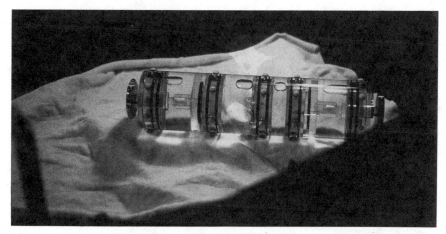

Figure 1. Photograph of Apparatus used for Crystal Growth

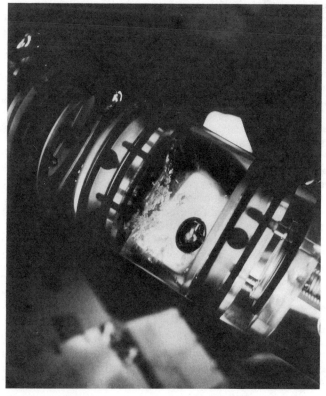

Figure 2. Photograph of Lead Iodide Crystals Grown Aboard the Discovery Shuttle Mission

two laboratory experiments, crystals grew only in the lower half of the vertical membrane; the upper half of the membrane was free of crystals. Crystals gradually formed a shelf out from the horizontal centerline of the membrane and also formed a curved support of crystals under the shelf and a beard of crystals hanging down from the shelf as shown in Figure 3. Late in the growth period some crystals grew on the top half of the membrane. Much of the shelf and its support fell to the bottom of the apparatus when the solutions were removed. Fast forwarding of the videotape of the laboratory experiments shows that crystals are falling off the shelf after about 20 minutes of crystal growth whereas this is not observed in the shuttle experiment.

Ten different samples of crystals were available for analyses; those on and off the membrane from the three different iodide concentrations in the shuttle and those on and off the membrane from the high and low iodide concentrations on earth.

CHARACTERIZATION OF CRYSTALS. Space grown crystals that fell off the membrane and corresponding crystals grown in the laboratory were both analyzed with Auger electron spectroscopy and scanning Auger microscopy methods. An Auger electron spectrum for the shuttle grown lead iodide crystals is shown in Figure 4. Peaks are observed for lead, iodide, and carbon. The lead transition occurs near 100 eV, the carbon transition occurs near 272 eV, and the iodide transitions occur as a doublet near 500 eV and 520 eV. Auger peak to peak height analysis gives a ratio of I/Pb of 1.9. Similar Auger electron spectra collected for the earth grown lead iodide crystals show the same peaks as observed in Figure 5. In this case the I/Pb ratio is 1.5. In addition, the relative amount of carbon on the earth grown sample is larger than that on the space grown sample. Finally, the signal to noise ratio for the shuttle grown crystals is much worse than that of the earth grown sample as shown in Figures 4 and 5, respectively.

Scanning Auger micrographs on the shuttle grown crystals show that large platelets of about 10 microns width and 30 microns long are formed. A scanning Auger micrograph of the shuttle grown crystals is given in Figure 6. The long flat lead iodide crystal on the right hand part of this photo is somewhat covered with smaller crystals that are resting on top of this platelet although there is some intergrowth observed at the top of the photo of this crystal. For the most part this platelet is quite uniform. Auger analyses on different parts of this sample did not show any differences in chemical composition from that of Figure 4. No visual observation of beam damage was observed unless the beam current was greater than about 10 nanoamperes.

DISCUSSION

SYNTHESIS. Crystals of lead iodide growth on earth were formed only on the lower half of the membrane indicating that gravitational effects are definitely important. The exact reason for the absence of crystals

Figure 3.   Photograph of Earth Grown Crystals, Showing Half
Coverage of Membrane

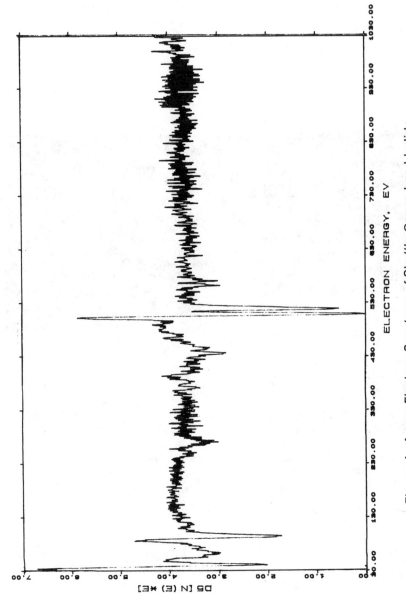

Figure 4. Auger Electron Spectrum of Shuttle Grown Lead Iodide Crystals

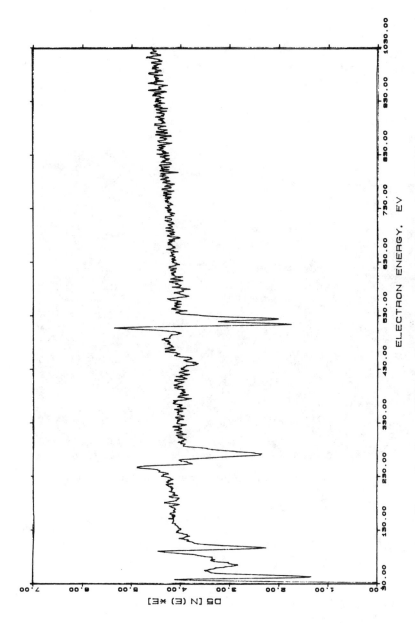

Figure 5. Auger Electron Spectrum of Earth Grown Lead Iodide Crystals

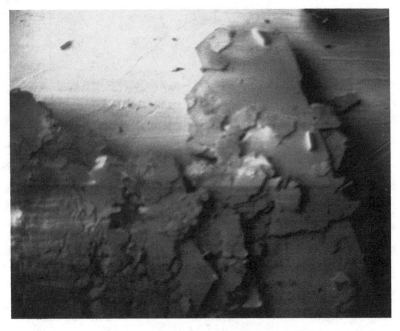

Figure 6. Auger Image of Shuttle Grown Lead Iodide Crystals

on the upper half of the membrane is not known although density and convection (3 − 4) gradients may be present during mixing of the lead and iodide ions. It is also possible that hydrodynamic pressures influence the growth mechanism.

Another interesting observation of the earth grown crystals is that nucleation only occurs on the membrane or on crystals that are attached to the membrane. This may have to do with intergrowth and twinning of the original crystals. In time, some of the crystals fall to the bottom of the reactor due to gravitational effects.

Crystal growth in the shuttle leads to nucleation all over the membrane as shown in Figure 2. Crystals in all experiments only grow on the side of the membrane initially containing the lead ions, therefore, iodide ions are migrating through the cellulose membrane. It is well known that iodide ions can diffuse through several substrates such as through silicate gels in the formation of PbI_2 crystals (17). One significant difference between space and earth grown nucleation processes is that space grown crystals do not need to be attached to the membrane or to other crystals as is the case with the earth grown crystals.

The appearance of a shelf of crystals with the earth grown lead iodide crystals may indicate that the different ions are mixing near this level of the tube. It may be possible that crystals are nucleating above this level and then falling down to this level where they attach to other crystals and to the membrane.

It is clear then that there is a great difference between the growth mechanism of crystals grown in space and those on earth. More nucleation sites are available in the space grown materials since crystallization can occur not only on all parts of the membrane but also in solution. This observation has important implications for growth of other materials like molecular sieves (15) such as the use of lower than normal concentrations in order to decrease the number of nucleation sites. This procedure may in fact lead to larger crystals than by using larger concentrations.

CHARACTERIZATION OF LEAD IODIDE CRYSTALS. The AES data of Figure 4 suggest that very pure crystals of lead iodide were grown in the shuttle. The amount of carbon contaminant is appreciably lower than the earth grown crystals (Figure 5) and the ratio of iodide to lead is much closer to the ideal ratio of 2 for the shuttle grown crystals than the earth grown materials.

In addition, the signal to noise ratio for the shuttle grown sample is much worse than the earth grown crystals. This suggests that the crystals grown in the shuttle are better insulators than the earth grown materials. This is consistent with the observation that the earth grown materials have more impurities than the shuttle grown crystals. Similar results have been found for proteins (12 − 13). The exact reason why shuttle grown crystals are more pure than earth grown crystals is not entirely clear although it would appear that the mechanism of crystal growth in space does not rely as heavily on foreign ions as nucleation centers.

CONCLUSIONS

Results of the shuttle and laboratory experiments suggest that the mechanism of nucleation of lead iodide crystals on membranes depends on gravitational effects. The crystals grown in space are more pure and better insulators than earth grown crystals. The observation that earth grown crystals only grow on the lower half of the membrane with a shelf of crystals growing out from the membrane suggests that secondary nucleation is occurring on the surface of the originally formed crystals. These data may provide insight for the preparation of other crystals grown in space such as molecular sieves.

ACKNOWLEDGMENTS

The efforts of Roland Pierson and Joseph O'Rourke toward machining and constructing the Plexiglas crystal growing apparatus as well as the financial support to SRC from the Internal Education Foundation and to CWJS from the Faculty Research Fund at Union College are gratefully acknowledged. SLS acknowledges the Department of Energy, Office of Basic Energy Sciences, Division of Chemical Sciences for support of this work. We thank NASA for its support to SRC through the Shuttle Student Involvement Program and the shuttle photographs and G. Nelson for initiating the shuttle crystal growth experiments.

Literature Cited

1. Kelter, R. B.; Snyder, W. B.; Buchar, C. S. J. Chem. Ed., 1987, 64, 228-231.
2. Kelter, R. B.,; Snyder, W. B.; Buchar, C. S. J. Chem. Ed., 1987, 64, 60-62.
3. Favier, J. J. Bull. Soc. Fr. Phys., 1986, 62, 22-23.
4. Dressler, R. F. U.S. Patent 4,615,760, 1986.
5. Nordine, P. C.; Fujimoto, G. T.; Greene, F. T. NASA Contract Rep., NASA, CR172027, 1987.
6. Clifton, K. S.; Owens, J. K. Appl. Optics, 1988, 27, 603-609.
7. Lal, R. B.; Trolinger, J. D.; Wilcox, W. R.; Kroes, R. L. Proc. SPIE Int. Soc. Opt. Eng., 1987, 788, 62-72.
8. Nigli, S.; Chadha, G. K.; Trigunayat, G. C.; Bagai, R. K. J. Cryst. Growth, 1986, 79, 522-526.
9. Singh, N. B.; Glicksman, M. E. Materl. Lett., 1985, 5, 453-456.
10. Van den Berg, L.; Schepple, W. F. Int. Sample. Tech. Conf., 1987, 19, 754-760.

11. Citterio, O.; Bonelli, G.; Conti, G.; Hattaini, E.; Santambrogio, E.; Sacco, B.; Lanzara, E.; Brauninger, H.; Buckert, W. Appl. Optics, 1988, 27, 1470-1475.
12. DeLucas, L. J.; Bugg, C. E.; Suddath, F. L.; Snyder, R.; Naumann, R.; Broom, M. B.; Pusey, M.; Yost, V.; Herren, B. Polym. Prepr. ACS, Divi. Polym. Chem., 1987, 28, 383-384.
13. DeLucas, L. J.; Bugg, C. E.; Suddath, F. L.; Snyder, R.; Naumann, R.; Broom, M. B.; Pusey, M.; Yost, V.; Herren, B.; Carter, D. J. Cryst. Growth, 1986, 76, 681-693.
14. Bromberg, L. E.; Rudman, A. R.; Eltseton, B. S. Vysokomol. Soedin, Ser. A, 1987, 29, 1669-1675.
15. Sand, L. B.; Sacco, A.; Thompson, R. W.; Dixon, A. G. Zeolites, 1987, 7, 387-392.
16. Cioffi, E. A.; Willis, W. S.; Suib, S. L. Langmuir, 1988, 4, 697-702.
17. Suib, S. L. J. Chem. Ed., 1985, 62, 81-82.
18. Madjid, A. H.; Vaala, A. R.; Anderson, W. F.; Pedulla, J. U. S. Patent 3 788 818, 1974.

RECEIVED May 9, 1990

Chapter 2

Templates in the Transformation of Zeolites to Organozeolites

Cubic P Conversions

Stacy I. Zones and Robert A. Van Nordstrand

Chevron Research Company, 576 Standard Avenue, Richmond, CA 94802–0627

This study concerns the conversion of Cubic P zeolite to organozeolites as influenced by a variety of quaternary ammonium templates. In general, when tri-methyl ammonium is attached to a cyclohexane, or more elaborated cyclic or polycyclic structure, high silica zeolites composed of four- and six-ring sili-cate subunits are produced. Conversion rates are proportional to organocation size with adamantane the fastest and cyclohexane the slowest. With symmetric tetraalkyl ammonium or linear diquaternary ammonium templates, organozeolites with predominantly five-ring subunits are produced instead. In one instance, a linear diquat gave no conversion of P and demon-strated an inhibitory effect when a subsequent tem-plate was introduced into the reaction.

This paper presents results of a continuation of our study (1,2) of the conversion of the small-pore zeolite, Cubic P, into various organozeolites--large-, medium-, and small-pore varieties. The main focus here is on the influence of structure of the quaternary amine template molecule on the rates of conversion and on the organo-zeolite produced.

Recently, we described the synthesis of an all-silica molecular sieve, SSZ-24 (isostructural with AlPO-5) composed of four- and six-ring silicate subunits (3). This result was unexpected because most high silica syntheses produce structures rich in five rings (pentasils). This new sieve is produced in a reaction using N,N,N-trimethyladamantammonium cation. This same cation can, under different synthesis conditions, produce a high silica form of chaba-zite, SSZ-13 (4,5). And, again, this structure is devoid of five rings. However, using the simple adamantane amine, clathrate sili-cates are produced, mostly five-ring structures (6-8). This raises questions about how the template, the silica-alumina ratio, and

0097–6156/90/0437–0014$06.00/0
© 1990 American Chemical Society

other factors control the features of crystal structure. Studies by ^{29}Si NMR (9) have shown the silicate solutions have such a multitude of silica forms that it is not clear how crystal structure control is achieved.

We have found a zeolite synthesis reaction which is at the boundary between the five-ring structure and the four- and six-ring structure production. The product depends on the organocation used. A second feature of this synthesis reaction is that no amorphous solid is involved or produced. This reaction derives silica and all of its alumina from the low silica zeolite, Cubic P. This zeolite, organic-free, had been encountered as a transient phase in the discovery of SSZ-13 (1).

Templates represented by the formula Me_3N^+R have been used in this work to study the influence of the R group, which is referred to as the nonpolar end of the cation. More complex cations have been used shedding some light on the influence of the polar end. One cation examined was not effective in converting the Cubic P but was effective in blocking the surface of the Cubic P so that other templates were ineffective. The study opens questions regarding how the polar and the nonpolar ends interact with Cubic P and how they relate to the synthesis of SSZ-13 and other zeolites (including void-filling).

<u>Experimental</u>

The conversion of Cubic P to organozeolites is observed by following the rise in pH (measured at room temperature) (1). X-ray diffraction was used to measure extent of conversion. Zeolite synthesis reactions were carried out at 135°C with 30 rpm tumbling in Parr 4745 reactors with Teflon cups. Unless otherwise stated, the synthesis mix consisted of 0.50 g Cubic P, 2.5 millimols quaternary ammonium halide, 5.0 g Banco "N" silicate solution (38.5% solids, $SiO_2/Na_2O = 3.22$), and 12 mL water. Cubic P was prepared by reacting "N" silicate, water, $Al_2(SO_4)_3$, $18H_2O$, and NaOH at 140°C for six days unstirred. The resultant Cubic P has $SiO_2/Al_2O_3 = 6$.

	Mix for Cubic P Synthesis	For Subsequent Synthesis
Al/Si	0.06	0.06
Na/Si	1.05	0.72
OH/Si	0.95	0.65
N+/Si	0.00	0.10
H_2O/Si	32	32

Organic syntheses were performed by quaternization of amines according to a new peptide procedure (10). In some instances, the amine was not available but the dimethyl amino derivative could be produced from the cyclic carbonyl using the Leukhart reaction (11).

The dimethyl amine derivative produced was quaternized directly by methyl iodide, added slowly to a chilled solution of the ternary amine in ethyl acetate. When other ring substituents are present, the reductive amination leads to isomeric configurations, as noted in Table I. Quaternary ammonium halide salts were characterized by melting point, TLC, or microanalytical data and ^{1}H and ^{13}C NMR.

ESCA analyses were performed on an HP 5950-A instrument. Elemental compositions were determined using ICP. SEM photos were obtained on a Hitachi S-570.

Results

Rates of Conversion of P to SSZ-13. Table I is a list of quaternary ammonium iodides used in this study together with their abbreviations used here and their carbon-to-nitrogen ratios. These templates all may be viewed as derivatives of Me_3N+-cyclohexane by further elaboration of the cyclohexane ring. (The salt obtained from cis-myrtanyl amine actually contains a methylene bridge from the cyclohexyl ring to the charged nitrogen. This salt does not lead to SSZ-13 but converts the P very slowly to mordenite. This template will not be considered further.)

A comparison of synthesis of the conventional sort, proceeding via a silica-alumina gel intermediate, and the synthesis used here starting with Cubic P zeolite, is shown in Figure 1. The synthesis with Cubic P is at least five times as fast as the conventional synthesis. The template used was 1-Ada.

The varying rates of conversion of Cubic P to the organozeolite SSZ-13 are shown in Figure 2. The data show that the large adamantane derivatives produced the most rapid conversion of P to SSZ-13, 1-Ada being more rapid than 2-Ada. (The 2-Ada derivative is "salted out" of solution to a greater extent than the 1-Ada. This may account for the slower rate by 2-Ada. The 9-Non template also has this "salting out" tendency.) The rates follow roughly in order of C/N+ ratio, the cyclo C_6 template showing the slowest rate.

The adamantane derivatives fill the chabazite cavity most completely, even to the point of some lattice distortion (1). Other factors favoring the adamantanes include the high C/N+ with consequent high hydrophobicity and rigidity. An interesting comparison is that of the norbornane templates and the 3 Me cyclo C_6. These all have the same C/N+, but the conversion rates for the norbornanes were faster and the conversion was more selective. The product from the 3 Me cyclo C_6 conversion contained some mordenite.

Figure 3 shows the change of pH with time for the same templates. The relative rates are similar. The rise in pH is due to release of NaOH from the buffer, sodium silicate, as the silica is incorporated into the organozeolite.

The SiO_2/Al_2O_3 values of the SSZ-13 produced were in the range 11 to 13. Na/Al values of the solid products dropped from 1.00 (for P) to 0.50 (for SSZ-13). Conversion to SSZ-13 as measured by XRD is proportional to the nitrogen content of the solid product, as shown in Figure 4. Data in this figure are from various times for each of the templates in this study. The same cation population is in the SSZ-13, regardless of the template used.

Table I. Quaternary Ammonium Iodides Used in the Conversion
of Cubic P to SSZ-13 R-N$^+$(CH$_3$)$_3$ I$^-$ Structures

R Group and Substitution Position	Conformational Isomers	C/N$^+$	Abbreviation in This Work
	1	9	Cyclo C$_6$
CH$_3$	2	10	2 Me Cyclo C$_6$
CH$_3$	2	10	3 Me Cyclo C$_6$
EXO	1	10	2-exo-Nor
ENDO	1	10	2-endo-Nor
	2*	11	2-Octo
	1	12	9-Non
	1	13	1-Ada
	1	13	2-Ada
CH$_3$ CH$_3$ CH$_2$	1	14	Cis Myrt.

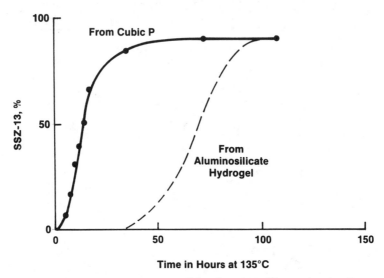

Figure 1. Crystallization profile for SSZ-13 synthesis from Cubic P zeolite and from conventional aluminosilicate hydrogel.

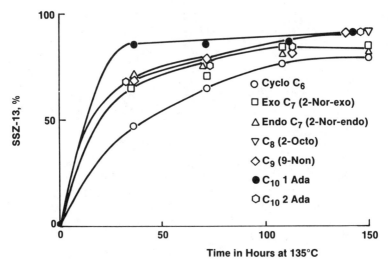

Figure 2. Crystallization of SSZ-13 from Cubic P zeolite in the presence of various organocations.

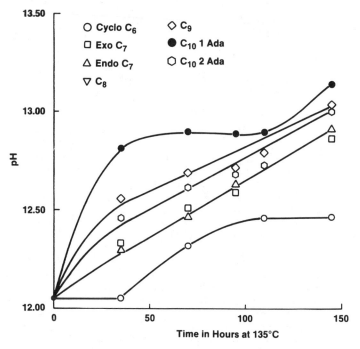

Figure 3. pH change versus reaction time for the crystallization of SSZ-13 from Cubic P zeolite in the presence of various organocations.

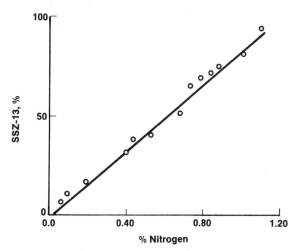

Figure 4. % crystalline SSZ-13 versus nitrogen content in zeolite product. Data obtained for conversion of Cubic P zeolite using a variety of organocations.

Conversion to Other Zeolites. A key question regards the significance of the trimethylammonium group on the template for the conversion of Cubic P. Table II contains the templates studied in this
connection. TMA causes conversion to the organozeolite sodalite and
at a rate comparable to most of the conversions to SSZ-13 shown in
Figure 2. The rates for this second portion of the study are shown
in Figure 5, based on pH measurements. Products in this portion are
generally of the pentasil class. Rates of the templates other than
TMA and TEA were relatively slow. TEA produced mostly mordenite
plus a little beta.

Diquat 6 produced mordenite, but at a slow rate in spite of
having trimethylammonium groups. In conventional synthesis and at
Al/Si of 0.02 to 0.07, Diquat 6 is quite efficient in producing
EU-1 (12); but in the present synthesis starting with Cubic P, the
conversion was slow and no EU-1 was made.

ZSM-5 Formation. In earlier work (1), it was found that with lower
concentrations of TPA, Cubic P was not converted to other zeolites
over several days. However, when TPA was used at higher concentrations of this study, slow production of ZSM-5 was observed. SEM
photos of the product showed large crystals of ZSM-5 and small
aggregates of Cubic P. When these two types of crystals were
separated by sedimentation from an ultrasonic bath, the electron
microprobe showed the ZSM-5 was only slightly enriched in silica.

The slow conversion of Cubic P brought about by TPA may be
attributed to the greater shielding of the charged nitrogen by the
propyl group compared to the methyl and ethyl groups.

Inhibition. Diquat 4 produced no change in Cubic P during several
days at 135°C, and the pH did not rise above the initial value.
Examination of this P sample by ESCA showed considerable ion
exchange had occurred, the sodium being replaced by nitrogen. The
Cubic P from a TPA conversion experiment showed much less ion
exchange.

An experiment was conducted which showed that Diquat 4 effectively blocked the surface of Cubic P, preventing its conversion by
a normally effective template. A corresponding experiment showed
that TPA did not block the surface. The Cubic P was heated at 100°C
for several days with either Diquat 4 or TPA, then washed and sub-
jected to the normal synthesis reaction using 2-exo-Nor as template.
Both XRD and pH (Figure 6) showed that Diquat 4 greatly retarded the
conversion to SSZ-13 and that TPA did not retard the conversion.
Two factors may be responsible for this difference in blocking conversion. The difference in shielding of the charged nitrogen by the
methyl versus the propyl groups should favor stronger retention of
the Diquat 4. Also, the Diquat 4 may have the advantage of a
"bidentate ligand" in its exchange retention on the Cubic P surface.

Discussion

A number of templates having the trimethylammonium group attached to
a cyclohexane ring (and derivatives thereof) produce the high silica
chabazite, SSZ-13. The rate of conversion of Cubic P to SSZ-13
increases as the nonpolar group increases in size up to C_{10} group,

Table II. Quaternary Ammonium Compounds Used in the
Conversion of Cubic P Zeolite to Organozeolites

Compound	C/N$^+$	Abbreviation in This Work
$(CH_3)_4 \, N^+ \, Br^{-(a)}$	4	TMA
$(CH_3CH_2)_4 \, N^+ \, Br^{-(a)}$	8	TEA
$(CH_3CH_2CH_2)_4 \, N^+ \, Br^{-(a)}$	12	TPA
$(CH_3)_3 \, N^+ - (CH_2)_4 - N^+ (CH_3)_3 \, 2I^{(-)}$	5	Diquat 4
$(CH_3)_3 \, N^+ (CH_2)_6 - N^+ (CH_3)_3 \, 2Br^{-(a)}$	6	Diquat 6
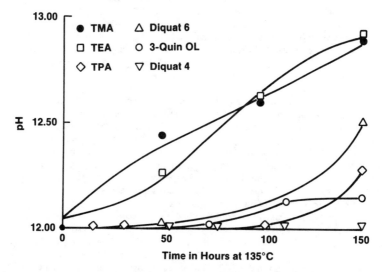	8	3-Quin-OH

^{a}Commercial material from Aldrich Chemical Company, used as
received.

Figure 5. pH change versus reaction time for the
crystallization of organozeolites from Cubic P zeolite in the
presence of various organocations. In this series (Table II),
none of the products are SSZ-13.

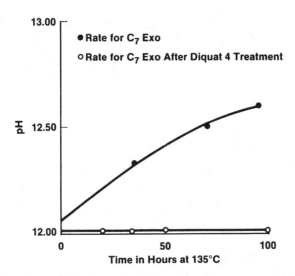

Figure 6. The inhibition of Cubic P zeolite conversion to SSZ-13 when the Cubic P has been pretreated with Diquat 4. TPA Br has no inhibitory effect.

adamantane. This template is probably the upper size limit for filling the chabazite cage.

Rigid, space-filling characteristics may be an important factor for the nonpolar part of the template molecule. For example, the norbornane and methylcyclohexyl organocations have the same C/N+ ratio. But the latter converted P much more slowly and less selectively to SSZ-13. In Figure 7, a strained, high energy conformation of the 3-methylcyclohexyl group can mimic the norbornane. If the norbornane shape gives a much better fit for the chabazite cage, NMR study might show if the methylcyclohexyl group is forced into similar configuration when the chabazite is synthesized in the presence of this template.

Other shapes of organocations do not produce SSZ-13 and, instead, give a slower synthesis of pentasil zeolites (mordenite, ZSM-5, beta) or no conversion product at all. This was observed for linear substituents on charged nitrogen. Two groups were considered: (1) homologous tetraalkyl ammonium compounds and (2) trimethyl ammonium diquats separated by a straight chain spacer. So the size, shape, and rigidity on the nonpolar part of the organocation all seem to be important factors in the rate and selectivity for converting P to chabazite.

The conversion of P to chabazite proceeds at a superior rate compared with the conventional synthesis from an alumia-silicate hydrogel. Our ESCA studies have shown that a superficial cation exchange can take place even in the absence of subsequent conversion. Shielded organocations like tetrapropyl ammonium give poorer surface ion-exchange. This may be related to the slower rates of organozeolite synthesis observed in this study. The ion-exchange which occurs on the surface may be a preliminary step in the conversion.

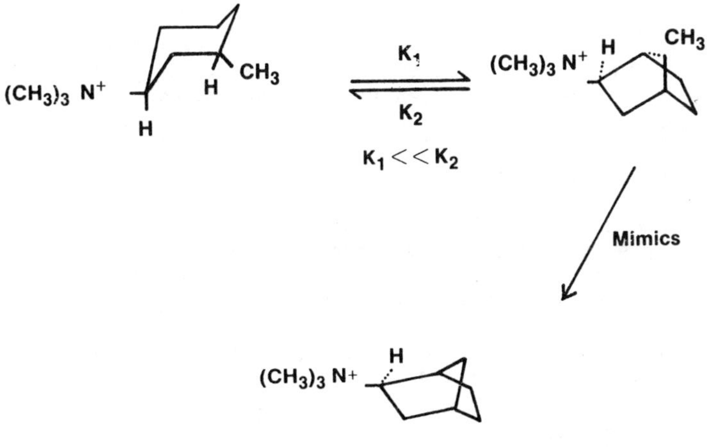

2-Exo-Norbornane Template

Figure 7. A representation of how a high energy conformation of the 3 Me cyclo C_6 template might mimic the isomeric 2-ExoNor template.

Some molecules are capable of interacting with the surface of P leading to an inhibition for further organozeolite synthesis. The special features of this inhibition are a template which does not produce a subsequent organozeolite in its own use and a stronger surface affinity through a "bidendate" behavior. If this is what happens with the Diquat 4, then spatial considerations are, again, important because the larger Diquat 6 does not give this same complete inhibition.

The general rules for transformation of one zeolite to another under hydrothermal conditions, as discussed recently by R. M. Barrer (13), must include the role of the organic template molecule in adding stability. This role is particularly significant in transformations of zeolite to organozeolites as described in this paper.

Acknowledgments

We thank Lun Teh Yuen for technical expertise in the synthesis work. Louis D. Scampavia also contributed in the organic synthetic work. ESCA analyses were carried out by Dr. C. L. Kibby. We thank the Process Research Department of Chevron Research Company for support over the course of this study.

Literature Cited

1. Zones, S. I., and Van Nordstrand, R. A., Zeolites 1986, 8, 166.
2. Chan, I. Y., and Zones, S. I., Zeolites 1989, 9, 3.
3. Van Nordstrand, R. A.; Santilli, D. S.; Zones, S. I.; ACS
    Symposium Series 368, Flank, W. H.; Whyte, T. E.; Eds.;
    "Perspectives in Molecular Sieve Science," 1988, pp 236-245.
4. Zones, S. I.; Van Nordstrand, R. A.; Santilli, D. S.; Wilson,
    D. M.; Yuen, L. T.; Scampavia, L. D.; "Proc. 8th International
    Zeolite Conference," Amsterdam, 1989, p 299.
5. Zones, S. I., U.S. Patent 4 544 538, 1985.
6. Stewart, A.; Johnson, D. W.; Shannon, M. D.; "Innovation in
    Zeolite Materials," Elsevier, 1987, p 87.
7. McClusker, L. J., Appl. Cryst. 1988, 21, 305.
8. Melville, J. B., and Kaduk, S. D., "Proc. 8th International
    Zeolite Conference," 1989, p 57.
9. Groenen, E. J. J.; Kortbeek, A. T. G. T.; MacKay, M.;
    Sudmeijer, O.; Zeolites, 1986, 6, 403.
10. Chen, F. C. M., and Benoiton, N. L., Can. J. Chem. 54, 3310.
11. Bach, R. D., J. Org. Chem., 1968, 33, 1647.
12. Casci, J. L.; Lowe, B. M.; Whittam, T. V.; Eur. Pat. Appl.,
    1981, 42 226.
13. Barrer, R. M., Hydrothermal Chemistry of Zeolites; Academic,
    New York, 1982; pages 161, 174, 210.

RECEIVED May 9, 1990

Chapter 3

Inelastic Neutron Scattering from Non-Framework Species Within Zeolites

J. M. Newsam[1], T. O. Brun[2,3], F. Trouw[2], L. E. Iton[2], and L. A. Curtiss[2]

[1]Exxon Research and Engineering Company, Route 22 East, Annandale, NJ 08801
[2]Materials Science Division, Argonne National Laboratory, Argonne, IL 60439

Inelastic and quasielastic neutron scattering have special advantages for studying certain of the motional properties of protonated or organic species within zeolites and related microporous materials. These advantages and various experimental methods are outlined, and illustrated by measurements of torsional vibrations and rotational diffusion of tetramethylammonium (TMA) cations occluded within zeolites TMA-sodalite, omega, ZK-4 and SAPO-20.

In many heterogeneous catalyst systems we are interested in the interface at a molecular level between an organic component and an inorganic matrix. An attractive probe of this margin would be one that would yield structural, bonding and dynamical data for the organic (or, in some circumstances, the inorganic) component, with an interpretation uncomplicated by a contribution from the inorganic (organic) phase. As introduced below, neutron scattering can, in a number of cases, provide this disproportionately large sensitivity to the organic phase. Neutron scattering cross-sections are quite small, rendering the technique essentially bulk sensitive, but requiring large numbers of scattering centers for effective measurement with the relatively modest fluxes available from today's neutron sources. In catalyst studies this can be a major problem, for active sites are generally restricted to external surfaces, and are therefore relatively dilute. Such is not the case for zeolites and related microporous materials (1–8), for which the active surface is internal.

---

[3]Current address: Manual Lujan, Jr., Neutron Scattering Center, Los Alamos National Laboratory, Los Alamos, NM 87545

Up the 50% of the total crystal volume represents pore space, and high effective concentrations of organic - substrate couples can thus be generated.

Neutron scattering has already been applied to a number of topical zeolite research areas. Early work ([9], [10]) focussed on opportunities for applications of inelastic scattering techniques for studying vibrational modes; inelastic scattering experiments have continued steadily over the past two decades and represent a large portion of the work discussed and cited here. Single crystal neutron diffraction measurements on a number of natural zeolite crystals have appeared ([11]) (although mineral samples often contain crystals up to many mm in size, crystals of synthetic zeolites are almost always much smaller, ~5μm or less being typical). Powder neutron diffraction has been applied to a number of zeolite problems ([12], [13]), including complete studies of a small number of zeolite – hydrocarbon complexes ([14–19]). Small angle neutron scattering has been used to study benzene molecule aggregation in sodium zeolite Y ([20]), and in preliminary studies of template-assisted crystallization ([21], [22]). Quasielastic neutron scattering measurements of, for example, methane in zeolite A ([23–25]) and benzene in mordenite ([26]) have been reported. In the present paper we focus on the advantages that inelastic and quasielastic neutron scattering provide, outlining experimental aspects and discussing measurements of torsional vibrations and rotational diffusion of tetramethylammonium (TMA) cations occluded within a number of zeolites.

<u>The Special Advantages of Neutron Scattering</u>

In systems that do not contain magnetic species, neutrons are scattered by the nuclei. Atomic (nuclear) scattering cross sections are, as above, small on an absolute scale. Rather large samples, typically several g, are therefore generally required, but apparatus for controlling the atmosphere, temperature and/or pressure of the sample environment can entail only minimal attenuation of the incident and scattered neutron beams, facilitating studies under controlled non-ambient conditions. The energy of a thermal neutron (~80meV or 645 cm^{-1}) is comparable to the energies characteristic of atomic and molecular vibrations. The energy changes that the neutron can undergo on scattering can be measured with reasonable precision (resolution) enabling inelastic neutron scattering to be used to measure such vibrational spectra.

The incoherent neutron scattering cross–section for hydrogen is extremely large compared to that of other elements (Figure 1 – arising from its $I = \pm 1/2$ nuclear spin; these two spin states – which occur without correlation from one crystallographic proton site to the next – have differing neutron scattering cross sections). This incoherent scattering does not contribute information to measurements that require coherence from one site to the next (such as those of small angle scattering, diffraction or phonon spectra) and, rather, gives rise to a troublesome background in powder diffraction measurements on protonated species (alleviated by using perdeuterated analogs ([14–19])). However, for the single particle excitations sampled by incoherent inelastic neutron scattering (IINS) techniques it determines that modes involving hydrogen atom motion (which additionally have large amplitudes because of the low proton mass) will dominate the measured spectra.

The neutron scattering process is not subject to the selection rules that govern the observability of modes in optical spectroscopies, and, if a detailed model of the atomic or molecular motions responsible for the IINS spectra is available, full quantitative

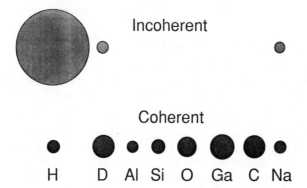

Figure 1. Relative coherent and incoherent scattering cross-sections for some zeolite atomic species.

treatment of both the peak positions and intensities in the IINS spectra is possible. Facilities for performing IINS measurements (see below) are accessible at most of the major neutron scattering centers.

A complementary form of scattering arises when the scattering proton is undergoing diffusive motion, the energy of the scattered neutron being subject effectively to a small doppler shift. This *quasielastic* scattering involves small energy transfers, $< \sim 1 \text{meV}$ (8cm^{-1}) and is observed as a broadening of the purely elastic peak in the spectrum. The signal measures the Fourier transform of the self-self correlation function (the probability of finding the proton at point **r** at time t, when it was originally located at the origin at time t=0). For a single diffusional process, the measured width of the line is related directly to the correlation time, or the residence time between jumps (which are assumed to occur instantaneously), and hence inversely to the diffusion constant for the motion. An Arrhenius plot of the logarithm$_e$ of the width versus reciprocal temperature yields the activation energy for the motion. The window of observability of proton diffusion by quasielastic neutron scattering techniques corresponds to diffusion constants in the approximate range 10^{-7} to 10^{-5} cm^2 s^{-1} (the lower limit determined by the available instrumental resolution - some 0.03 μeV for a spin-echo spectrometer and the upper limit dictated by the requirement of distinguishing a very broad Lorenztian peak from background). The observation of a quasielastic component is, however, a signature that proton diffusional motion is occurring on this time scale. Analysis of the form of the quasielastic scattering enables inferences to be made of the time-scale and also the geometry of the motional process. The geometrical information is conveyed by the elastic incoherent structure factor (EISF), that is the manner in which the normalized quasielastic intensity (or, equivalently, the difference from unity of the non-Bragg elastic component) varies with scattering angle. When a model for the diffusional process is available, the shape of the EISF can be calculated and compared with experiment. A first indication as to an appropriate model can frequently be inferred from the large Q ($Q=4\pi\sin\theta/\lambda$) limit of the EISF. Where diffusion entails hopping between adjacent sites, the limiting value of the EISF is determined by the limiting probability of the proton being at its original position. Thus, for example, for a proton hopping backwards and forwards between two equivalent sites, this limiting probability, and the limiting value of the EISF will be 1/2.

<u>Measurement Methods And Instrumentation</u>

Conceptually the simplest way of performing an inelastic scattering experiment is to permit a monochromatic (monoenergetic) beam of neutrons to impinge on the sample, and to record the scattered intensity at a fixed scattering angle as a function of the Bragg angle, 2θ, of diffraction from a suitable analyzer crystal. As the resolution, $\Delta E/E$ varies as $\Delta\theta\cot\theta$, the scan is best performed in backscattering mode (large 2θ). A monochromatic incident beam can be extracted from the broad incident spectrum also by Bragg diffraction from a suitable monochromator, or alternatively by velocity selection in a chopper (in contrast to the constant (*in vacuo*) velocity of electromagnetic radiation, the neutron's velocity, v, scales with its kinetic energy, E, as $E = mv^2/2$, its wavelength being inversely related to its momentum, $\lambda = h/mv$). The correspondence between velocity and energy (or wavelength) can be exploited by pulsing the incident, monochromatic beam and recording the scattered intensity as a function of the neutron arrival time at a detector after scattering from the sample. All the neutrons in a given pulse arrive at the sample simultaneously, but those neutrons which gain energy through an inelastic scattering process ('down scattering') then arrive at the detector sooner than those that are scattered elastically. The time of flight becomes a measure of the energy transfer that has occurred within the sample.

The measurements of the dynamics of templating TMA$^+$ cations discussed below were made on an inverted geometry time-of-flight spectrometer installed at the Intense Pulsed Neutron Source (IPNS) of Argonne National Laboratory (Figure 2). In this spallation source, the neutrons are generated by stopping a 500MeV proton beam pulsed at 30Hz in a uranium target. Neutrons are 'boiled' out of the target nuclei with a continuous spectrum of high energies. This pulse of hot neutrons is allowed partially to thermalize in a moderator that is viewed by the experimental beam pipes. All of the neutrons generated in a single pulse depart from the target/moderator assembly within a narrow time window. The neutrons that, following scattering by the sample, have a defined energy are detected following Bragg diffraction from pyrolytic graphite analyzer crystals (the Be filter selectively scatters out the higher order harmonics diffracted by the 004 etc. planes). In any given time-of-flight spectrum, those neutrons detected first are those which were the first (most energetic) to reach the sample and hence lost most energy in the sample to attain the selected final energy. This configuration offers the advantage of being able to record, simultaneously, the time-of-flight powder diffraction pattern from the sample by using a detector without an energy-selecting analyzer crystal.

Some topical examples

Water, hydroxyl groups and ammonium cations.   Several IINS measurements of hydrated zeolites have been reported (9, 27–36). The influence of the zeolite host structure on the (dis-)ordering and dynamical properties of sorbed water is reflected in the frequencies and widths of the translational (centre of mass vibrations) bands at ~10 – 40meV (80 – 320 cm^{-1}) and the librational bands (restricted rotations) at ~60 – 100meV (480 – 800 cm^{-1}). In larger pore zeolites, the degree of definition in the IINS spectra is more limited and the spectra more closely ressemble those for water itself (although with less order than is developed in ice). Additionally, proton motion associated with riding motion of the water molecules on non-framework cations and, particularly in the smaller pore systems, coupled framework - water molecule motion is suggested. The degree of quantitative interpretation of these various measured data remains limited. The vibrational characteristics of bridging hydroxyl groups (37) and ammonium cations (38) in zeolite rho have also been examined by IINS. A related application has been the use of IINS to probe hydride formation in small particles of Pd within the cages of zeolite Y (39).

Hydrocarbon sorbate vibrations.   IINS spectra have been recorded for a number of simple sorbate molecules within aluminosilicate zeolites, including hydrogen in A (40, 41), acetylene in X (42), ethylene in A (43) and X (44–46), and p-xylene (47) in X type materials. In addition to intramolecular modes, where interaction between the sorbate and the non-framework cations is strong (for example in the ethylene – silver zeolite A system (43)), vibrational transitions associated with sorbate motion with respect to the zeolite's internal surface can be observed. The latter modes, and the dependence of their frequencies on loading, structure and composition are of particular interest as they convey detailed information about the character of the zeolite – sorbate

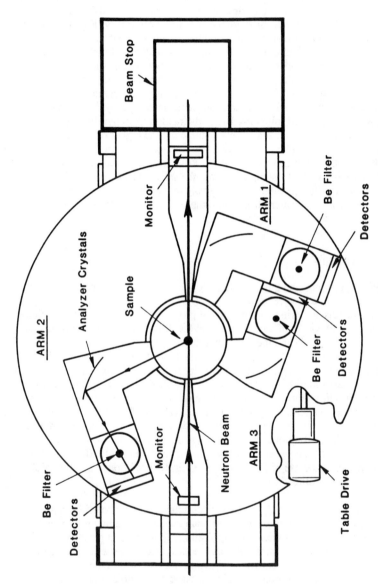

Figure 2. Schematic diagram of the QENS spectrometer at the Intense Pulsed Neutron Source (IPNS), Argonne National Laboratory.

interaction. Unfortunately, quantitative interpretations of such spectra are rarely straightforward, but substantial insight can be gained when IINS data are considered in combination with complementary techniques such as diffraction, nmr, and infra-red and Raman spectroscopies.

Torsional vibrations of occluded template cations.   Of equal interest to the zeolite - organic interaction that determines the utility of zeolites in technological sorptive and catalytic applications, is the role of the inorganic - organic interface during the process of zeolite crystallization. Prior to embarking on a detailed program of *in situ* studies of the role of templating molecules in zeolite synthesis, we are examining the interaction between occluded templating species and the fully-formed zeolite (Figure 3). This interaction is manifested in a number of ways. For example, the temperature of template weight loss features in thermogravimetric analysis scans depend on the character of the template and of its containing cage, and the magnitudes of the ^{13}C chemical shifts observed for TMA$^+$ cations in zeolites vary with the dimensions of the pores within which they are housed (48–50). Certain of the motional properties of the TMA$^+$ cations are conveniently studied by neutron scattering techniques.

The vibrational spectrum of the tetramethylammonium cation in the region 150 - 550 cm^{-1} contains both torsional and vibrational modes. The ν_8 and ν_{19} vibrational modes of E and T$_2$ symmetry involve C-N-C bond angle bending. These modes are Raman active and have been studied for TMA$^+$ in several zeolite environments, although little change in frequency is observed (51). The ν_4 and ν_{12} torsional modes involve partial rotation about C - N bonds and form respectively a singlet (A$_2$) and a triplet (T$_1$) which are both Raman inactive. These torsional modes are directly observed in the IINS spectra and prove to be sensitive to the character of the TMA$^+$ cation (see Table 1) environment(52).

TABLE 1
Torsional and Bending Mode Frequences (cm^{-1}) for TMA$^+$ Cations in Various Environments

environment	Torsion		Bending		
	Singlet	Triplet	Singlet	Triplet	
	A$_2$ (ν_4)	T$_1$ (ν_{12})	E (ν_8)	T$_2$ (ν_{19})	ref.
TMA-Cl	301		371	456	(53)
TMA-Br	294	363		456	(53)
TMA-Br	290	370		455	(52)
TMA-I	265	344		451	(53)
TMA-sodalite	224	310	373	462	(57)
TMA-SAPO-20	226	311	375	466	(57)
TMA-ZK-4 (β)	226	"	"	"	(57)
TMA-ZK-4 (β)	245	325	372	460	(52)
TMA-ZK-4 (α)	210	(310)	371	454	(57)
TMA-omega	209	298	362	455	(60)
TMA-omega	206	293	365	454	(52)
TMA-montmorillonite	221	306	369	461	(61)
TMA-Cl in D$_2$O solid soln	244	310	–	–	(62)
TMA-Cl in D$_2$O soln	230	–	370	–	(62)
TMA-solution			370	455	(51)
TMA free ion (calc)	199	287	385	485	(52)

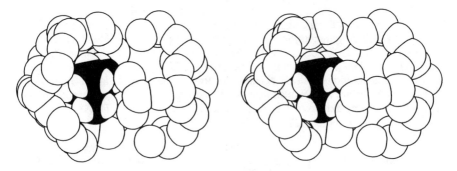

Figure 3. 'Exploded' strereoview representation of a TMA⁺ cation in the sodalite (β) cage in TMA-sodalite.

In the halide salts, TMAX, X = Cl, Br, I, the torsional mode frequencies evolve smoothly to lower force constants in the series Cl - Br - I (53). The crystal structures are similar, and the change in torsional frequencies correlates with the anion polarizabilities. The TMA$^+$ torsional frequencies observed in zeolites (Table 1) imply a weaker interaction between the TMA$^+$ cation and its environment. The IINS spectra, $150 \leq v \leq 550$ cm^{-1}, for TMA-sodalite and SAPO-20 and are similar (Figure 4), with lower torsional frequencies observed for TMA$^+$ cations occluded in the slightly larger gmelinite cages of zeolite omega, and in the supercage of zeolite ZK-4 (Table 1). A dependance of the torsional frequencies on the entrapping zeolite cage size is further suggested by estimates of torsional frequencies computed for the free TMA$^+$ ion based on a Hartree-Fock *ab initio* treatment (52), which indicate still lower torsional mode frequencies. The dependence of the torsional frequencies on the non-framework and framework compositions of the zeolite is still being explored, but, in principle, this sensitivity of the torsional modes to environment might be used to explore the stage at which complete cages are formed around the 'templating' TMA$^+$ cations during synthesis. Although minimum measurement times of several hours for each spectrum are currently necessary, instrument optimization for such a synthesis experiment and/or appropriate control of the synthesis conditions should permit such experiments to be performed.

Diffusional motion.   Many rotational and translational diffusion processes for hydrocarbons within zeolites fall within the time scale that is measurable by quasielastic neutron scattering (QENS). Measurements of methane in zeolite 5A (24) yielded a diffusion coefficient, D= 6×10^{-6} cm^2 s^{-1} at 300K, in agreement with measurements by pulsed-field gradient nmr. Measurements of the EISF are reported to be consistent with fast reorientations about the unique axis for benzene in ZSM-5 (54) and mordenite (26), and with 180° rotations of ethylene about the normal to the molecular plane in sodium zeolite X (55). Similar measurements on methanol in ZSM-5 were interpreted as consistent with two types of methanol species (56).

Quasielastic scattering is observed from TMA-sodalite and omega for T ≥ ~80K (57) (Figure 5). A quasielastic component is, in contrast, not observed for the bromide salt below 300K (58). In each case, molecular translation is prevented and the quasielastic scattering indicates that rotational diffusion is occurring. Arrhenius plots of the logarithm$_e$ of the quasielastic width versus reciprocal temperature indicate small activation barriers, 1.8(5) and 1.5(5) kJ mol^{-1} for the sodalite and omega samples respectively. These barriers are much smaller than that measured for methyl group rotation by nmr methods (59), estimated from the torsional mode frequencies (53) or, indeed, calculated using *ab initio* methods for TMA$^+$ in a range of possible model environments (57). The broadening is therefore interpreted as arising from whole body TMA$^+$ cation reorientational motion. In constrained environments this motion has a high activation barrier (59). The freer TMA$^+$ environment in the zeolites (that is reflected in the approach of the torsional mode frequencies towards the free-ion values) apparently results in a dramatic reduction in this activation barrier, such that whole body can occur, even down to relatively low temperatures ≥ 80K. Further experiments, combined with detailed dynamical simulations are currently underway to confirm and further quantify these findings. The geometrical information conveyed by the EISF for the TMA-sodalite and omega systems has also been examined. Although detailed inferences about the character of the reorientational motion are difficult in these relatively complicated cases, the present data do suggest different modes of

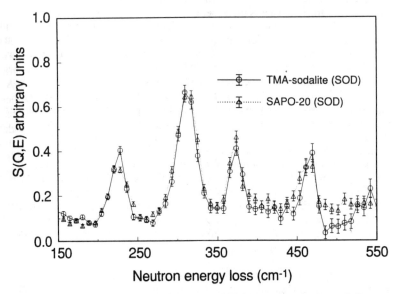

Figure 4. Comparison of the INS spectra of TMA+ cations in the sodalite cage of the aluminosilicate zeolite TMA-sodalite, and the silicoaluminophosphate molecular sieve SAPO-20.

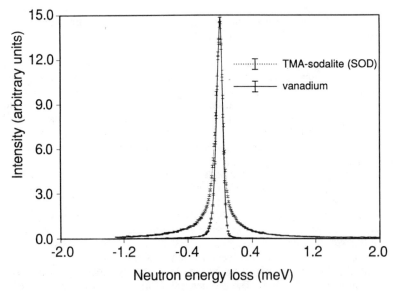

Figure 5. Quasielastic neutron scattering spectrum of TMA+ cations in the sodalite cage of the aluminosilicate zeolite TMA-sodalite compared with the instrumental resolution function.

reorientation in the two cases (57), not inconsistent with the differing crystallographic environments of the TMA$^+$ cation in the two systems.

Conclusion

Inelastic and quasielestic neutron scattering offer a special perspective on the vibrational and diffusional properties of hydrogen-containing non-framework species within zeolites. Results for a number of systems have already appeared, including IINS measurements of water, hydroxyl groups and ammonium cations, and a small number of simpler hydrocarbons in a variety of aluminosilicate zeolites. Interpretations have to date generally been qualitative. Measurements of torsional and bending mode frequencies of TMA$^+$ cations in a number of zeolites have indicated sensitivity in the former (but not the latter) to environment, a sensitivity that may prove exploitable in studies of zeolite synthesis phenomena. Quasielastic scattering measurements have probed translational diffusion of methane in zeolite A and hydrocarbon rotational diffusion has been observed in a number of systems. Quasielastic scattering that is interpreted as indicating whole body TMA$^+$ cation reorientation is observed from TMA-sodalite and omega for T $\geq$ ~80K.

Acknowledgments

We thank those who have contributed to our own IINS and QENS studies: R. A. Beyerlein, S. K. Sinha, D. E. W. Vaughan, and the staff of the Intense Pulsed Neutron Source at Argonne National Laboratory. (This work supported in part by the U.S. Department of Energy, BES-Materials Sciences, contract W-31-109-ENG-38.)

Literature Cited

1. Breck, D. W. Zeolite Molecular Sieves: Structure, Chemistry and Use; Wiley and Sons (reprinted R. E. Krieger, Malabar FL, 1984): London, 1973.
2. Barrer, R. M. Zeolites and Clay Minerals as Sorbents and Molecular Sieves; Academic Press: London, 1978.
3. Barrer, R. M. Hydrothermal Chemistry of Zeolites; Academic Press: London, 1982.
4. Olson, D.; Bisio, A. Eds. Proceedings of the Sixth International Zeolite Conference (Butterworths, Surrey, UK, 1984).
5. Murakami, Y.; Iijima, A.; Ward, J. W. Eds. New Developments in Zeolite Science and Technology (Kodansha - Elsevier, Tokyo - Amsterdam, 1986).
6. Jacobs, P. A.; van Santen, R. A. Eds. Zeolites: Facts, Figures, Future (Elsevier, Amsterdam, 1989).
7. Newsam, J. M. Science 1986, 231, 1093-1099.
8. Kerr, G. T. Scientific American 1989, 100-105.
9. Boutin, H.; Safford, G. J.; Danner, H. R. J. Chem. Phys. 1964, 40, 2670-2679.
10. Egelstaff, P. A.; Stretton Downes, J.; White, J. W. In Molecular Sieves; Barrer, R. M. Ed.; Society of Chemical Industry: London, 1968; pp. 306-318.
11. Kvick, Å. Trans. Amer. Cryst. Assoc. 1986, 22, 97-106.
12. Newsam, J. M. Physica 1986, 136B, 213-217.
13. Newsam, J. M. Materials Science Forum 1987, 27/28, 385-396.
14. Kahn, R.; Cohen de Lara, E.; Thorel, P.; Ginoux, J. L. Zeolites 1982, 2, 260-6.

15. Wright, P. A.; Thomas, J. M.; Cheetham, A. K.; Nowak, A. K. Nature (London) 1986, 318, 611-614.
16. Fitch, A. N.; Jobic, H.; Renouprez, A. J. Phys. Chem. 1986, 90, 1311-1318.
17. Newsam, J. M.; Silbernagel, B. G.; Garcia, A. R.; Hulme, R. J. Chem. Soc. Chem. Comm. 1987, 664-666.
18. Taylor, J. C. Zeolites 1987, 7, 311-18.
19. Czjzek, M.; Vogt, T.; Fuess, H. Angew. Chem. 1989, 786-787.
20. Renouprez, A. J.; Jobic, H.; Oberthur, R. C. Zeolites 1985, 5, 222-224.
21. Brun, T. O.; Epperson, J.; Iton, L. E.; Trouw, F.; Henderson, S.; White, J. private communication 1989,
22. Henderson, S. J.; White, J. W. J. Appl. Crystallogr. 1988, 21, 744-50.
23. Cohen de Lara, E.; Kahn, R. J. Physique (Orsay, France) 1981, 42, 1029-1038.
24. Cohen de Lara, E.; Kahn, R.; Mezei, F. J. Chem. Soc., Faraday Trans. I 1983, 79, 1911-1920.
25. Stockmeyer, R. Zeolites 1984, 4, 81-86.
26. Jobic, H.; Bee, M.; Renouprez, A. Surf. Sci. 1984, 140, 307-320.
27. Belitskii, I. A. Geol. Geofiz 1970, 26-29.
28. Belitskii, I. A.; Gabuda, S. P.; Joswig, W.; Fuess, H. Neues Jahrb. Mineral. Monatsh 1986, 541-551.
29. Bogomolov, V. N.; Zadorozhnii, A. I.; Plachenova, E. L.; Pogrebnoi, V. I. Zh. Strukt. Khim. 1978, 19, 259-263.
30. Fuess, H. Ber. Bunsenges. Phys. Chem. 1982, 86, 1049-54.
31. Fuess, H.; Stuckenschmidt, E.; Schweiss, B. P. Ber. Bunsenges. Phys. Chem. 1986, 90, 417-21.
32. Pechar, F.; Schweiss, P.; Fuess, H. Chem. Zvesti 1982, 36, 779-783.
33. Pechar, F.; Fuess, H. Acta Montana 1983, 64, 59-67.
34. Pokotilovskii, Y. N. Zh. Strukt. Khim. 1968, 9, 1079-81.
35. Ramsay, J. D. F.; Lauter, H. J.; Tompkinson, J. J. Phys., (Colloq. C7) 1984, 73-79.
36. Stuckenschmidt, E.; Fuess, H. Ber. Bunsenges. Phys. Chem. 1988, 92, 1083-1089.
37. Wax, M. J.; Cavanagh, R. R.; Rush, J. J.; Stucky, G. D.; Abrams, L.; Corbin, D. R. J. Phys. Chem. 1986, 90, 532-534.
38. Udovic, T. J.; Cavanagh, R. R.; Rush, J. J.; Wax, M. J.; Stucky, G. D.; Jones, G. A.; Corbin, D. R. J. Phys. Chem. 1987, 91, 5968-5973.
39. Jobic, H.; Renouprez, A. J. Less Comm. Metals 1987, 129, 311-316.
40. Braid, I. J.; Howard, J.; Nicol, J. M.; Tomkinson, J. Zeolites 1987, 7, 214-218.
41. Nicol, J. M.; Eckert, J.; Howard, J. J. Phys. Chem. 1988, 92, 7117-7121.
42. Howard, J.; Robson, K.; Waddington, T. C. Zeolites 1981, 1, 175-80.
43. Howard, J.; Robson, K.; Waddington, T. C.; Kadir, Z. A. Zeolites 1982, 2, 2-12.
44. Howard, J.; Waddington, T. C.; Wright, C. J. J. Chem. Soc. Chem. Commun 1975, 775-776.
45. Howard, J.; Waddington, T. C.; Wright, C. J. J. Chem. Soc., Faraday Trans. II 1977, 73, 1768-1787.
46. Howard, J.; Nicol, J. M.; Eckert, J. Springer Ser. Surf. Sci. 1985, 2 (Struct. Surf.), 219-224.
47. Dimitrova, R.; Natkaniec, I. Proc. VIth Int. Symp. Heterog. Catal. 1987, 210-215.
48. Jarman, R. H.; Melchior, M. T. J. Chem. Soc. Chem. Comm. 1984, 414-415.

49. Hayashi, S.; Suzuki, K.; Shin, S.; Hayamizu, K.; Yamamoto, O. Chem. Phys. Lett. 1985, 113, 368-371.
50. Newsam, J. M.; Silbernagel, B. G.; Melchior, M. T.; Brun, T. O.; Trouw, F. In Proc. Vth. Int. Symp. Inclusion Phen. Molec. Recognition; Atwood, J. L. Eds.; Plenum: New York, 1989; in press.
51. Dutta, P. K.; Del Barco, B.; Shieh, D. C. Chem. Phys. Lett. 1986, 127, 200.
52. Brun, T. O.; Curtiss, L. A.; Iton, L. E.; Kleb, R.; Newsam, J. M.; Beyerlein, R. A.; Vaughan, D. E. W. J. Amer. Chem Soc. 1987, 109, 4118-4119.
53. Ratcliffe, C. I.; Waddington, T. C. J. Chem. Soc. Faraday Trans. II 1976, 72, 1935-1956.
54. Jobic, H.; Bee, M.; Dianoux, A. J. J. Chem. Soc., Faraday Trans. 2 1989, 85, 2525-2534.
55. Wright, C. J.; Riekel, C. Mol. Phys. 1978, 36, 695-704.
56. Jobic, H.; Renouprez, A.; Bee, M.; Poinsignon, C. J. Phys. Chem. 1986, 90, 1059-1065.
57. Brun, T. O.; Curtiss, L. A.; Trouw, F.; Iton, L. E.; Newsam, J. M. 1989, to be submitted.
58. Albert, S.; Gutowsky, H. S.; Ripmeester, J. A. J. Chem. Phys. 1972, 56, 3672.
59. Sato, S.; Ikeda, R.; Nakamura, D. J. Chem. Soc., Faraday Trans. 2 1986, 82, 2053-2060.
60. Bradley, K. F.; Chen, S.-H.; Brun, T. O.; Kleb, R.; Loomis, W. A.; Newsam, J. M. Nucl. Instr. Meth. 1988, A270, 78-89.
61. Neumann, D. A.; Rush, J. J.; Nicol, J. M.; Wada, N. In NIST Tech. Note #1257; Eds.; Nat. Inst. Standards and Testing: Maryland, 1989; pp. 12-14.
62. Brown, A. N.; Newbery, M.; Thomas, R. K. J. Chem. Soc., Faraday Trans. 2 1988, 84, 17-33.

RECEIVED May 9, 1990

Chapter 4

Silicoaluminophosphate Molecular Sieves

X-ray Photoelectron and Solid-State NMR Spectroscopic Results Compared

David F. Cox and Mark E. Davis

Department of Chemical Engineering, Virginia Polytechnic Institute
and State University, Blacksburg, VA 24061

The local environments of T-atoms in SAPO materials were examined using solid-state NMR, a bulk
probe, and XPS, a surface sensitive probe. T-atom 2 p binding energies in XPS were found to
vary in a predictable fashion with changes in NMR
chemical shifts. The comparison demonstrates
that XPS is sensitive to variations in the second
coordination sphere for T-atoms in SAPO molecular
sieves. XPS was also found to give a reasonable,
quantitative measure of superficial (surface)
T-atom fractions thus providing information about
elemental homogeneity by comparison to bulk chemical analysis.

Silicoaluminophosphates (SAPO's) (1) are molecular sieves which
contain tetrahedra of oxygen surrounded silicon, aluminum, and
phosphorus. These microporous solids not only exhibit properties
characteristic of zeolites but also show unusual physiochemical
traits ascribable to their unique chemical compositions (1,2).

We have been investigating the nature of these novel solids
in order to ascertain information which may lead to new catalytic
applications. Two important questions to be addressed are: (i)
what are the local T-atom (tetrahedral atom) arrangements
(silicon, aluminum, and phosphorus), and (ii) what is the
elemental homogeneity? Questions (i) has been addressed mainly
through the use of solid-state NMR while the answer to question
(ii) is most appropriately obtained from surface to bulk elemental
analyses.

The purpose of our work is to show that the local
environments of Si, Al, and P in SAPO's can be probed via
solid-state NMR and XPS and that their results give a consistent
picture of the second coordination shell for all the T-atoms.
Also, of secondary importance, we will illustrate that XPS can be
useful in ascertaining whether elemental homogeneity is achieved
in SAPO molecular sieves.

0097–6156/90/0437–0038$06.00/0
© 1990 American Chemical Society

Experimental Section

The samples used in this study are those that have been described
previously (3-5).

Magic angle spinning ^{29}Si, ^{27}Al and ^{31}P NMR spectra were
recorded on a variety of spectrometers. The data reported here
are from results obtained at 200 MHz proton. Thus, the ^{29}Si, ^{27}Al
and ^{31}P frequencies are 39.7 MHz, 52.15 MHz and 81.0 MHz,
respectively. Chemical shifts for aluminum are reported relative
to $Al(NO_3)_3$ in aqueous solution at infinite dilution and are not
corrected for second-order quadrupole effects. The phosphorus and
silicon chemical shifts are reported relative to 85 wt% H_3PO_4 and
Me_4Si, respectively.

Bulk chemical analyses of the solid was performed by
Galbraith Laboratories, Knoxville, TN.

Chemical analysis of the superficial regions of the samples
was performed by X-ray photoelectron spectroscopy (XPS) with a
Perkin-Elmer Phi 5300 ESCA, employing Mg $K\alpha$ x-rays. Binding
energies are referenced to the gold $4f_{7/2}$ transition at 83.8 eV
(gold was sputtered onto a small spot of the sample prior to
analysis).

Results and Discussion

Bulk and Surface Compositions. The chemical compositions of the
molecular sieves used in this study are given in Table I in terms
of tetrahedral atom (T-atom) fractions, and are grouped according
to structure type. The bulk compositions of $AlPO_4$-5, $AlPO_4$-20 and
VPI-5 show the ideal 1:1 ratio of Al and P characteristic of
aluminophosphate molecular sieves. The SAPO materials have
frameworks consisting of Si, Al and P T-atoms.

The two SAPO-20 materials have very different bulk
concentrations. SAPO-20A has a composition consistent with a
"silicon-substituted aluminophosphate", while the high bulk
concentration of Si in SAPO-20B is more consistent with a
"phosporous-substituted aluminosilicate". The other SAPO samples
all have sufficiently low concentrations of Si to be considered as
silicon-substitued $ALPO_4$'s.

Superficial (surface) compositions are also listed in Table
I as determined by XPS with atomic sensitivity factors
characteristic of the electron energy analyzer used for these
studies. The superficial compositions agree reasonably well with
the bulk values for all samples with the exceptions of SAPO-5 and
the two Si-VPI-5 samples which show a significant enrichment of Si
at the surface. Note that samples which were examined with and
without deposited gold (the reference for the binding energy
scale) showed little difference in measured compositions with XPS.

The generally good agreement between bulk and superficial
compositions for homogeneous samples provides some assurance that
XPS, despite its surface sensitivity, samples a sufficient portion
of the framework structure to give reasonable T-atom compositions.
The composition determinations are aided by the fortuitous
occurrence of the 2p photoelectron peaks within a fairly narrow
range of values at low binding energies (Si 2p, Al 2p and P 2p at

TABLE I.   Superficial and Bulk Compositions of Molecular Sieves

Sample	Bulk Comp.	Superficial Comp.
SAPO-37	$Si_{0.12}Al_{0.5}P_{0.38}$	$Si_{0.13}Al_{0.54}P_{0.33}$
AlPO$_4$-5	$Al_{0.5}P_{0.5}$	$Al_{0.52}P_{0.48}$
SAPO-5	$Si_{0.08}Al_{0.47}P_{0.45}$	$Si_{0.20}Al_{0.51}P_{0.29}$
AlPO$_4$-20	$Al_{0.5}P_{0.5}$	$Al_{0.52}P \cdot _{0.48}$
SAPO-20A	$Si_{0.13}Al_{0.47}P_{0.40}$	$Si_{0.10}Al_{0.55}P_{0.35}$
SAPO-20B	$Si_{0.50}Al_{0.38}P_{0.12}$	$Si_{0.41}Al_{0.44}P_{0.15}$
VPI-5	$Al_{0.51}P_{0.49}$	$Al_{0.54}P_{0.46}$
Si-VPI-5A	$Si_{0.06}Al_{0.49}P_{0.45}$	$Si_{0.17}Al_{0.54}P_{0.29}$
Si-VPI-5B	$Si_{0.09}Al_{0.48}P_{0.43}$	$Si_{0.31}Al_{0.37}P_{0.32}$

approximately 102, 75 and 135 eV, respectively). These low
binding energies correspond to high values of photoelectron
kinetic energy and represent the least surface-sensitive portion
of the XPS spectrum. The similarities in photoelectron kinetic
energies gives rise to similar inelastic mean-free paths. Thus,
the signal attenuation with respect to depth in the sample is
essentially constant for all three types of T-atoms.

The agreement between XPS and bulk analysis for the majority
of the samples supports the conclusion that significant variations
between the measured bulk and superficial compositions are indeed
due to inhomogeneity in the sample. The compositions measured by
XPS in these cases also give relative T-atom fractions which agree
with known mechanisms of silicon substitution in aluminophosphate
materials (see discussion below). The SAPO-5 composition data in
Table I indicate an enrichment of silicon at the surface, and the
composition suggests silicon substitutes primarily for phosphorus.
The two Si-VPI-5 samples prepared by different synthetic routes
both show an enrichment of silicon at the surface, but in
Si-VPI-5A (synthesized from a single phase synthesis gel (5)) the
substitution appears to be primarily for phosphorus. In contrast,
surface silicon in Si-VPI-5B appears to substitute for
aluminum-phosphorus pairs. The two phase synthesis used to
prepare Si-VPI-5B (5) appears to yield a significantly less
homogeneous crystal with more a silicon-rich surface than the
corresponding single phase synthesis for Si-VPI-5A.

XPS-NMR Comparisons. While the physics associated with NMR and
XPS measurements are very different, both techniques are sensitive
to changes in the local environment and electronic structure of
the individual atomic species. The observation of a correlation
between bulk T-atom environments probed by NMR and surface T-atom
environments probed by XPS would indicate that XPS truly provides
information about the framework structure and not just a defective
region at the surface of the crystals.

Phosphorus. Previous ^{31}P solid-state NMR studies of SAPO
materials have detected only a single type of local environment
for phosphorus T-atoms: four aluminum second nearest neighbors as
in a pure aluminophosphate (ALPO$_4$) material (3,4). This
environment is illustrated in Figure 1. The NMR chemical shifts
from this environment typically range from -25 to -35 ppm.
    Figure 2 compares the P 2p binding energies from XPS with
the ^{31}P chemical shifts from solid-state NMR. Chemical shifts are
reported as single points except for the VPI-5 materials which
show two ^{31}P NMR peaks associated with different crystallographic
locations in the VPI-5 topology (6). The two different chemical
shift values for the VPI-5 materials are joined by vertical lines
in Figure 2, but each is consistent with the local environment
illustrated in Figure 1. The XPS binding energies vary over a
range of only 0.3 eV. Since the typical accuracies for
determining binding energies are ± 0.1 eV, the spread in P 2p
binding energies is essentially that expected from experimental
uncertainty. Also, the narrow range of binding energies for P 2p
core levels in similar local environments but different molecular
sieves demonstrates that the binding energy scale referenced to Au
4f$_{7/2}$ reproducibly accounts for differences in the extent of
charging for the different samples.
    The XPS results are in agreement with NMR in that no
significant variations in the binding energies are measured for
phosphorous T-atoms which are confined to a single type of local
(2nd coordination sphere) environment. These results agree with a
previous XPS study by Suib et al. of oxygen polarizabilities in
SAPO's (7). Those studies found no variation in the
polarizability of oxygen ions surrounding P^{5+} sites (i.e., no
local electronic variations) as a function of silicon
incorporation (7).

Aluminum. Previous ^{27}Al NMR studies have demonstrated four
possible local environments for Al in SAPO materials (3,4). These
environments are illustrated in Figure 3, and may be classified as
either phosphorous rich (i.e., ALPO$_4$-like) with a chemical shift
ranging from 30 to 40 ppm, or silicon rich (i.e., zeolite-like)
with a chemical shift greater than 48 ppm. Both types of
environments are characteristic of a substitution mechanism
involving silicon substitution for phosphorus. A fifth
possibility for an Al environment involves two Si and two P second
nearest neighbors. However, no such environment has yet been
identified by NMR, either because the Al chemical shift is similar
to that for the silicon- or phosporous-rich environments, or
because materials with an appropriate level of Si to give rise to

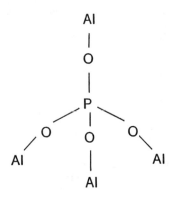

$\delta = -25$ to -35 ppm

Figure 1.   Phosphorus environments determined from ^{31}P solid-state NMR.

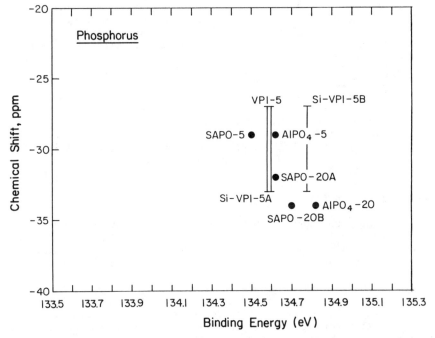

Figure 2.   Comparison of P 2p binding energies from XPS and ^{31}P chemical shifts from NMR.

such environments in a particular framework structure have not been synthesized.

The comparison between Al 2p binding energies from XPS and ^{27}Al NMR chemical shifts is shown in Figure 4. The data points group into two regions on the plot consistent with the ideas of two different classes of environments found in NMR. The general trend shows that the Al binding energy decreases with increasing chemical shift (i.e., increasing Si concentration). The binding energies vary over a range of 1.4 eV, with the points for the individual groupings varying by about 0.5 eV, clearly larger than experimental uncertainty.

The decrease in Al 2p binding energies with increasing Si content can be rationalized by considering the effect of changing the second coordination sphere of aluminum from P^{5+} to Si^{4+}. The charge associated with the first coordination sphere of O^{2-} anions around the Al centers will be less strongly polarized towards the second coordination sphere as the ionic charge on the second-nearest-neighbor cations decreases. Thus, the increased charge density associated with the anions in the first coordination sphere appears as a net negative change on the Al^{3+} centers and decreases the binding energy. This explanation is in agreement with previous observations of a strong effect of P^{5+} substitution in zeolites on the polarizability of oxygen surrounding Al^{3+} centers (7).

The result given for Si-VPI-5B appears to be anomalous. However, notice that the superficial composition of this sample gives Al/P $\simeq$ 1. Thus, the aluminum environment near the surface of Si-VPI-5B is more like an $AlPO_4$ rather than a SAPO.

Silicon. Two Si substitution mechanisms have been identified previously for aluminophosphate structures by NMR (3,4). The first mechanism, Si substitution for P, was mentioned above with regard to changes in the second coordination sphere for Al cations, and gives rise to isolated silicon cations with four aluminum second nearest neighbors (i.e., a zeolite-like environment). The second mechanism involves the substitution of Si atoms for an aluminum-phosphorus pair, and gives rise to regions of the sample where silicon is incorporated with four silicon nearest neighbors (i.e., silica-like domains). These two types of environments may be distinguished easily with ^{29}Si NMR (3,4,8). A third, aluminum-rich, environment (one Si and three Al second-nearest neighbors) is also possible, but the NMR chemical shift is not significantly different than that observed for isolated Si atoms with four Al second nearest neighbors. The three possible Si environments are illustrated in Figure 5.

The comparison between Si 2p binding energies from XPS and ^{29}Si NMR is given in Figure 6. NaA (Si:Al = 1:1), ZSM-5 (Si:Al = 28) and sodalite (Si:Al = 6) samples are also included for comparison. The vertical lines connect the values of the chemical shift associated with the two distinct, NMR-resolvable Si environments discussed above. NaA and SAPO-37 are two exceptions; they exhibit only single peaks, and the vertical lines represent the broadness of the resonances.

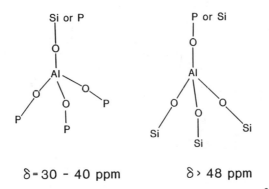

Figure 3. Aluminum environments determined by ^{27}Al solid-state NMR.

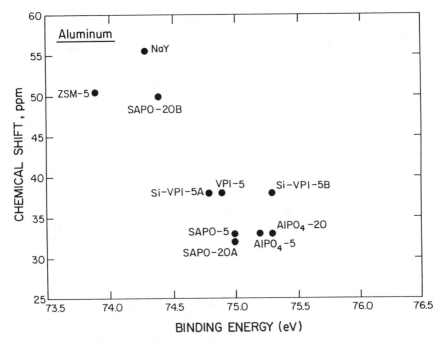

Figure 4. Comparison of Al 2p binding energies from XPS and ^{27}Al chemical shifts from NMR.

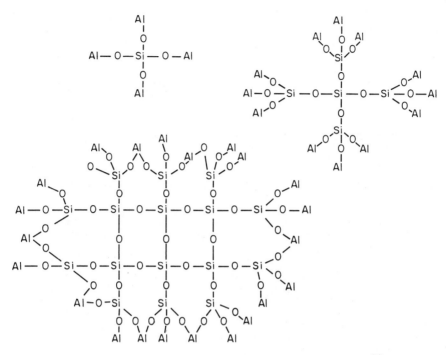

Figure 5. Silicon environments determined from ^{29}Si solid-state NMR.

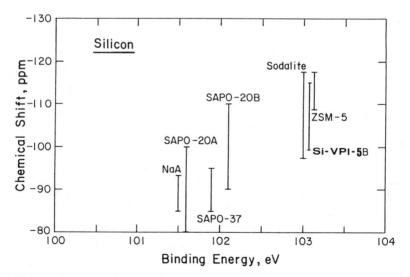

Figure 6. Comparison of Si 2p binding energies from XPS and chemical shifts from NMR. No ^{29}Si data is available for the Si-VPI-5A sample.

The trend observed in Figure 6 is an increase in binding energy with increasing magnitude of the chemical shift. More generally, this trend corresponds to an increase in binding energy as the silicon environments change from aluminum-rich to silicon-rich, which typically means with increasing Si concentration. The data for SAPO-37 and NaA represent two special cases in this respect. The SAPO-37 used in this study is the same used in previous NMR studies where it was demonstrated conclusively that silicon substitutes solely for phosphorus and hence contains only aluminum rich environments (3). The NaA sample, despite a 50% Si T-atom fraction, falls in the region of the plot associated with Al rich environments because of the ideal 1:1 ratio of Si:Al. Hence, despite the high silicon content in the NaA sample, all Si atoms have four Al second nearest neighbors and an equivalent local environment to that of isolated Si atoms in an ALPO$_4$ framework (substitution for phosphorus). The silicon binding energies in XPS therefore change in a reproducible manner with changes in the second coordination sphere in agreement with NMR observations.

The increase in silicon binding energies on changing from aluminum-rich to silicon-rich environments may be understood using a similar line of reasoning to that reported above for aluminum. As the second-nearest-neighbor Al^{3+} cations are replaced with more positively charged Si^{4+}, the charge density associated with the O^{2-} anions in the first coordination sphere will be more strongly polarized towards the second coordination shell and appear as a net decrease in charge density with respect to the isolated Si atom case. The upper limit in binding energy therefore approaches the 103.4 eV value for pure SiO$_2$ (9).

Conclusions

The correlations observed between XPS and solid-state NMR indicate that XPS binding energies are sensitive to changes in the local environments (second coordination shell) of T-atoms in silicoaluminophosphate (SAPO) materials. These results demonstrate that XPS truly samples enough of the framework near the surface to be a useful characterization tool for SAPO materials much as it has proven to be a useful tool for zeolite characterization (10-13). Having demonstrated the capability of XPS for providing framework information, the (semi)quantitative nature of XPS composition determinations is recognized by the similar bulk and surface compositions observed for homogeneous framework compositions. XPS therefore provides a means of determining superficial compositions and can be used as a check for surface enrichment of specific T-atoms. For all the samples examined in this study, the surface compositions measured by XPS were consistent with known mechanisms for silicon substitution into ALPO$_4$ materials.

Literature Cited

1. Flanigen, E. M.; Patton, R. L.; Wilson, S. T. Stud. Sur. Sci. Catal. 1988, 37, 13-28.
2. Martens, J. A.; Janssens, C.; Grobet, P. J.; Beyer, H. K.; Jacobs, P. A. Stud. Sur. Sci. Catal., 1989, 49A, 215-226.
3. Sierra de Saldarriaga, L.; Saldarriaga, C.; Davis, M. E. J. Am. Chem. Soc. 1987, 109, 2686-2691.
4. Hasha, D; Sierra de Saldarriaga, L.; Saldarriaga, C.; Hathaway, P. E.; Cox, D. F.; Davis, M. E. J. Am. Chem. Soc. 1988, 110, 2127-2135.
5. Davis, M. E.; Montes, C.; Hathaway, P. E.; Garces, J. M. Stud. Sur. Sci. Catal., 1989, 49A, 199-214.
6. Davis, M. E.; Montes, C.; Hathaway, P. E.; Arhancet, J. P.; Hasha, D. L.; Garces, J. M. J. Am. Chem. Soc. 1989, 111, 3919-3924.
7. Suib, S. L.; Winiecki, A. M.; Kostapapas, A. Langmuir 1987, 3, 483-488.
8. Martens, J. A.; Mertens, M.; Grobet, P. J.; Jacobs, P. A. Stud. Surf. Sci. Catal. 1988, 37, 97.
9. Wagner, C. D.; Riggs, W. M.; Davis, L. E.; Moulder, J. F.; Muilenberg, G.E. Handbook of X-Ray Photoelectron Spectroscopy; Perkin-Elmer Physical Electronics: Eden Prairie, MN, 1978; p 52.
10. Barr, T. L. Appl. Surf. Sci. 1983, 15, 1-35.
11. Barr, T. L.; Lishka, M. A. J. Am. Chem. Soc. 1986, 108, 3178-3186.
12. Wagner, C. D.; Passoja, D. E.; Millery, H. F.; Kinisky, T. G.; Six, H. A.; Jansen, W. T.; Taylor, J. A. J. Vac. Sci. Technol. 1982, 21, 933-944.
13. Winiecki, A. M.; Suib, S. L.; Occelli, M. L. Langmuir, 1988, 4, 512-518.

RECEIVED May 9, 1990

Chapter 5

Crystallization of Aluminophosphate VPI-5 Using Magic Angle Spinning NMR Spectroscopy

Mark E. Davis[1], Brendan D. Murray[2], and Mysore Narayana[2]

[1]**Department of Chemical Engineering, Virginia Polytechnic Institute and State University, Blacksburg, VA 24061**
[2]**Shell Development Company, Westhollow Research Center, 3333 Highway 6 South, Houston, TX 77082**

Variable temperature Magic Angle Spinning Nuclear Magnetic Resonance spectroscopy was used for the first time to study the crystallization process of the aluminophosphate, VPI-5. The results of these in-situ NMR experiments were compared to results obtained for VPI-5 synthesized by conventional methods in teflon lined autoclaves. The excellent agreement between the experiments demonstrate that in-situ Magic Angle Spinning Nuclear Magnetic Resonance spectroscopy can be used to monitor the entire crystallization process of VPI-5 and possibly other molecular sieves.

The use of Magic Angle Spinning Nuclear Magnetic Resonance, (MASNMR), spectroscopy to study molecular sieves has been widely reported (1). MASNMR spectroscopy has been used to elucidate a wealth of structural information about molecular sieves. The technique has also been used to study chemical properties, sorption and numerous chemical processes. Another area that has been exploited is the use of MASNMR spectroscopy to identify species in dilute wet gels and in solids extracted from zeolite synthesis mixtures (2,3,4,5).

Although MASNMR spectroscopy has helped identify some of the precursors to molecular sieves, the technique has not been used to monitor in situ the entire process of molecular sieve synthesis.

In this report we will describe the use of variable temperature MASNMR spectroscopy to follow the crystallization of the aluminophosphate, VPI-5. The results of the in-situ experiments are compared to those obtained for VPI-5 synthesized in autoclaves by traditional methods.

0097–6156/90/0437–0048$06.00/0

Experimental Section

The x-ray powder diffraction patterns of all the samples were
recorded on a Siemens I2 automated diffraction system using Cu K$_\alpha$
radiation. Argon adsorption isotherms were obtained at liquid argon
temperatures using an Omnisorp 100 analyzer. Argon adsorption
capacities and x-ray diffraction patterns were used to ascertain
phase purity.

Bulk chemical analyses of the solid samples (after dissolution
into HCl solutions) were performed using an inductively coupled
plasma (ICP) spectrometer. The ICP system consisted of a
Jarrell-Ash ICAP 9000 and a Jarrel-Ash Atomscan 2400.

Magic angle spinning ^{31}P NMR spectra were recorded on a Bruker
CXP 200 spectrometer. The ^{31}P spectra were taken at a frequency of
80.4 MHz and with sample spinning rates of 4-5kHz. Chemical shifts
are reported relative to 85 wt% H$_3$PO$_4$. Magic angle spinning ^{27}Al
NMR spectra were recorded on a Bruker CXP-200/400 spectrometer. The
^{27}Al spectra were taken at frequencies of 104.3 MHz and 52.2 MHz
with a rotation rate of 3-5 kHz. The ^{27}Al chemical shifts are
reported relative to Al(NO$_3$)$_3$ in aqueous solution at infinite
dilution and are not corrected for second order quadrupole effects.
Chemical shifts downfield of the standard are expressed as positive
shifts. All of the samples were sealed inside special zirconia
rotors. The temperature of the gel inside the spectrometer was
precisely controlled during all of the in-situ experiments by a
variable temperature control unit. Failure to duplicate the
conditions of the autoclave experiments was found to result in
impure products.

Pseudoboehmite alumina (Catapal-B) and 85 wt% H$_3$PO$_4$ were used
exclusively as the aluminum and phosphorus starting materials.
Aqueous (55 wt%) tetrabutylammonium hydroxide (TBA) was purchased
from Alfa.

VPI-5 Synthesis

Fifty five grams of pseudoboehmite was slurried in 150 g of water.
In a separate beaker, 100 g of water was added to 90 g of phosphoric
acid. The phosphoric acid solution was then added to the alumina
slurry to form a precursor mixture. This mixture was aged for two
hours at 25°C with no agitation. 186 g of 55 wt% TBA was added to
the mixture and the resulting gel was vigorously agitated for
approximately 2 hours. The gel composition was TBA•Al$_2$O$_3$•P$_2$O$_5$•50
H$_2$O. The reaction mixture was divided into parts and then charged
into 15 mL teflon-lined autoclaves which were statically heated to
150°C at autogenous pressure in forced convection ovens. At
specific times, the autoclaves were removed from the oven,
immediately quenched in cold water, and the pH of the contents
measured. The products were recovered by slurrying the autoclave
contents in water, decanting the supernatant liquid, filtering the
white solid, and drying the solids in air.

The dilute nature of the synthesis gel described above makes
sample spinning at very high speeds difficult inside the NMR
spectrometer. Because of this problem, it was necessary to use a
gel that contained less water. The gel used to study the synthesis

of VPI-5 inside the NMR spectrometer was prepared by first filtering
and drying the starting gel described above and then adding enough
water to obtain a gel composition of 0.24 TBA•1.3 Al_2O_3•P_2O_5•15 H_2O.
No free liquid was evident in this paste-like solid. This gel
composition has been shown previously to crystallize VPI-5 (3).

Results and Discussion

Autoclave Synthesis of VPI-5. Table I shows how the properties of
the gel change after various heating times. The TBA that is present
in the starting gel is rapidly expelled into solution upon heating
and is not found in any of solid samples collected. The pH was
found to rise sharply after 45 minutes and reaches a final pH of 7.0
after 23 hours. The solid Al/P ratio decreases to a ratio of 1.06
over the course of the reaction while the liquid Al/P ratio rises to
a value of 0.9.
    The powder x-ray diffraction pattern of VPI-5 is shown in
Figure 1 (6). The most intense reflection, (I/I_o =100%)in the
pattern appears at 5.38 degrees 2θ, (d-spacing of 16.43Å). The
powder x-ray diffraction patterns of the three samples collected
after 30, 45 and 60 minute heating times are seen in Figure 2. The
characteristic pattern of VPI-5 begins to appear between 45 and 60
minutes. The structure of VPI-5 is known to contain two
crystallographically unique aluminum and phosphorus sites. These
sites are located in the 6-membered rings (S1) (see [001])
projection in Figure 3 and in the two adjacent 4-membered rings
(S2), and are in an atomic ratio of 2 to 1, respectively. The ^{31}P
and ^{27}Al MASNMR spectrums of well synthesized VPI-5 are shown in
Figures 3 and 4. The ^{31}P spectrum of hydrated VPI-5 exhibits three
resonances with approximately equal areas. The resonance at -33 ppm
is assigned to phosphorus occupying the S2 sites (7). The resonances
at -23 ppm and -27 ppm have been assigned to phosphorus in the S1
sites. The resonance at -23 ppm disappears upon evacuation of the
sample. The spectrum of a partially dehydrated sample exhibits only
the resonances at -27 ppm and -33 ppm in a ratio of 2 to 1. The
location of water in VPI-5 at differing levels of hydration is
discussed in recent articles (8,9). The effect of occluded template
molecules has been previously ruled out since samples calcined in
air at 550°C display approximately the same ^{31}P MASNMR spectrum.
    The ^{27}Al MASNMR spectrum of VPI-5 has also been reported
earlier (7). The spectrum contains a major resonance near 37 ppm, a
neighboring resonance at 28 ppm, and a small resonance below 0 ppm.
The chemical shifts of these resonances are consistent with other
hydrated aluminophosphate molecular sieves that contain
tetrahedrally coordinated aluminum that are linked to four
phosphorus atoms through bridging oxygen atoms (10,11).
    Figure 5 shows the changes in the ^{31}P and ^{27}Al MASNMR spectrum
of gel as it is heated. The main ^{27}Al resonance in the starting gel
is centered at 7 ppm. As the reaction progresses the signal at 7
ppm decreases in intensity while a signal at approximately 40 ppm
becomes the dominant resonance after 60 minutes. The ^{31}P MASNMR
spectrum of the starting gel exhibits a resonance at -16 ppm that
shifts upfield to -23.5 ppm after 45 minutes and than splits into
two peaks at -27 ppm and -34 ppm at 60 minutes. Upon controlled

Table I. Properties of Samples Collected
As a Function of Heating Time

Heating Time (min.)	pH	Wt. of Solid/Wt. of Reaction mixture Charged	Solid (Al/P)	Liquid (Al/P)
0	4.4	0.361	1.3 (S0)	--
30	4.2	0.080	2.1 (S30)	0.2
45	4.4	0.049	2.0 (S45)	0.4
60	5.6	0.061 (0.233) *	1.7 (S60)	0.3
75	6.2	0.056 (0.212) *	1.5 (S75)	0.3
23 Hours	7.0	--	1.06 (--)	0.9

* Wt. of Oxide Recovered/Wt. of Oxide Charged

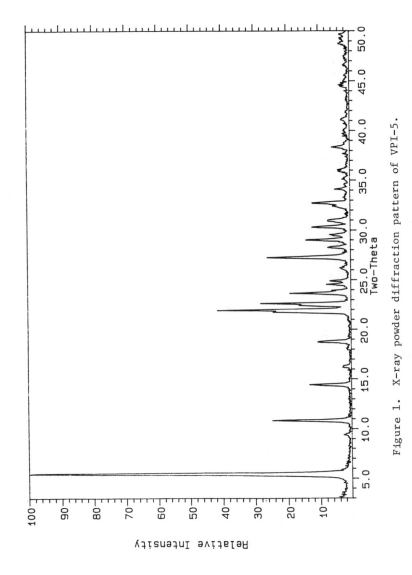

Figure 1. X-ray powder diffraction pattern of VPI-5.

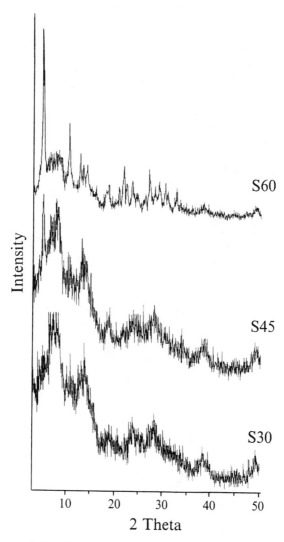

Figure 2. X-ray powder diffraction data of solid phases obtained
during the crystallization of VPI-5.

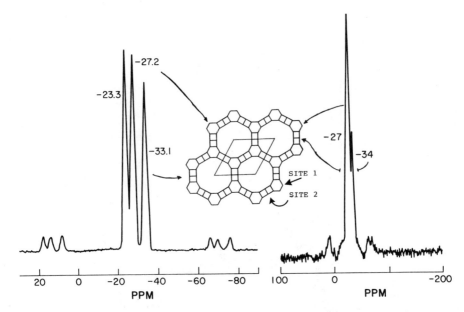

Figure 3. The ^{31}P MASNMR spectrum of hydrated VPI-5, (left) partially dehydrated, (right). [001] projection of VPI-5.

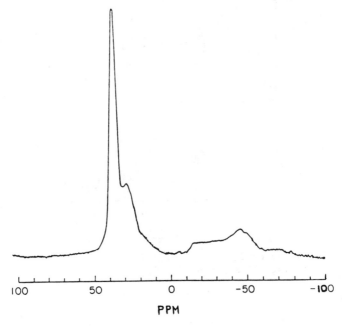

Figure 4. The ^{27}Al MASNMR spectrum of as-synthesized VPI-5.

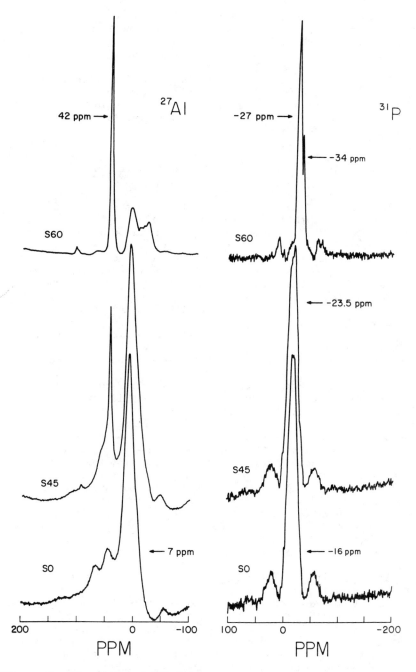

Figure 5. The ^{27}Al and ^{31}P MASNMR spectra of the solids obtained during the crystallization of VPI-5. The solids were dried prior to analysis.

hydration, three peaks are seen at -23 ppm, -27 ppm and -33 ppm as expected for VPI-5. Reaction times between 4 and 24 hours were found to produce the highest yield of product.

The argon adsorption capacities of the starting gel and samples heated for 30, 45, 60 and 75 minutes were determined. The percent crystallinity was defined as a sample's argon adsorption capacity divided by the argon adsorption capacity of pure VPI-5. Samples were evacuated overnight at 350°C prior to argon adsorption. The results are listed in Table II. It is apparent that the formation of VPI-5 occurs after 45 minutes.

Synthesis of VPI-5 Inside a NMR Spectrometer. The gel used to study the synthesis of VPI-5 inside the NMR spectrometer was obtained by adding enough water to the dried starting gel to obtain a gel composition of 0.24 TBA•1.3 Al$_2$O$_3$•P$_2$O$_5$•15 H$_2$O. No free liquid was evident with this "past-like" solid. Figure 6 shows that the ^{31}P MASNMR spectrum of this starting gel is nearly identical to that of the spectrum of the starting gel used in the autoclave studies. A very broad resonance centered at -17 ppm is present along with a narrower resonance at 1 ppm. If water is added to the starting gel the peak at -17 ppm disappears and the resonance at 1 ppm narrows as seen in Figure 7. Evacuation of the starting gel reduces the intensity of the resonance at 1 ppm.

Upon heating the starting gel at 150°C for 45 minutes the resonance at -17 ppm slowly shifts upfield to -25 ppm,(see Figure 8). The ^{31}P spectrum obtained after 60 minutes is shown in Figure 9. The characteristic three peak pattern of VPI-5 is clearly evident along with some amorphous material that is seen between 0 to -20 ppm.

The ^{27}Al MASNMR spectrum of the sample after 60 minutes is shown in Figure 10. The spectrum is essentially identical to the spectrum of the sample heated in the autoclave for 60 minutes. The large resonance near 39 ppm is due to tetrahedrally coordinated framework aluminum and the small resonance near 6 ppm is due to octahedrally coordinated aluminum that has not yet reacted (3).

After three hours the sample was removed from the spectrometer, washed and reanalyzed. The ^{31}P MASNMR spectrum of this sample is presented in Figure 11. Washing the sample removes the amorphous material that is detected between 0 and -20 ppm. The excellent agreement between the results obtained from the autoclave experiments and the in-situ MASNMR spectroscopy experiments suggests the synthesis of VPI-5 can be studied directly inside a MASNMR spectrometer.

Possible Mechanism of Crystallization

The steps involved in the crystallization of VPI-5 from the starting gel can be described by interpreting the experimental results. The ^{31}P and ^{27}Al MASNMR spectra suggests that in the starting gel, the alumina is coated with phosphoric acid. The acid slowly moves into the layers of the alumina. Upon heating, aluminum and phosphorus are expelled into the liquid phase and a near steady state solid weight is achieved after approximately 30 minutes. After 30 minutes the solid Al to P ratio begins to drop and the pH starts to rise.

Table II. Percent Crystallinity of Solids

Sample	% Crystallinity[*]
S0	0
S30	0
S45 **	0
S60	80
S75	80

[*] Samples evacuated overnight at 350°C prior to argon adsorption. Percent crystallinity defined as argon adsorption capacity divided by the adsorption capacity of pure VPI-5.

** Sample also evacuated at room temperature prior to adsorption. The reflection at 5.38 degrees two-theta is removed upon evacuation.

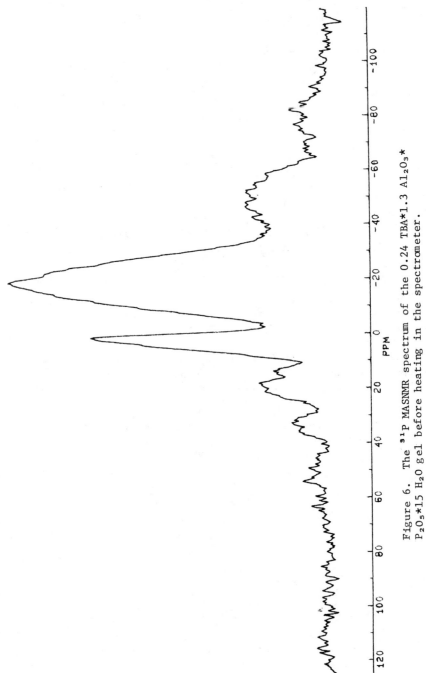

Figure 6. The ^{31}P MASNMR spectrum of the 0.24 TBA*1.3 Al_2O_3*
P_2O_5*15 H_2O gel before heating in the spectrometer.

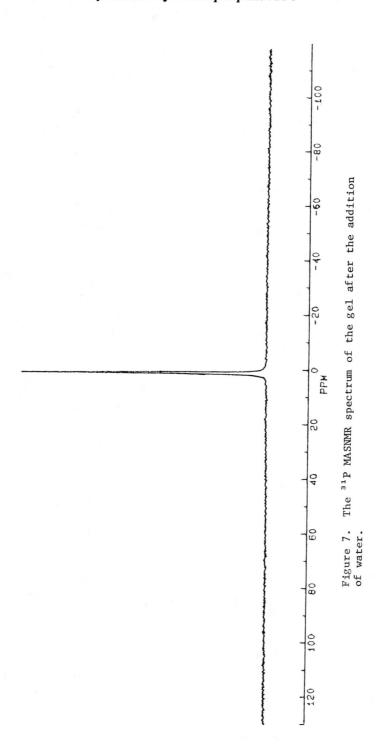

Figure 7. The ^{31}P MASNMR spectrum of the gel after the addition of water.

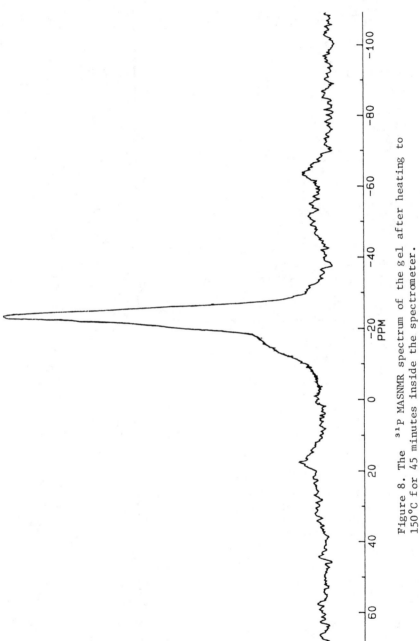

Figure 8. The ^{31}P MASNMR spectrum of the gel after heating to 150°C for 45 minutes inside the spectrometer.

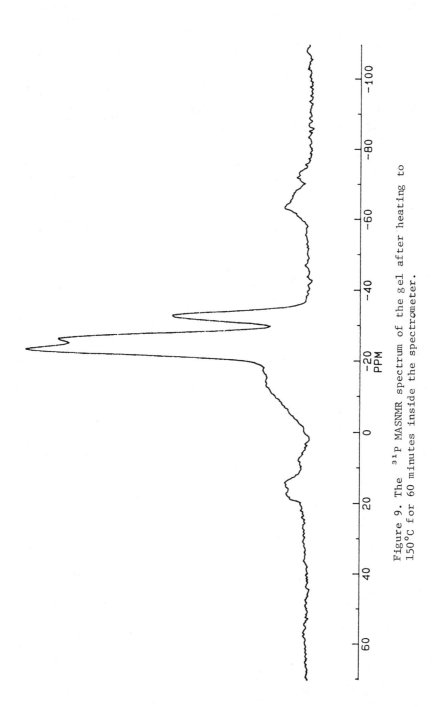

Figure 9. The ³¹P MASNMR spectrum of the gel after heating to 150°C for 60 minutes inside the spectrometer.

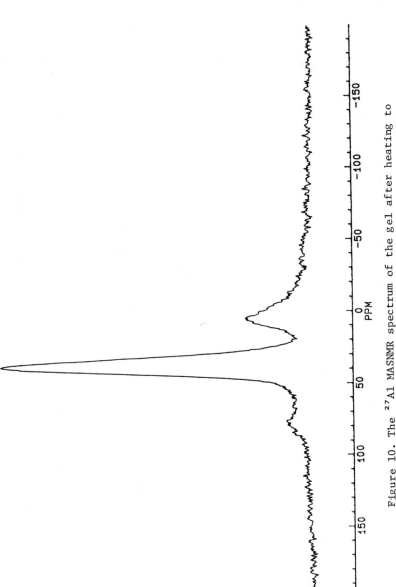

Figure 10. The ^{27}Al MASNMR spectrum of the gel after heating to 150°C for 60 minutes inside the spectrometer.

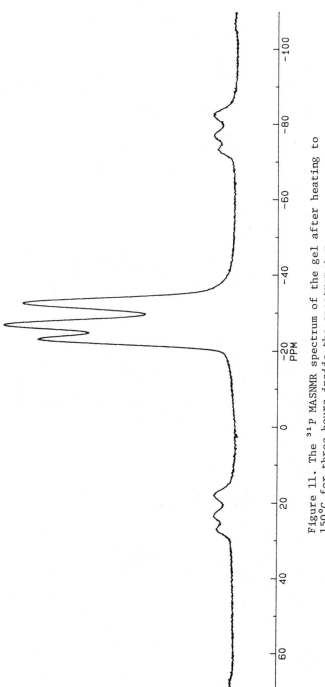

Figure 11. The ^{31}P MASNMR spectrum of the gel after heating to 150°C for three hours inside the spectrometer.

The NMR spectra suggest the presence of $O_3Al-O-PO_3$ linkages. After 45 minutes, the x-ray powder diffraction pattern displays a reflection at a 2θ of 5.38 which corresponds to a d-spacing of 16.43Å. This reflection is removed by evacuating the sample. Samples evacuated after 45 minutes maintain this reflection in their x-ray powder diffraction pattern. The argon adsorption results shown in Table II are consistent with a layered structure being formed at 45 minutes. After 45 minutes, the layers may quickly cross link to form VPI-5.

The TBA does not appear to act as a template or space filler. TBA is quickly expelled into solution and is not detected in any of the heated samples. (It is not detected during the rapid VPI-5 crystallization that occurs between 45 and 60 minutes). The possible role of TBA as a catalyst for the formation of the VPI-5 structure in concentrations below the detection limit can not be ruled out, however TBA seems to mainly serve as an effective pH moderator.

The crystallization of VPI-5 most likely does not involve aqueous phase ion transport because the crystallization time is fast and the analytical data indicate that short and long range order appear simultaneously (3). Further evidence for these speculations comes from the observation that the crystallization of VPI-5 from a paste like gel requires the same time as when a dilute gel was used. This is consistent with a solid state transformation.

Conclusions

Variable temperature Magic Angle Spinning Nuclear Magnetic Resonance spectroscopy was used for the first time to study crystallization process of the aluminophosphate, VPI-5. The products derived from the in-situ experiments were similar to those obtained for VPI-5 synthesized by conventional autoclave methods. The excellent agreement between these experiments demonstrates that in-situ Magic Angle Spinning Nuclear Magnetic Resonance spectroscopy can be used to monitor the entire crystallization process of VPI-5 and possibly other molecular sieves.

Literature Cited

1. Englehardt, G.; Michel, D. High Resolution Solid-State NMR Of Silicates And Zeolites. J. Wiley, Chichester, 1987.
2. Ginter, D. M.; Radke, C. J.; Bell, A. T. Stud. Sur. Sci. Catal., 1989, 49A, 161-168.
3. Davis, M. E.; Montes, C.; Hathaway, P. E.; Garces, J. M. Stud. Sur. Sci. Catal., 1989, 49A, 199-214.
4. McCormick, A. V.; Bell, A. T. Catal. Rev. Sci. Eng., 1989, 31, 97-127.
5. Clague, A. D. H.; Alma, N. C. M. Analytical NMR, 1989, 115-156.
6. Crowder, C.; Garces, J. M.; Davis, M. E. Advances In X-ray Analysis. 1989, 32, 503.
7. Davis, M. E.; Montes, C.; Hathaway, P. E.; Arhancet, J. P.; Hasha, D.L.; Garces, J. M. J. Am. Chem. Soc., 1989, 111, 3919-3924.

8. Wu, Y.; Chmelka, B.S.; Grobet; P.J.; Jacobs, P.A.; Davis, M.E.; Pines, A.; <u>Nature</u>, submitted for publication.

9. Davis, M.E.; Montes, C.; Hathaway, P.E.; Garces, J.M.; <u>Zeolites: Facts Figures Futures; Proceedings of the Eighth International Zeolite Conference</u>, Elsevier, 1989, 199-214.

10. Blackwell, C. S.; Patton, R. L. <u>J. Phys. Chem.</u>, 1984, **88**, 6135-6139.

11. Blackwell, C. S.; Patton, R. L. <u>J. Phys. Chem.</u>, 1988, **92**, 3965-3970.

RECEIVED May 9, 1990

Chapter 6

Redox Catalysis in Zeolites

J. A. Dumesic[1] and W. S. Millman[2]

[1]Laboratory for Surface Studies, University of Wisconsin, P.O. Box 413, Milwaukee, WI 53201
[2]Department of Chemical Engineering, University of Wisconsin, 1415 Johnson Drive, Madison, WI 53706

The active sites for redox reactions carried out using zeolitic catalysts are the exchangeable cations introduced into the zeolite matrix. These cations may occupy a number of different exchange sites, and in general, the catalytic activity of a cation depends upon its site location. This leads to the possibility of altering the catalytic activity by changing the site population of the exchangeable cations. We discuss two methods which have been used to alter the sites occupied by Fe^{2+}/Fe^{3+} cations in Y zeolite structures: increasing the silicon-to-aluminum ratio via silicon-substitution and co-exchanging Fe together with a second cation (Eu).

The effect on the catalytic activity for N_2O decomposition on increased silicon content of the zeolite is to provide nearly 2 orders of magnitude increase in the turn-over-frequency (TOF) of the Fe cations. The increased TOF is accompanied by a change in the Mössbauer spectra of the Fe^{2+} cations. These cations exhibit spectra which are indicative of cations moving from sites of high coordination (Site I) to sites of lower coordination (Sites I'/II' and/or II). In addition, the rate of oxidation of the cations decreases with increasing Si content of the lattice. When Eu and Fe are coexchanged into a Y zeolite, which hasn't had its silicon-to-aluminum ratio increased, the Fe^{2+} Mössbauer spectrum is intermediate between those of the Fe exchanged Y-zeolites having silicon-to-aluminum ratios of 4.5 and 6.2 as determined by ^{29}Si NMR. The catalytic activity for N_2O decomposition, while greater than normal Fe-Y, is less than expected from correlation with Mössbauer spectra. This observation is discussed in terms of interaction between Eu and Fe. When Eu is present alone it does not undergo redox until the temperature is increased to 873 K. When both cations are present Eu can undergo redox half reactions at 723 K. EPR spectra show that different sites may be involved when both cations are present as opposed to when Eu is present alone.

0097–6156/90/0437–0066$06.00/0

Catalytic oxidation-reduction (redox) reactions in zeolites are generally limited to reactions of molecules for which total oxidation products are desired. One important class of such reactions falls under the category of emission control catalysis. This encompasses a broad range of potential reactions and applications for zeolite catalysts. As potential catalysts one may consider the entire spectrum of zeolitic structural types combined with the broad range of base exchange cations which are known to carry out redox reactions.

Transition metals exchanged into Y-zeolite offer a basis upon which to build an understanding of the important parameters involved in designing zeolitic redox catalysts. Y-zeolite was chosen for this study because it is the most thoroughly characterized catalytic zeolite. Thus, one can address such questions as what non-framework cation sites are occupied, whether the cations move between sites, whether interactions between the cations themselves are important and how these factors relate to the kinetics of catalytic reactions.

The choice of exchange cation is limited to those which do not undergo reduction to the metal, as redox reactions using these systems are generally irreversible (1). A criterion for long term catalytic stability of exchangeable cations in zeolites is the ability to undergo repeated oxidation and reduction cycles and to reach a reproducible stable state after each cycle. This criterion provides the possibility of a catalyst not only showing good activity, but also of having a long catalytic life. Two redox couples which will be discussed here are Fe^{3+} - Fe^{2+} and Eu^{3+} - Eu^{2+}.

The reversibility of the redox cycle involving H_2 and O_2 was established for Fe by Boudart and co-workers (2) using Mössbauer spectroscopic techniques. They proposed that the oxygen was held between two Fe^{3+} cations. Fu et al. (3) showed that Fe-Y acted as a redox catalyst for reactions of CO with NO, CO with O_2, and N_2O with CO. The ability of Fe-Y to decompose N_2O into its elements was established in the work of Hall and co-workers (4), who also showed that Fe-Mordenite was as active as Fe-Y, despite containing only 16% as much Fe as its Y-zeolite counterpart. This difference in catalytic activity was thought to result from differences in the environments of the Fe within the zeolite structures. The objective of the present study was to alter the cation environment and relate that environment to the catalytic activity; this was accomplished by varying the silicon-to-aluminum ratio of Y-zeolite and by co-exchanging Fe with Eu.

EXPERIMENTAL

The starting material for preparation of the samples was Linde Na-Y (SK-40, lot 1280-133). Before ion-exchange, the samples were washed with a pH 5.0 buffer solution. The procedures used to introduce Fe^{2+}, Eu^{3+} and other ions have been described earlier (5). The iron containing samples were stored under vacuum after preparation and drying under N_2 at 400 K for 5 h followed by oxidizing in 25% O_2 in He at a final temperature of 773 K for 2 hr. Gases were purified by standard means (6).

Spectroscopic characterization of the zeolites was carried out using a variety of techniques. The equipment is listed below:

1) Austin Science Associates Model S-600 Mössbauer Spectrometer equipped with a Tracor-Northern Model N6-900 multi-channel analyzer.
2) Varian Model E-115 ESR spectrometer equipped with variable temperature accessories, covering a range from 2.4 K to 600 K.

3) A Spectrascan IV plasma spectrophotometer used for elemental analysis in determining the unit cell compositions.

4) A Phillips X-ray powder diffractometer equipped with scintillation counter and computer interface was used for crystallinity determinations.

Redox reactions and corresponding half-reactions were carried out in a Cahn Model C-2000 electrobalance operated in a flow mode to determine the oxidation state from weight changes. Catalytic activity was determined using a single pass differential flow reactor. Products were analyzed by a UTI 100-C quadrupole mass spectrometer using continuous sampling. This system is described more completely elsewhere (6).

The catalysts used in this study are described in Table 1 by their unit cell compositions.

Table 1. Unit Cell Composition of Catalysts

Sample[a]	Unit Cell Composition[b]
FeY(2.4)	$H_{5.2}Na_{22.6}Fe_{13.6}(AlO_2)_{55.0}(SiO_2)_{137}$
FeY(3.5)	$H_{4.2}Na_{13.5}Fe_{10.8}(AlO_2)_{38.3}(SiO_2)_{153.7}$
FeY(4.5)	$H_{10.0}Na_{6.6}Fe_{7.4}(AlO_2)_{31.4}(SiO_2)_{160.6}$
FeY(6.2)	$H_{2.8}Na_{5.4}Fe_{8.0}(AlO_2)_{23.8}(SiO_2)_{168.2}$
EuY	$Na_{11}Eu_{13}(AlO_2)_{50}(SiO_2)_{137}$
EuFeY	$H_{0.2}Na_{8.7}Eu_{7.7}Fe_{10}(AlO_2)_{52}(SiO_2)_{137}$

a) The value in parentheses is the Si/Al ratio from ^{29}Si NMR.
b) Cell compositions based on elemental analysis. The Si substituted samples are normalized to Si + Al = 192.

The location of the exchangeable cations within the zeolite matrix has been the subject of study by different techniques. First the use of xray diffraction has been sucessfully applied for single cations in some cases (10). Another technique which has had some recent sucess to univalent and divalent cations is Far IR(11). Mid-IR has been used to follow the change in location of cations as a function of time when exposed to different gas phase environments (9), however this method is a rather indirect method for determining the location of the cations. The technique which we will rely on here is principally Mössbauer spectroscopy. While this technique is an indirect method it provides coordination environment information and oxidation state information in the short range (next nearest neighbor). Particularly with the silicon substituted zeolites this is the only information which can be used to infer cation location because the substitution of Si for Al is random (12) and no long range order is present either within the ß-cages of a unit cell or between unit cells, thus making it difficult to extract reliable information from techniques which rely on long range ordering. Thus, Mössbauer together with ESR spectroscopy will be used there to indirectly infer changes in cation location.

RESULTS AND DISCUSSION

Mössbauer spectra of the ^{57}Fe in these samples following oxidation at 725 K are shown in Figure 1. The spectra show that as the silicon-to-aluminum ratio increases the spectra change from two doublets having isomer shifts of 0.38 and 0.33 mm/s to three doublets having isomer shifts of 0.36, 0.34 and 0.95 mm/s and associated quadrupole splittings of about 1.85, 1.08 and 0.68 mm/s. The first two doublets are associated with Fe(III) while the third having higher isomer shift is associated with Fe(II) which has not been completely oxidized, as will be seen in the spectra of the reduced samples. It is evident that the contribution of the Fe(II) species increases with the silicon-to-aluminum ratio. The spectrum of the EuFe-Y sample (Fig. 1E) is similar to that of the normal Fe-Y (Si/Al = 2.4).

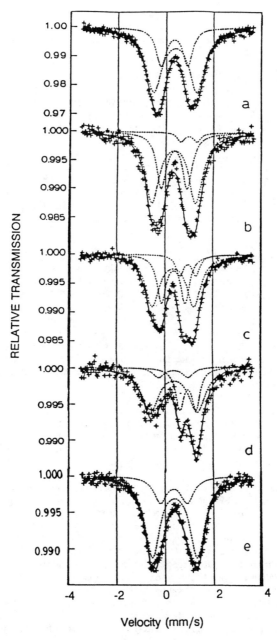

Figure 1. Mössbauer Spectra of Fe in oxidized samples. a) FeY (2.4); b) FeY (3.05); c) FeY (5.0); d) FeY (6.2); e) EuFeY. All samples oxidized in flowing O_2 at 725 K diluted 75% with He.

Hydrogen reduction of the above samples results in the Mössbauer spectra shown in Figure 2. It is apparent that as the Si/Al ratio increases the spectral contribution of the Fe with the smaller quadrupole splitting (QS) increases at the expense of the signal with the larger QS. These signals are denoted as the inner and outer doublets, respectively. As shown in Fig. 2E, incorporation of Eu together with Fe results in spectra nearly identical to that in Fig. 2C, which has a Si/Al ratio of 6.2.

Because quadrupolar interactions depend on the coordination environment of the ion, changes in Mössbauer quadrupole splittings are sensitive to the sites at which the cations are located, whereas differences in the isomer shift are sensitive to the oxidation state of the ion. For reduced samples, the ferrous cations in site I have been assigned to the outer doublet, while those in sites I', II' and/or II have been assigned to the inner doublet (2).

The changes in the Mössbauer spectra as the Si/Al ratio increases are interpreted as iron cations moving out of site I into the ß-cages and possibly the supercages. Thus, at the higher Si/Al ratios the fraction of Fe in more accessible sites is greater. This change with Si/Al ratio is also consistent with what has been observed for the exchangeable ions in going from X-zeolite to Y-zeolite. It is also consistant with calculations of the charge density at the different locations as a function of increasing Si content (13), ie., the charge density at Site I decreases with increasing Si content.

The changes in the location of iron cations as one changes the Si/Al ratio or introduces a large, high-valent cation discussed above are reflected in the ability of the zeolites to carry out redox catalysis. Figure 3 shows the turnover frequency (TOF) as a function of the silicon-to-aluminum ratio of the zeolite (as determined by ^{29}Si NMR) for the decomposition of N_2O into its elements. The EuFe-Y sample (shown as a solid square) clearly has a TOF corresponding to a sample having a Si/Al ratio of about 3, i.e., its TOF is about 10 times greater then normal Fe-Y. Yet, a comparison of the Mössbauer spectrum for the EuFe-Y sample with spectra for the silicon-substituted Fe-Y samples suggests that the activity of the former sample should have been about 100 times greater than normal Fe-Y, based on the catalytic activity results for the silicon-substituted Fe-Y series of samples. Thus, there is clearly some other factor involved in addition to location of the cations.

Eu-Y shows essentially no activity for N_2O decomposition, and ^{151}Eu Mössbauer spectroscopy indicates that the Eu does not undergo reduction in 1 atm of flowing H_2 at 700K for 5 h. However, when Fe is present, reduction of Eu does occur. Unfortunately, ^{151}Eu does not have a large quadrupolar splitting so that no information concerning the location of these cations is available from the Mössbauer spectroscopy.

Iton (7) has shown that ESR spectroscopy provides information on the location of Eu cations in Y-zeolite. The ESR spectra of Eu^{2+} in EuFe-Y are shown in Figure 4A, following reduction in CO for 50h at 770 K. The spectra are characterized by three effective g-values at 6.0, 4.9, and 3.0. The first two resonances are associated with Eu at sites I', II' and II (7). Treatment in a 3:1 mixture of CO and O_2 leads to a significant, and rapid decrease in the intensity of the resonance at $g_{eff} = 4.9$, indicating a rapid conversion to Eu^{3+}. This is in agreement with the changes in EPR spectra observed upon treatments of reduced samples with O_2. Of the sites Iton assigned to this resonance, site II is the most accessible; therefore, we interpret the resonance at 4.9 to Eu^{2+} cations at this site. Unlike site II', Eu cations at site II do not have the possibility of interacting with another ion at site I'. This leaves the resonance at 6.0 to be associated with sites I' and II'. Recalling the Mössbauer spectra of EuFe-Y, Fe was also shown to be located at Sites I, and II, and thus the possibility exists for interactions between the two different cations.

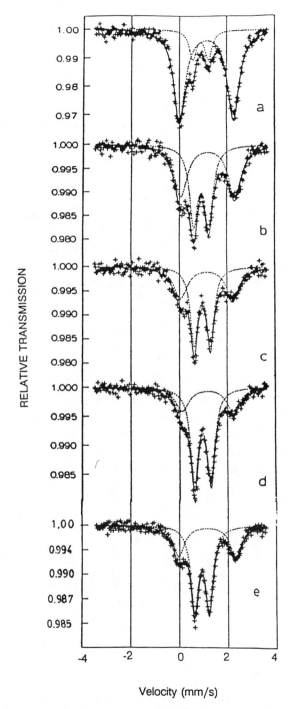

Figure 2. Mössbauer Spectra of reduced samples. a) FeY (2.4); b) FeY (3.5); c) FeY (4.5); d) FeY (6.2); e) EuFeY. All samples reduced in flowing H_2 at 700K diluted 75% in He.

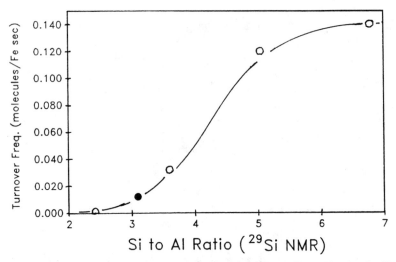

Figure 3. Turnover frequency for Fe containing samples of different Si to Al ratios for N_2O decomposition (o). The EuFeY data (●) represents where the turnover frequency (all transition metal cations) of N_2O decomposition fall on the line for samples of varying Si to Al ratios.

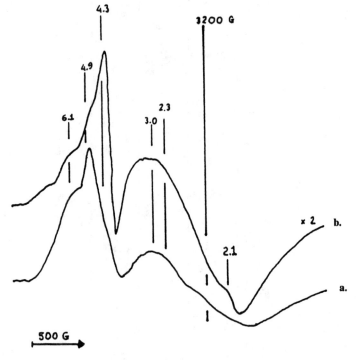

Figure 4. X band ESR spectra of EuFeY. a) spectrum following reduction at 770K for 50 in 1 atm of CO b) spectrum of a following 1 minute treatment in .8 atm of a 3:1 mixture of CO to O_2.

Comparing the rates of oxidation and reduction between the high-silica Fe-Y and EuFe-Y, one finds that the Fe-Y is difficult to oxidize while the EuFe-Y is difficult to reduce. This is illustrated in Figure 5, where the EuFe-Y sample is in a completely oxidized state with a $CO:O_2$ gas ratio of 10:1 at 773 K, while Fe-Y is not completely oxidized in pure O_2 at 773 K (6). Thus, considering that the decomposition of N_2O is limited by the rate of reduction of the metal cation (4), the EuFe-Y sample would be expected to have a lower TOF than the Fe cation distribution would predict. This is likely due to the influence of Eu-O-Fe pairs in sites I' and II' which are more difficult to reduce than Fe-O-Fe pairs. One could further predict that the EuFe-Y sample would be more active than the Fe-Y with high Si/Al ratio for reactions which are limited by rate of oxidation of the cations.

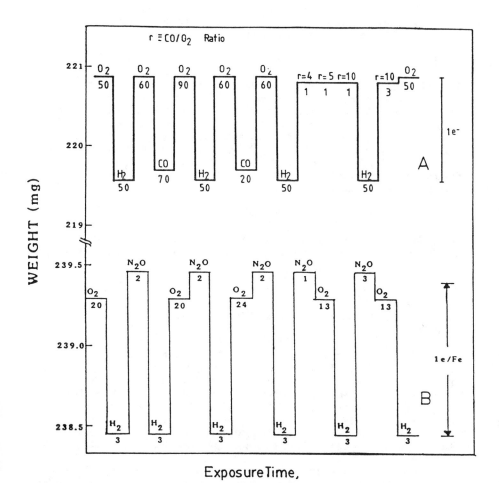

Exposure Time,

Figure 5. Oxidation-reduction cycles for EuFeY (A) and FeY (6.2) (B). The additional weight in N_2O represent the added ease with which the two electron process occurs relative to the four electron process in O_2. The data indicate the treatment gas on top of the horizontal lines and the time of treatment in hrs below the line.

An explanation of the difficulty of oxidizing the high-silica Fe-Y samples can be related to the zeolite structure and the nature of the oxidizing gas. Although the Fe in the high-silica Y-zeolite samples is not completely oxidized in molecular oxygen, it is converted to Fe^{3+} when the oxidizing agent is N_2O. One must consider that in the first case four electrons are involved in the reaction, while in the second case only 2 electrons are involved. An oxidation process which requires 4 electrons needs 4 cations, if each cation can undergo a one electron change in oxidation state. If we take for example, a pair of cations at site I' and II', then we have a requirement of two sets of these sites for reaction with O_2. However, as the Si/Al ratio is increased, the probability of having two sets of these sites in close proximity decreases. Accordingly, one could still accommodate a 2-electron process such as reaction with N_2O, but a 4-electron process becomes more difficult to accommodate. Because these reactions occur at elevated temperatures one could expect cations to migrate, and in fact complete oxidation in O_2 can be accomplished with Fe-Y at 850 K for 80h. However, the rate at which the last 10 % of the iron is oxidized is very slow, occurring over about 75 h.

CONCLUSION

Altervalent metal cations exchanged into zeolites can function as oxidation-reduction centers. Importantly, the redox properties of these cations can be controlled by altering the structure of the zeolite, the Si/Al ratio of the zeolite and by co-exchanging the zeolite with a second metal cation. The factors responsible for controlling the redox properties of the exchange cations are (i) the site-locations of the cations in the zeolite, (ii) the formation of cation-oxygen-cation bridges in oxidized samples, and (iii) the separation between exchange cations. The present paper has documented the above phenomena for Fe-Y and EuFe-Y with respect to N_2O decomposition. More generally, we suggest that zeolites offer the possibility of formulating redox catalysts with controlled catalytic properties.

ACKNOWLEDGMENTS

We acknowledge support from the National Science Foundation for this work. We also acknowledge the critical participation of the following colleagues in this work: Luis Aparicio and Maria Ulla.

REFERENCES

1. Jacobs, P.A., Utterhoeven, J.B. and Beyer, H.K., *J. Chem. Soc., Faraday Trans. 1*, **73**, 1755 (1977)
2. a) Garten, R.L., Delgass, W.N., and Boudart, M., *J. Catal.*, **18**, 90 (1970)
   b) Delgass, W.N., Garten, R.L., and Boudart, M., *J. Phys. Chem.*, **73** 2970 (1969)
3. Fu, C.M., Deeba, M., and Hall, W.K., *Ind. Eng. Chem., Prod. Res. Dev.*, **19**, 229 (1980)
4. Leglise, J., Petunchi, J.O., and Hall, W.K., *J. Catal.*, **86**, 392 (1984)
5. Aparicio, L.M., Ulla, M.A., Millman, W.S., and Dumesic, J.A., *J. Catal.*, **110**, 330 (1988)
6. Aparicio, L.M., Dumesic. J.A., Fang, S-M., Long, M.A., Ulla, M.A., Millman, W.S. and Hall, W.K., *J. Catal.*, **104**, 381 (1987)
7. Iton, L.E., and Turkevich, J., *J. Phys. Chem.*, **81**, 435 (1977)
8. Ulla, M.A., Aparicio, L.M., Balse, V.R., Dumesic, J.A., and Millman, W.S., *J. Catal.*, **123**, 195 (1990)
9. Ulla, M.A., Aparicio, L.M., Dumesic, J.A., and Millman, W.S., *J. Catal.*, **117**, 237 (1989)
10. Pearce, J.R., Mortier, W.J., Uytterhoeven, J.B. and Lundsford, J.H., *J. Chem. Soc., Faraday Trans. I*, **77**, 937 (1981)
11. Ozin, G.A., Baker, M.D., Godbar, J. and Shihua, W., *J. Amer. Chem. Soc.*, **107**, 1195 (1985); Baker, M.D., Godbar, J. and Ozin, G.A., *ibid*, **107**, 3033 (1985)
12. Klinowski, J., Ramdas, S., and Thomas, J.M., *J. Chem. Soc., Faraday Trans. II*, **78**, 1025 (1982)
13. Dempsey, E., *J. Phys. Chem.*, **73**, 3660 (1969)

RECEIVED May 9, 1990

Chapter 7

Propylene Conversion over Silicoaluminophosphate Molecular Sieves

Robert T. Thomson[1], Norazmi Mat Noh, and Eduardo E. Wolf

Department of Chemical Engineering, University of Notre Dame, Notre Dame, IN 46556

The catalytic activity of silicoaluminophosphate molecular sieves (SAPO-5, SAPO-11, SAPO-34) has been studied during propylene conversion. During the reaction, SAPO-34 and SAPO-5 yielded C_2-C_7 hydrocarbons but both catalysts deactivated severely during reaction. The initial activity of SAPO-34 which contained sites of stronger acidity was higher than SAPO-5. SAPO-11, showing lower activity than SAPO-5 and SAPO-34 as well as rapid deactivation, yielded only C_6 hydrocarbons. Differences in the product distribution observed during both reaction studies arise from the different acidity, pore structure and pore size of the SAPO molecular sieves.

Syngas conversion to methanol has been shown to take place on supported palladium catalyst [1]. Methanol can in turn be converted to gasoline over ZSM-5 via the MTG process developed by Mobil [2]. In recent work we have reported syngas (CO/H_2) conversion to hydrocarbon products on bifunctional catalysts consisting of a methanol synthesis function, Pd, supported on ZSM-5 zeolites [3]. Work on syngas conversion to hydrocarbon products on Pd/SAPO molecular sieves has been published elsewhere [Thomson et. al., J. Catal., in press].Therefore, this paper will concentrate on propylene conversion.

Conversion of low molecular weight olefins to longer chain aliphatic products, a reaction of industrial importance, has traditionally been performed over amorphous solid acid catalysts, such as silica-alumina and silica treated with phosphoric acid [4]. Recently, shape-selective molecular sieves, most notably the medium-pore zeolite ZSM-5 [5-8], have been investigated as catalysts for the conversion of light olefins to heavier species. At high temperatures or long contact periods, gasoline or diesel range fuel can be selectively formed from a mixture of C_3 and C_4 olefins, with the product distribution being sensitive to both temperature and

[1]Current address: Mobil Research and Development Corporation, Paulsboro Research Laboratory, Paulsboro, NJ 08066

olefin partial pressure [5]. Over ZSM-5, Tabak et. al.[5] reported that equilibrium limitations on the product distribution were found to be dominant above 625 K. Researchers at Union Carbide have examined olefin oligomerization over SAPO molecular sieves, with several of these catalysts displaying activity for the synthesis of higher olefins from propylene [9]. Narrow-pore SAPO-34 was quite active in the synthesis of light olefins, but wide-pore SAPO-5 deactivated quickly. SAPO-11 and SAPO-31, both having medium-sized pores, selectively converted propylene to liquid aliphatic hydrocarbons, thus providing an attractive route from light olefins to gasoline-range products [10].

The reaction pathway for propylene conversion is believed to consist of three acid-catalyzed steps [5]; (1) oligomerization of C_3H_6 to hexenes, nonenes, etc., (2) cracking of long chain olefins to smaller olefins, and (3) copolymerization of propylene with product olefins. The initial step in this route accounts for the high yields of C_6, C_9, C_{12} observed by Tabak [5] and Wilshier et. al. [8] at low reaction temperatures (ca. 450 K). Successive cracking (step 2) and repolymerization (step 3) within the molecular sieve pores leads to the equilibration of the olefin mixture [5,6]. High temperatures and long contact times, by increasing the number of reactions, would therefore be expected to lead to a product distribution closer to equilibrium. The properties of the acid sites, as expected, play an important role in determining final products. The extent to which the original oligomerization reactions occurs and the rate at which long chain olefins are cracked will ultimately determine if the product distribution approaches equilibrium predictions. Deviations from equilibrium observed in zeolite-catalyzed systems reflect the shape-selective properties of the catalyst.

Propylene conversion over three SAPO molecular sieves (SAPO-5, SAPO-11, and SAPO-34) was conducted at a variety of operating conditions. Catalyst behavior was correlated with the physical and chemical properties of the SAPO molecular sieves. The objective of this work was to determine the relative importance of kinetic and thermodynamic factors on the conversion of propylene and the distribution of products. The rate of olefin cracking compared to the rate of olefin polymerization will be addressed to account for the observed trends in the product yields. The processes responsible for deactivation will also be addressed.

Experimental Methods.

Catalysts Preparation. The silicoaluminophosphate (SAPO) molecular sieves employed in this study were synthesized in the laboratory of Professor Mark Davis in the Department of Chemical Engineering of the Virginia Polytechnic Institute, following the methods reported in U.S. Patent 4,440,871. The three different samples, distinguished by their microscopic structure, were the wide-pore SAPO-5, medium-pore SAPO-11, and the narrow-pore SAPO-34. Verification of their microscopic structure (through x-ray diffraction) and micropore diameters (by argon adsorption measurements) was performed at VPI. The SAPO molecular sieves were provided in the ammonium cation form. *Ex situ* calcination at 873 K for one hour in oxygen was performed on the SAPO samples prior to their use as catalysts for the propylene conversion.

Catalysts Characterization. Following pretreatment of the SAPO molecular sieves, the catalysts were characterized by temperature programmed desorption (TPD) of ammonia and infrared spectroscopy. To assess the acidity of the samples, the desorption of ammonia from the catalysts was performed in a manner similar to that described by van Hooff et. al. [11]. For the   ammonia TPD experiments, typically 0.1 gram of the molecular sieve sample was supported on quartz wool inside a 9 mm O.D. quartz reactor equipped with axial thermowell which contacted the top of the sample bed. The reactor was heated by a 300 watt heating jacket (Briskheat), and temperature was monitored by a Eurotherm 211 programmable temperature controller. Analysis of the reactor effluent stream was performed using a thermal conductivity cell detector.

The samples were initially outgassed in 30 cc/minute of helium flow (Linde Ultra High Purity) at 323 K for 30 minutes. Following outgassing, the temperature was kept constant as 60 cc/minute of nitrogen (Linde Prepurified) was introduced as carrier gas for ammonia. Shortly thereafter, ammonia was introduced into the nitrogen stream and, hence, into the reactor. After 15 minutes of this flow pattern, the ammonia inlet valves were closed and nitrogen was passed over the sample for ten minutes. Thereafter, the carrier gas was switched to helium for use with the TC cell. Helium outgassing (30 cc/minute) continued as the sample was held at 323 K for one hour, then heated at a rate of 5 K/minute to 373 K. The reactor temperature was held at this higher level for one hour, then cooled to 323 K. Following outgassing, helium was then allowed to flow through the TC cell for ammonia detection.

During ammonia TPD the sample was heated from 323 K to 823 K at 10 K/minute, followed by a ten minute hold at this upper temperature, after which time the setpoint was dropped to 323 K. Evaluation of the sample was repeated using the same procedure until reproducible results were achieved, which required three to four cycles.

Characterization of the acid sites on molecular sieve catalysts was also performed by infrared spectroscopy. FTIR analysis of the molecular sieve catalyst in the form of wafers placed in a glass cell was performed *in vacuo*. The samples were prepared by compressing 30 mg of catalyst into 1.9 cm diameter wafers. After placing the sample in the holder and evacuating the cell down to approximately 10^{-2} torr, the wafer was moved to the furnace section of the cell, heated at a rate of 6 K/minute to 650 K, and maintained at this upper temperature for 30 minutes. Following pretreatment, the sample was moved to the path of IR beam and allowed to cool to 300 K. After collection of reference spectra under vacuum, the cell was filled with 760 torr of ammonia (Matheson Anhydrous Grade). After five minutes of exposure to ammonia, the sample was again evacuated to 10^{-2} torr. The spectra collected after NH_3 adsorption, which was found to be stable within ten minutes, was compared to that obtained immediately following pretreatment. The nature of the acid sites on the molecular sieve was determined by examining the features in the 3400 cm^{-1} to 3900 cm^{-1} range.

Supported palladium catalysts for syngas conversion were characterized by CO chemisorption, ammonia TPD and SEM. Compositional characterization using XRD, MAS NMR and argon adsorption were done at VPI.

Activity Studies. Propylene conversion over the specified catalysts was performed in a 9 mm o.d. quartz tubular reactor. The amount of catalyst charged into the reactor was 0.1 gram of SAPO-5, while only 0.05 gram of SAPO-11 and SAPO-34 was used to ensure constant contact time. The catalyst powders in the form of fine particles (100-200 µm) were supported on quartz wool in the reactor, equipped with an axial thermowell which contacted the top of the catalyst bed. Reaction temperature was monitored by an Omega Model 4002 temperature controller and gas flows regulated by Brooks mass flow controllers.

All product analysis of effluent gas streams was performed by on-line gas chromatography. Two different gas chromatographs were employed, each with heated sample valve connected to the reactor effluent stream. Analysis of light (C_1-C_6) hydrocarbons and dimethyl ether was performed by a Varian 1400 GC equipped with flame ionization detector. Separation of the products was accomplished by a 20' column (1/8" O.D.) packed with Porapak Q. Analysis of hydrocarbons in the C_5 to

C_{12} range was accomplished by use of a Supelco SPB-1 capillary column, which is 60 meters long with an inside diameter of 0.75 mm. In the course of sample analysis, the column temperature was held at 308 K for 12 minutes following sample injection, then increased at 4 K/minute to 398 K, where the temperature was held constant for 10 minutes. This heating pattern provided good separation characteristics, allowed products up to C_{12} to elute before cooling, and permitted capillary column analysis of the reactor effluent streams every 60 minutes.

After outgassing for thirty minutes with nitrogen (Linde Prepurified) at ambient temperature, all catalyst samples were pretreated at 650 K in nitrogen for one hour. Following pretreatment, the samples were cooled to reaction temperature, then exposed to a stream of propylene (Matheson C.P. Grade) diluted in nitrogen. Reaction temperatures ranged from 550 K to 650 K, and C_3H_6 inlet partial pressures of 16.2 kPa to 35.8 kPa were used. Diluted streams of 1-butene (Matheson C.P. Grade) were also used. Total inlet flow rate was held constant at 100 cc/minute, and all experiments were performed at atmospheric pressure.

Results.

Catalyst Characterization. The three SAPO molecular sieves employed in this study represents the three pore sizes of molecular sieves, ranging from 0.4 nm to 0.8 nm. While SAPO-5 and SAPO-11 have unidimensional pores, SAPO-34 has a multidimensional pore system with supercages. The chemical composition and total ammonia uptake of the three SAPO molecular sieves are listed in Table I.

Table I. Framework Composition and Quantitative Analysis of Ammonia Desorption
Experiments for SAPO Molecular Sieves

Desorbed Catalyst	Composition	Total NH_3 (moles/gcat) $\times 10^4$
SAPO-5	$(Si_{.05}Al_{.5}P_{.45})O_2$	6.1
SAPO-11	$(Si_{.06}Al_{.49}P_{.44})O_2$	5.6
SAPO-34	$(Si_{.07}Al_{.52}P_{.41})O_2$	15.7

The patterns observed during ammonia desorption from SAPO-5 (Figure 1a) and SAPO-11 (Figure 1b) exhibited similar qualitative and quantitative acid character, while SAPO-34 showed peaks at both low and high temperatures during ammonia desorption (Figure 1c). Table I lists the total amount of ammonia desorbed (total area beneath curves given in figure 1) from each sample .

Characterization of the catalyst acid sites by infrared spectroscopy correlated well with the results of ammonia desorption experiments. The transmission spectra of HZSM-5 (Figure 2a), and SAPO-34 (Figure 2b) following in vacuo pretreatment at 650 K all showed absorbance bands near 3610 cm^{-1}. Since these two molecular sieves were the only samples to show high-temperature NH_3 desorption peaks (Figure 1), the acid sites which generate the 3610 cm^{-1} band are relatively strong. This conclusion for HZSM-5 has also been reached by other investigators [13]. The amount of infrared radiation which passed through the SAPO-5 and SAPO-11 wafers was not sufficient to permit evaluation of the acid sites by this method.

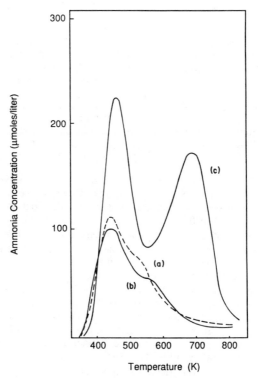

Figure 1. Patterns of Ammonia Desorption During Programmed Heating of Molecular Sieve Samples (T= 50 C - 550 C at 10 C/min) a) SAPO-5 b) SAPO-11 c) SAPO-34

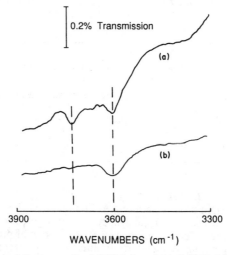

Figure 2. Infrared Spectra of a) ZSM-5 and b) SAPO-34 catalysts Wafers Following *in vacuo* Pretreatment.

<u>Activity Studies.</u>

Propylene Conversion. Catalyst activity studies for the three SAPO samples were initially conducted at a temperature at 550 K and propylene partial pressure of 16.2 KPa. Since the observed activity of the SAPO catalysts was low (conversion never exceeding 15%, and most often below 3%), rates of reaction are reported assuming differential reactor behavior. Deactivation occurred in all cases, with the loss in activity of SAPO-5 and SAPO-34 being particularly rapid (Figure 3). The severity of SAPO-34 deactivation at 550 K made estimation of its initial activity (the rate of propylene converted at time-on-stream= 0 hours) difficult. However, it is clear that the narrow-pore SAPO-34 displayed the highest initial catalyst activity, with the medium-pore SAPO-11 and the wide-pore SAPO-5 having almost equal initial activities.

At the initial sampling point (time-on-stream= 1/2 hour for SAPO-5 and SAPO-34, and time-on-stream= 1 hour for SAPO-11), propylene conversion over the three SAPO molecular sieves yielded different product distributions (Figure 4). Both SAPO-5 and SAPO-34 produced C_2-C_7 olefins, with the latter catalyst also synthesizing a significant amount of hydrocarbons in the C_8 range. The wide-pore molecular sieve showed the highest yield of hexenes, while the narrow-pore SAPO-34 produced primarily linear butenes. In contrast, reaction over the medium-pore SAPO-11 led to only hexenes, with no evidence of the C_9 trimer or other products. The isomeric distribution among the C_6 olefins from the three SAPO samples also depended on the choice of catalyst (Table II). The identification of the C6 isomers were approximated from the boiling point data. Analysis of the products from SAPO-34 showed that primarily straight-chained hexene isomers were formed by the narrow-pore molecular sieve, with only three distinct peaks present in the trace from the capillary column. Both SAPO-5 and SAPO-11 yielded a broader range of C_6 olefins, with the distribution from the medium-pore catalyst showing many similarities to the distribution found from the medium pore ZSM-5 [17]. SAPO-5 appeared to have greater selectivity for highly branched C6 olefins than SAPO-11.

Table II. Chromatographic Analysis of Major Hexene Products From Propylene
Conversion Over SAPO Catalysts
(Propylene Inlet Pressure= 16.2 KPa)

Reaction		Possible Isomers (tentative assignments)					
Catalyst	Temp. (K)	methyl pentenes	1-hexene	trans-2 hexene	cis-2 hexene	3-hexene	other hexenes
SAPO-5	550	0.87	0.43	1.0	0.13	0.72	0.36
SAPO-11	550	0.28	0.43	1.0	0.22	0.74	0.06
SAPO-34	550	0.0	0.11	1.0	0.25	0.0	0.0

The activity of the SAPO catalysts was also studied at different operating temperatures. The rapid deactivation of SAPO-5 and SAPO-34 required the use of fresh catalyst samples and limited catalyst evaluation to just two temperatures at 550 K and 650 K. The rate of deactivation, as well as the rate of propylene conversion, depended on the operating temperature (Figure 5). In contrast to reaction over the ZSM-5 catalysts [17], propylene conversion increased at higher temperature over

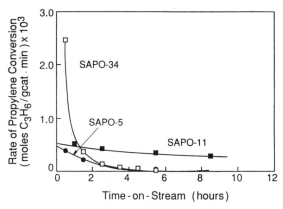

Figure 3. Comparison of Deactivation Trends over SAPO Molecular Sieves (Propylene Inlet pressure= 16.2 kPa, Temp.= 550 K)

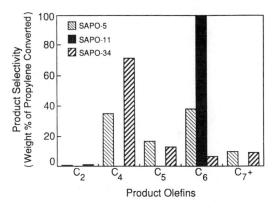

Figure 4. Distribution of Non-Propylene Olefins Products Over SAPO Catalysts (Propylene Inlet Pressure= 16.2 kPa, Temp.=550 K, time= 30 min)

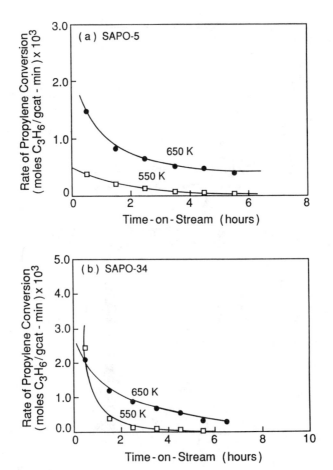

Figure 5. SAPO Activity and Deactivation Trends During Propylene Conversion at Various Temperatures   (a) SAPO-5 (b) SAPO-34

SAPO-5, with the estimated initial activity being five-fold higher at 650 K than at 550 K (Figure 5a). The narrow-pore SAPO-34 displayed very different trends. Estimated activity at time-on-stream= 0 hours was much higher at the lower temperature, but the catalyst deactivated at a slower rate at the higher temperature (Figure 5b). The loss of SAPO-34 activity was so rapid that less than 10% of its estimated initial activity at 550 K remained after just ninety minutes of elapsed time.   The product distribution from the wide-pore and narrow-pore SAPO's at 650 K changed as the propylene conversion decreased. At the lower temperature (550 K), the relative yields of olefins other than propylene remained fairly constant during the entire exposure period over both SAPO-5 and SAPO-34. At higher temperature, the yield of hexenes from SAPO-5 increased with decreasing propylene conversion, going from 38% at time-on-stream of 1/2 hour to 67% after 7.5 hours (Figure 6a). The narrow-pore SAPO-34 showed increased preference of ethylene and hexene products at greater exposure times, with these olefin products accounting for nearly 40% of the converted propylene after 7.5 hours (Figure 6b). The color of the catalysts, both white before before reaction, changed after propylene conversion , with SAPO-5 ending up black and SAPO-34 becoming greenish-blue.

The rate of SAPO-11 deactivation was sufficiently slow to permit the use of a single sample for evaluation at various temperatures. Over SAPO-11, *only* hexene products were synthesized at all conditions employed in this study. Increased temperature led to decreased propylene conversion (Figure 7). Use of n-butene feed over SAPO-11 resulted in the selective formation of C_8 range products. Analysis of the reaction products indicated that despite the exclusive formation of dimers from propylene or butene feeds, the effluent stream contained a large number of isomers. The variety of isomers resulting from the dimerization reaction suggests that skeletal rearrangement of olefins is possible without olefin cracking occurring on the SAPO-11 acid sites. (*See* Figure 8.)

Discussion.

Olefin oligomerization were found to occur on SAPO molecular sieves, though their activity was far less than the of zeolite ZSM-5[17]. While showing very different initial activity , the wide-pore SAPO-5 and the narrow pore SAPO-34 both deactivated severely (Figure 3). Both of these catalysts yielded a wide spectrum of products presumably following the pathway described by Tabak et. al. [5], in which numerous olefin polymerization and scission reactions take place. Strangely, medium pore SAPO-11 showed complete selectivity for olefin dimers

The relative rates of olefin combination and scission influence the severity of SAPO-34 deactivation. Formation of large hydrocarbon molecules in the supercages of this narrow-pore molecular sieve is the most likely reason for its loss in activity with time on stream (Figure 5b). The rate of "coke" formation within the SAPO-34 framework therefore depends on the rate at which large olefins are cracked and isomerized to products which can diffuse out of the micro pores (i.e. linear olefins). The rate of SAPO-34 deactivation at 650 K is slower than at 550 K, which may be due to the faster cracking of large molecules, more rapid isomerization, and faster diffusion through the 4.3 Å pores at the higher temperature.

In addition to the rates of olefin reactions, mass transfer also plays an important role in determining the extent of propylene conversion and the product distribution from SAPO molecular sieves. Restrictions on molecular movement may be severe in the SAPO catalysts, due to pore diameters (4.3 Å for SAPO-34) and structure (one-dimensional pores in SAPO-5 and SAPO-11). The deactivation of SAPO-5 and SAPO-11 catalysts may be more directly related to mass transfer than the coking of SAPO-34. Synthesis of large or highly-branched products, having low diffusivities, inside the pores of SAPO-5 or SAPO-11 essentially block internal acid

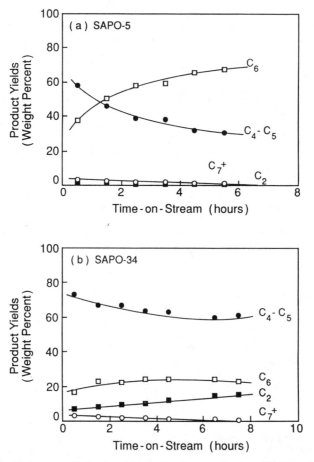

Figure 6. Influence of Catalysts Deactivation on Product Distribution over SAPO Catalysts (Propylene Inlet Pressure- 16.2 kPA, Temp.= 650 K) (a) SAPO-5 (b) SAPO-34

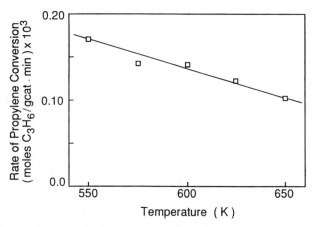

Figure 7.   Temperature Dependence of Propylene   Conversion Rate over SAPO-11

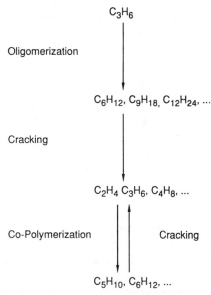

Figure 8.   Route of Acid Catalyzed Conversion of Propylene [6]

sites. These "coke" molecules limit the entry of reactants and the exit of smaller products from the molecular sieve channels. The increase in activity of SAPO-5 at higher temperature (Figure 5a) may result from higher diffusivities of light olefins as well as faster movement of coke molecules out of the micropores. This temperature effect on coking rates is similar to that reported by Rollman [14]. It is interesting to note that the activation energy of intrazeolite diffusion tends to be higher for bulky molecules [15], implying that higher temperatures will have a greater effect on the movement of coke precursors.

The effect which deactivation has on the product distribution from SAPO-5 and SAPO-34 can also be traced to diffusion considerations. The yield of hexenes at 650 K from SAPO-5 (Figure 5a) and SAPO-34 (Figure 5b) increased as the C_3H_6 conversion levels decreased, suggesting that the probability of the C_3H_6 dimer undergoing further reaction decreases at greater deactivation levels. This trend may result from the blockage of internal acid sites, increasing the fraction of C_6H_{12} formed near the ends of the pores or on the external SAPO surface. Formation of the dimer closer to the pore mouths decreases the likelihood of further oligomerization or cracking reactions, thus increasing the hexene yield while decreasing the final butene and pentene fractions.

The preferential formation of hexenes over SAPO-11 may be consequence of the diffusion patterns within the unidimensional micro pores. Reactants must diffuse into the channel "next to one another" for a bimolecular reaction to take place. Formation of C_9 olefins, which would yield C_4 and C_5 products after cracking, requires either (1) diffusion of three adjacent propylene molecules to an active site, or (2) formation of hexene, followed by diffusion to the end of the micropore and diffusion of adjacent hexene and propylene molecules back towards an acid site. The necessity of hexene transfer to the end of the pore via the second pathway increases the likelihood of C_6H_{12} desorption to the gas phase before further reaction. The selective formation of hexenes over medium-pore SAPO-11, which has a one-dimensional micropore system similar to the wide-pore SAPO-5, may be a consequence of steric restrictions, as will be discussed later.

While the rates of reactions and mass transfer exert a large influence on the catalytic conversion of propylene, the chemical and physical characteristics of the SAPO molecular sieves must also be addressed. SAPO-34, having the greatest initial activity among the SAPO catalysts (Figure 3) possesses strong acid sites (Figures 1c and 2b). In both forms of characterization, the strength of the acid sites on SAPO-34 appear to be equivalent to those on HZSM-5. Therefore, on ZSM-5 and SAPO catalysts, acid sites of greater strength lead to higher amounts of olefin products.

The acidity and micropore diameter of SAPO-11 may account for the absence of non-dimer products from propylene conversion over this catalyst. Comparison of the NH_3 desorption patterns from the wide-pore SAPO-5 (Figure 1a) and medium-pore SAPO-11 (Figure 1b) reveals nearly identical acidic features. However, lack of products other than hexenes from SAPO-11 indicates that, unlike SAPO-5, the medium-pore molecular sieve is not capable of further oligomerization and cracking to produce C_4 and C_5 products. The absence of nonenes in the reactor effluent may result from low intraphase diffusivity, which is expected with larger molecules [16]. If propylene trimers exist within the micropores, the absence of butenes and pentenes implies that SAPO-11 acid sites are unable to crack olefins. It has been speculated that the behavior of SAPO-5 and SAPO-11 for oxygenate conversion is related to the formation of tertiary carbon intermediates [17]. The importance of tertiary carbonium ions in the cracking of aliphatic hydrocarbons has been discussed by Pines [13]. Since SAPO-5 and SAPO-11 have only moderately acidic sites, formation of tertiary carbonium ions may be required to crack olefins, due to the added stability of such intermediates over secondary and primary species. The micro pores of SAPO-5, 8.0 Å in diameter, should easily accommodate a tertiary carbonium ions. However, the medium-sized, elliptical pores of SAPO-11 may not allow the formation of the highly-branched intermediate. Therefore, the absence of butenes and pentenes may indicate steric restrictions on reactions catalyzed by SAPO-11, a problem not encountered over SAPO-5, which yields a variety of olefins during C_3H_6 conversion. Selective formation of octenes from n-butene feed provides further evidence for the inability of the medium-pore SAPO to crack olefins. Propylene and pentenes should be found as products of acid-catalyzed C_8H_{16} scission [13], yet only hydrocarbons in the C_4 and C_8 ranges appeared in the reactor effluent.

The pore sizes of SAPO-5 and SAPO-34 affected the distribution of hexene products during the conversion of propylene. The major difference in the C_6 fraction synthesized over SAPO-5 and the medium-pore catalyst SAPO-11 at 550 K, as seen in Table II, lies in the component(s) responsible for the peak with a retention time of 6.8 minutes, which was significantly larger for the wide-pore SAPO-5 catalyst. If olefin retention times during capillary column separation follow the boiling point trend, dimethylbutenes will elute earlier than methylpentenes and linear hexenes. The peak which appears 6.8 minutes after sample injection represents one of the first peaks in the C_6 range, suggesting that it is generated by dimethylbutenes. Therefore, propylene conversion within the wide-pore SAPO-5 permits the formation of greater amounts of dimethylbutenes. The medium-sized micropores of SAPO-11, as with HZSM-5, inhibit the formation of dimethyl aliphatics when the branches are located on the same or adjacent carbons [18]. Only a limited number of hexene isomers are formed over the narrow-pore SAPO-34 molecular sieve (Table II). While consideration of the pore size leads to the assumption that all C_6 products are straight-chain, another possibility must also be addressed. Propylene dimers formed within the pores of SAPO-34 would be expected to react further and equilibrate rapidly, favoring lighter (C_3, C_4) olefins. However, reactions on the external surface of the crystallite would not experience micropore mass transfer limitations or secondary reactions. The initial product of propylene dimerization, 2-methylpentene [13,4], would be selectively formed and detected if C_6 products resulted primarily from reactions outside the micro pores.

Literature Cited.

1. Kikuzuno, Y., Kagami, S., Naito, S., Onoshi, T., Tamaru, K. Chem. Lett. 1981 1249
2. Weisz, P.B. Adv. Catal. 1962 137
3. Thomson, R.T.; Wolf, E.E. Appl. Catal. 1988 41 65
4. Oblad, A.G.; Mills,G.A.; and Heinemann, H. "Polymerization of Olefins", Catalysis VI 1958.
5. Tabak, S.A., F.J. Krambeck, and W.E. Garwood AIChE Journal , 1986 32(9), 1526.
6. Garwood, W.E., ACS Symp.Ser. 1983 218 383.
7. Bessell, S., and D. Seddon, J. Catal., 1987 105 270.
8. Wilshier, K.G., P. Smart, R. Western, T. Mole, and T. Behrsing, Appl. Catal., 1987 31, 339.
9. Pellet, R.J., J.A. Rabo, and G.N. Long, Proc. North. Amer. Catal. Soc., Poster 75.
10. Pellet, R.J., G.N. Long, and J.A. Rabo, Stud. Surf. Sci. Catal., 1986 28, 843.
11. Post, J.G., and J.H.C. van Hooff, Zeolites, 1984 4, 9.
12. Lok, B.M., B.K. Marcus, and C.L. Angell , Zeolites 1986 6, 185-194.
13. Pines, H., The Chemistry of Catalytic Hydrocarbon Conversions, Academic Press, New York, 1981.
14. Rollmann, L.D. J. Catal. 1977 47, 113.
15. Weisz, P.B. Pure Appl. Chem. 1980 52, 2091.
16. Hayhurst, D.T., and A.R. Paravar Zeolites 1988 8, 27.
17. Thomson, R.T. PhD. Thesis, University of Notre Dame, Indiana, 1988.
18. Chen, N.Y., and W.E. Garwood Catal. Rev.-Sci. Eng. 1986 28(2-3), 185.

RECEIVED May 9, 1990

Chapter 8

Lattice Concentrations of B-, Ga-, and Fe-Substituted Zeolite ZSM-5

R. J. Gorte, T. J. Gricus Kofke, and G. T. Kokotailo

Department of Chemical Engineering, University of Pennsylvania, Philadelphia, PA 19104

A method has been developed for determining the lattice concentrations of Fe, Ga, and B in high-silica ZSM-5 based on the observation of well-defined adsorption complexes in temperature-programmed desorption, thermogravimetric analysis (TPD-TGA) of 2-propanamine and 2-propanol. Following adsorption of 2-propanamine on the hydrogen form of these materials, weakly adsorbed molecules not associated with the Fe, Ga, or B lattice sites desorbed unreacted below ~450K. Following removal of these excess molecules, the coverage on each material corresponded to one 2-propanamine/lattice Fe, Ga, or B. On H-[Fe]-ZSM-5 and H-[Ga]-ZSM-5, the 1:1 species reacted to form ammonia and propene between 600 and 650K; on H-[B]-ZSM-5, it desorbed unreacted between 500 and 600K. TPD-TGA measurements on samples in which Fe and Ga were not in the ZSM-5 lattice failed to observe the adsorption complex. For 2-propanol, 1:1 complexes were again observed with H-[Fe]-ZSM-5 and H-[Ga]-ZSM-5. Excess 2-propanol desorbed unreacted on these materials, leaving a 1:1 complex which decomposed to propene and water at higher temperatures.

An important question in characterization of high-silica molecular sieves regards the location of trivalent species, whether these are part of the lattice structure or just present as an oxide which is not part of the lattice. A number of methods are available for determining the lattice concentration, each with its own advantages and disadvantages.

0097–6156/90/0437–0088$06.00/0

Solid-state NMR can distinguish between lattice and nonlattice species, but is difficult to use in some cases. Ion exchange capacity is simple; however, caution must be used since ion exchange capability has been observed for high-silica materials.[1] The size of the unit cell determined by x-ray diffraction is widely used to determine lattice concentrations, but may be unreliable if the unit cell size is affected by more than lattice concentrations.[2]

In this paper, we will discuss the use of temperature programmed desorption as a complementary technique for obtaining lattice concentrations for high-silica materials. While TPD of ammonia is commonly used as a measure of the number of acid sites in zeolites, this paper will discuss the use of the reactive probe molecules, 2-propanamine and 2-propanol. It has previously been demonstrated that well-defined adsorption complexes, with a stoichiometry of one molecule/Al atom could be observed in high-silica zeolites for a number of adsorbates.[3,4] These complexes were found to be independent of Si/Al_2 ratio for a series of H-ZSM-5 zeolites[5] and were unaffected by changes in zeolite structure in going from H-ZSM-5 to H-ZSM-12 and H-mordenite.[6] It was particularly easy to identify the stoichiometric complexes with 2-propanamine and 2-propanol, since these molecules reacted at the acid sites to desorb as propene and either ammonia or water. In this paper, we will discuss the extention of these ideas to molecular sieves which contain Fe, Ga, and B.[7,8]

EXPERIMENTAL TECHNIQUES

The samples were all prepared in a hydrothermal system by sequentially adding TPA-Br and Ludox® HS-30 to acidic solutions of either $Fe(NO_3)_3 \cdot 9H_2O$, $Ga(NO_3)_3$, or boric acid. A NaOH solution was then added to lower the pH; and crystallization was carried out in unstirred, teflon-lined autoclaves at $175^{\circ}C$ for 4-5 days. The samples were then rinsed, dried, washed with a 2 M NaOH solution to remove any amorphous material, calcined in air, and ammonium ion exchanged. The hydrogen form of each molecular sieve was formed by heating the ammonium form in vacuum to 700K. A more detailed description of the preparation conditions is given elsewhere.[7,8] A summary of the data obtained by x-ray diffraction, n-hexane adsorption, electron microscopy, and elemental analysis for each of the materials is given in Table I. In each case, the unit cell size is consistent with the Fe, Ga, and B being in the crystalline lattice. In addition to the above, a sample of high-silica ZSM-5, $Si/Al_2=880$, was obtained from the Mobil Oil Company for use as a reference.

To determine the effect of nonframework species on the adsorption properties, materials containing nonframework Fe and Ga were prepared. (Fe_2O_3)-ZSM-5 was obtained by heating H-[Fe]-ZSM-5 in air at 873K for 2 hours. As will be discussed in the results section, a fraction of the Fe is probably still present in the framework after this pretreatment; however, the sample turned from white to light brown in color and x-ray diffraction data showed a change in the unit cell volume toward that of a high silica zeolite. (Ga_2O_3)-ZSM-5 was prepared by impregnating a high-silica ZSM-5 (Si/Al_2=2000) with a $Ga(NO_3)_3$ solution. The high-silica ZSM-5 was prepared in hydrothermal solution with TPA-Br and Ludox® HS-30 as above, with the addition of ammonium fluoride for making large, regular-shaped crystals. $Ga(NO_3)_3$ in the sample was then decomposed to Ga_2O_3 by either heating in vacuum or by heating in air to 800K. No noticeable differences were observed in the adsorption properties following the two different pretreatment conditions.

The equipment and procedures used to measure adsorption were the same as those used in previous papers.[5-9] The TPD-TGA experiments were carried out simultaneously in a high vacuum chamber equipped with a Cahn 2000 microbalance and a Spectramass quadrupole mass spectrometer. This system could be evacuated with a turbomolecular pump to a base pressure below 1×10^{-7} Torr. Between 10 and 20 mg of zeolite were spread over the flat sample pan of the microbalance to avoid bed effects in adsorption and desorption.[10,11] The heating rate for the TPD-TGA experiments was maintained at 10K/min by a feedback controller, and a thermocouple placed near the sample was used for temperature measurement. During the desorption experiment, the sample weight was continuously monitored using the microbalance and the desorbing gases were observed using the mass spectrometer which was interfaced with a microcomputer to allow several mass peaks to be measured simultaneously. The mass peaks monitored were the most intense peaks in the fragmentation patterns of ammonia (m/e=17), 2-propanamine (m/e=44), 2-propanol (m/e=45), water (m/e=18), and propene (m/e=41). Other experimental details have been discussed elsewhere.

Prior to each TPD-TGA experiment, the samples were exposed to between 10 and 15 Torr of 2-propanol or 2-propanamine for approximately 5 min at 295K. This exposure was sufficient to fill a substantial fraction of the zeolite pore volume of each sample. Desorption measurements were performed following evacuation of the samples for 1 to 20 hrs to remove some of the weakly adsorbed species. While the evacuation time did affect the amount of weakly adsorbed species observed in TPD at lower temperatures, it had no affect on the well-defined, stoichiometric complexes observed in this study. Following an adsorption-desorption experiment, the sample weight returned to its original value for each adsorbate on each sample, indicating that no residual products were left undetected in the zeolite.

RESULTS AND DISCUSSION

Stoichiometric adsorption complexes between an adsorbate and framework Fe, Ga, and B were observed most clearly following 2-propanamine adsorption. Fig. 1 shows the TPD-TGA results for 2-propanamine on the H-[Fe]-ZSM-5 and (Fe$_2$O$_3$)-ZSM-5 samples and demonstrates the ability of the technique to distinguish between framework and nonframework Fe. On both samples, there is simultaneous desorption of propene and ammonia between 600 and 650K. That these decomposition products from 2-propanamine desorb simultaneously and well above their normal, individual desorption temperatures[7] implies that their formation is limited by the reaction of 2-propanamine. For the H-[Fe]-ZSM-5, the coverage of 2-propanamine corresponding to this desorption event is one molecule/Fe atom. For the (Fe$_2$O$_3$)-ZSM-5 formed by heating H-[Fe]-ZSM-5 in air at high temperatures, the coverage of 2-propanamine corresponding to desorption of ammonia and propene is decreased by more than 80% based on the relative areas under the propene peak.

The evidence suggests that 2-propanamine interacts with the protons associated with the framework Fe atoms to form 2-propyl ammonium cations which maintain 2-propanamine in the zeolite to high temperatures. Above approximately 600K, the decomposition rate for these cations to form propene and ammonia becomes appreciable. The decomposition reaction is very similar to the Hofmann elimination reaction found for quaternary ammonium salts and provides indirect evidence that ammonium ions are involved in the reaction. When Fe is removed from the framework of the molecular sieve, the associated proton site is lost, along with the capability for forming the ammonium ion and carrying out the reaction at that site.

The results for the H-[Ga]-ZSM-5 and (Ga$_2$O$_3$)-ZSM-5 samples are similar and are shown in Fig. 2. For H-[Ga]-ZSM-5, we again see formation of propene and ammonia between 600 and 650K, at a coverage corresponding to one 2-propanamine/Ga. That the decomposition reaction occurs in the same temperature range on H-[Ga]-ZSM-5 and H-[Fe]-ZSM-5 does not indicate that the acid sites in these two materials are of identical strength. Other evidence suggest that H-[Ga]-ZSM-5 has stronger acid sites than H-[Fe]-ZSM-5. Studies of propene adsorption have shown that the sites generated by Fe in the ZSM-5 framework are not capable of causing

Table I. Properties of zeolite samples used in this study

Sample Volume	SiO_2/T_2O_3	Particle size (µm)	Pore Volume $(cm^3/100g^*)$	Unit Cell Å^3
[Ga]-ZSM-5	95	20**	.17	5348
(Ga$_2$O$_3$)-ZSM-5	50	10x10x50	.16	-
[Fe]-ZSM-5	140	1x10	.15	5382
[B]-ZSM-5	85	30**	.17	5251
ZSM-5	880	1	.19	5337

* Obtained from n-hexane adsorption, assuming a liquid density for the adsorbate.
** Particles were agglomerates of smaller crystallites.

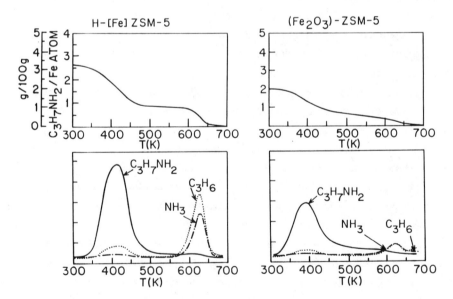

Figure 1.   TPD-TGA results for 2-propanamine on H-[Fe]-ZSM-5 and (Fe$_2$O$_3$)-ZSM-5.

oligomerization at room temperature, while those generated by Ga can, implying that H-[Ga]-ZSM-5 is capable of protonating propene at this temperature while H-[Fe]-ZSM-5 is not.[12] Furthermore, the decomposition of 2-propanamine on H-[Al]-ZSM-5, which appears to have stronger sites than either H-[Ga]-ZSM-5 or H-[Fe]-ZSM-5, also occurs between 600 and 650K during TPD in our apparatus.[5] This suggests that, so long as the Brønsted-acid site is capable of maintaining the ammonium ion up to the reaction temperature, the decomposition of 2-propanamine is independent of the acid strength.

Also shown in Fig. 2 are the TGA results for (Ga_2O_3)-ZSM-5. Only 2-propanamine was observed in desorption at ~400K, and this desorbed in very small quantities. It is apparent that the Ga_2O_3, which is outside the ZSM-5 framework in this sample, is not capable of reacting 2-propanamine. Of equal interest is the fact that most of the 2-propanamine was removed by evacuation at room temperature on this sample, even though all samples were initially saturated with 2-propanamine and similar evacuation procedures were used in each case. The difference cannot be due to diffusion since the (Ga_2O_3)-ZSM-5 had the largest average particle size of the samples we studied. Nor can it be due to the presence of additional adsorbate molecules at the acid sites, beyond the one/site coverage, since other high silica samples which we studied did show a substantial peak at 400K in the desorption of 2-propanamine. One possibility is that the peak at 400K is due to 2-propanamine interacting with silanols which terminate the zeolite structure or are present as defects in the framework. It has been reported that some of these silanols can give ion-exchange capabilities[1] and that their concentration can be affected by pretreatment and, presumably, by synthesis conditions.[13] The ZSM-5 sample used to make the (Ga_2O_3)-ZSM-5 in our study consisted of large, regular-shaped crystals and, therefore, likely had the lowest silanol concentration.

If the desorption feature at 400K is indeed due to 2-propanamine interacting with silanols in the zeolite, it could have important consequences in the use of spectroscopic techniques for determining the number of acid sites in zeolitic materials. While these sites may be important for some reactions under certain conditions,[14] they are unlikely to be important for more demanding reactions like hydrocarbon cracking. Since 2-propanamine is a relatively strong base, the species present at the silanols could be in the protonated, ammonium-ion form which would be spectroscopically difficult to distinguish from molecules present at the acid sites generated by framework Al, Fe, or Ga. It should be possible to distinguish between the two adsorption sites by heating the sample; however, it is important to consider that the conditions necessary for removing the more weakly adsorbed species will be strongly affected by the experimental configuration.[10,11] For example, we have found that NH_3 can be easily removed from even the Al sites in H-[Al]-ZSM-5 at 500K using our apparatus,[7] while much higher temperatures are needed when a carrier gas is used to remove the desorbing ammonia.[15]

Fig. 3 shows the TPD-TGA results for H-[B]-ZSM-5 and for the high-silica ZSM-5 sample. There is no reaction to form propene and ammonia on either sample; however, there is a second, high-temperature feature on H-[B]-ZSM-5, in addition to the 400-K state found on both samples, corresponding to a coverage of one molecule/B. It appears that the B sites are able to protonate 2-propanamine at room temperature, but that 2-propanamine desorbs below the temperature necessary for the 2-propyl ammonium ion to decompose to propene and ammonia. From this, it is apparent that the sites formed by B in the ZSM-5 lattice are much weaker than those formed by Fe or Ga.

We also studied the adsorption of 2-propanol on H-[Fe]-ZSM-5, H-[Ga]-ZSM-5, and H-[B]-ZSM-5, with the results having been discussed in detail elsewhere. On H-[B]-ZSM-5, we were unable to maintain 2-propanol on the sample during evacuation at room temperature, again indicating that the acid sites on this material are very weak. On H-[Fe]-ZSM-5 and H-[Ga]-ZSM-5, molecules in excess of one/Fe or Ga desorbed unreacted below 400K. The 2-propanol which remained had a coverage of one/Fe or Ga and reacted to form propene and water at ~430K on H-[Fe]-ZSM-5 and ~415K on H-[Ga]-ZSM-5. This difference in temperature of reaction appears to be a measure of the relative acid strengths of these materials. Acid-catalyzed dehydration of 2-propanol proceeds through an oxonium-ion intermediate, but ^{13}C NMR studies of $(CH_3)_2$[13]CHOH on H-[Al]-ZSM-5 indicate that proton transfer to the alcohol is not complete in zeolites and the adsorbed species are not true oxonium ions.[ref] The difference in reaction temperatures may be due to the degrees of protonation at the different acid sites.

SUMMARY

We have shown that TPD-TGA studies of 2-propanamine and 2-propanol on H-[Fe]-ZSM-5, H-[Ga]-ZSM-5, and H-[B]-ZSM-5 can provide a useful method for determining the framework concentrations of Fe, Ga, and B. For Ga and Fe containing

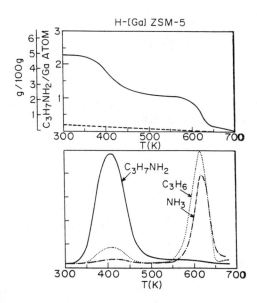

Figure 2.   TPD-TGA results for 2-propanamine on H-[Ga]-ZSM-5.
TGA results for (Ga$_2$O$_3$)-ZSM-5 are shown as a dashed line for comparison.

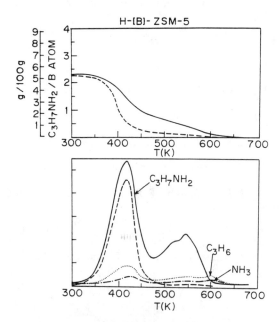

Figure 3.   TPD-TGA results for 2-propanamine on H-[B]-ZSM-5. The
dashed lines are results for the ZSM-5 sample with Si/Al$_2$=880.

ZSM-5, acid-catalyzed decomposition of both 2-propanol and 2-propanamine was observed only when the Ga and Fe were present as part of the ZSM-5 framework. Due to the fact that the B sites in H-[B]-ZSM-5 are much more weakly acidic, no reaction was observed for the probe molecules we studied; however, with 2-propanamine, a well-defined state corresponding to adsorption at the B sites could still be observed. Finally, these studies give insight into the relative strengths of the sites generated by the incorporation of various trivalent ions into the ZSM-5 framework.

REFERENCES

1.   G.L. Woolery, L.B. Alemany, R.M. Dessau, and A.W. Chester, Zeolites, 6 (1986) 14.

2.   C.A. Fyfe, H. Strobl, G.T. Kokotailo, G.J. Kennedy, and G.E. Barlow, JACS 110 (1988) 3373.

3.   M.T. Aronson, R.J. Gorte, and W.E. Farneth, J. Catal., 98 (1986) 434.

4.   M.T. Aronson, R.J.Gorte, and W.E. Farneth, J. Catal., 105 (1987) 455.

5.   T.J. G. Kofke, R.J. Gorte, and W.E. Farneth, J. Catal., 114 (1988) 34.

6.   T.J.G. Kofke, G.T. Kokotailo, R.J. Gorte, and W.E. Farneth, J. Catal., 115 (1989) 265.

7.   T.J.G. Kofke, G.T. Kokotailo, and R.J. Gorte, J. Catal., 116 (1989) 252.

8.   T.J.G. Kofke, G.T. Kokotailo, and R.J. Gorte, Appl. Catal., 54 (1989) 177.

9.   T.J.G. Kofke and R.J. Gorte, J. Catal., 115 (1989) 233.

10.   R.J. Gorte, J. Catal., 75 (1982) 164.

11.   R.A. Demmin and R.J. Gorte, J. Catal., 90 (1984) 32.

12.   T.J.G. Kofke, PhD Thesis, University of Pennsylvania, 1989.

13.   R.M. Dessau, K.D. Schmitt, G.T. Kerr, G.L. Woolery, and L.B. Alemany, J. Catal., 104 (1987) 484.

14.   W.R. Moser, C.-C. Chiang, and R.W. Thompson, J. Catal. 115 (1989) 532.

15.   C.T-W. Chu and C.D. Chang, J.Phys. Chem., 89 (1985) 1569.

RECEIVED May 9, 1990

Chapter 9

Controlled Pore Size, High Alumina Content Silica–Aluminas

Larry L. Murrell[1], N. C. Dispenziere, Jr., and K. S. Kim[2]

Exxon Research and Engineering Company, Route 22 East, Clinton Township, Annandale, NJ 08801

A new class of high alumina silica-aluminas has been prepared with high pore volume and with controllable pore size. For 20 wt % SiO_2 content materials steam treatment at 760°C serves to substantially <u>increase</u> the gas oil cracking activity. Steam stability at 870°C has also been demonstrated for these unique materials. These samples are prepared by recrystallization of co-gels of aluminum and silicon alkoxy compounds using a two-step procedure. Amorphous alumina shows similar physical properties to those of the silica-alumina co-gels when re-crystallized in an analogous manner.

Work at Chevron and Grace (1-2) showed that high alumina content, silica-alumina solid acids could be prepared with high paraffinic gas oil cracking activities. These silica-alumina co-gels had an average pore diameter of 3 nm with a surface area of 100-130 m^2/g (3). This is low surface area compared to high-silica silica-alumina gels, i.e., greater than 400 m^2/g, and is due to the low pore volume of ca. 0.2 cc/g in the case of the materials described in the Chevron and Grace patents (3). We have found that alumina or high-alumina-content silica-aluminas prepared from alkoxy precursor(s) can be prepared with pore volumes of 1.5-2 cc/g using a two step procedure. This two step procedure relies on gelation of the alkoxy precursor(s) with an amount of water stoichiometric with the number of alkoxy functions. After the gelation has occurred, addition of water in a second separate

[1]Current address: Engelhard Corporation, Menlo Park, CN 40, Edison, NJ 08818
[2]Current address: Polaroid Corporation, 750 Main Street, Cambridge, MA 02139

0097–6156/90/0437–0097$06.00/0
© 1990 American Chemical Society

step leads to recrystallization of the amorphous gel to form a
boehmite phase of considerable pore volume and, therefore, low
density. In the case of high-alumina silica-alumina gels prepared by this two-step procedure, acidities similar to those
described in the Chevron and Grace patents were obtained for 15-
20 % SiO_2 content. The amount of water used in the recrystallization step was found to have a major impact on the pore size
and pore volume obtained in the final product. This class of
wide pore high-alumina, silica-aluminas has high temperature
steam stability combined with high cracking activity after severe
steam treatment. In fact, steam treatment at 760°C substantially
increases the activity compared to the 500°C calcined co-gel.
High-silica silica-aluminas, in comparison, are completely deactivated by steam treatment at 870°C.

Experimental

Aluminum isobutoxide (AIB) or mixtures with tetraethoxysilane
(TEOS) were precipitated with a stoichiometric amount of water to
hydrolyze all of the aluminum-oxygen and silicon-oxygen bonds.
The dried gels were amorphous by x-ray powder diffraction. In
all of the experiments the undried gels were then recrystallized
in one of two ways. In one case water was added to the gel in a
manner similar to an incipient wetness impregnation to fill the
pores of the amorphous alumina or silica-alumina gel. In the
other case about 10 times this amount of water was added while
the mixture was stirred. The gel was then dried for 16 hours at
120°C. All preparations were decomposed in a N_2 purged muffle
furnace programmed to 500°C in ca. 4 hrs. After N_2 decomposition
all samples were calcined at 500°C for 16 hours. An example of
the two step procedure employed in this work is given for 20 wt %
SiO_2-content SiO_2-Al_2O_3. Thirty-two grams of H_2O were added to a
mixture of 162.74 g AIB (containing 32 g Al_2O_3) and 20.24 g TEOS
(containing 8 g SiO_2) to gel the sample. To the warm co-gel an
additional 20 g of H_2O was added. The sample was dried at 120°C
for 16 hours after a 2-hour period in contact with water. This
limited water addition was just sufficient to fill the pore voids
of the co-gel. In the case of the two step procedure where
excess water was employed, 16 g of H_2O was added to a mixture of
81.37 g AIB and 10.12 g TEOS to gel the sample. To the warm co-
gel 150 g H_2O was added and the mixture was stirred for 2 hours.
The co-gel was filtered and dried at 120°C for 16 hours. Quite
different materials were obtained from this seemingly modest
change in recrystallization conditions. The acidity of the
samples was estimated by gas oil cracking activity as employed in
other recent studies (4-5). Gas oil cracking activities were
found to track well with model compound acid characterization
studies in recent work of McVicker, et.al. (6)

Results and Discussion

We will first consider the case of a silica-free gel. Aluminum
isobutoxide is precipitated by addition of a stoichiometric

amount of water. If the gel was dried at 120°C and calcined at 500°C for 16 hours an amorphous alumina phase is formed of 600 m^2/g surface area and a pore volume of 1.4 cc/g. This amorphous gel had a broad distribution of pores below 6 nm diameter. If this amorphous, 500°C calcined gel is calcined at 800°C for 16 hours the amorphous phase converts to alpha-Al_2O_3 of surface area <5 m^2/g. This is very poor stability compared to transitional aluminas (4). If the amorphous gel phase is not dried, but impregnated with additional water to fill the pore voids similar to an incipient wetness impregnation; then recrystallization to form a wide-pore boehmite-like phase occurs. This boehmite-like material, after drying at 120°C and calcination at 500°C for 16 hours converts, at least partially, to a transitional gamma-alumina with a surface area of 494 m^2/g with a pore volume of 2.1 cc/g. The average pore diameter of this material is 17 nm. If ten times the amount of water is added to the undried amorphous gel then a boehmite-like material of much lower pore volume is formed than for the previous example with water added only to the pores of the gel. After drying at 120°C and calcination at 500°C for 16 hours, the surface area of this alumina was 294 m^2/g with a pore volume of 0.4 cc/g. The average pore diameter of this alumina was centered at 7 nm, quite distinct from that for the previous material. The amount of water added to an amorphous gel can clearly change markedly the properties of the boehmite-like alumina formed upon recrystallization. A recent publication (7) reports that it is possible to continuously shift the pores of alumina to larger size by a swing in pH of the slurry phase containing boehmite. In the case where water is added to fill the pores of the amorphous alumina gel in our work recrystallization occurs to give relatively large primary particles of a boehmite, which in turn forms the 17 nm pores within the calcined alumina structure. (7-8)

We will next consider the case of a low silica content co-gel. A 5% silica-content silica-alumina was prepared by precipitation of aluminum isobutoxide and tetraethoxy-silane as described for the silica-free gel. After gelation water was added just sufficient to fill the pore voids of the gel. The added water led to formation of a boehmite-rich phase during recrystallization. After drying at 120°C and calcination at 500°C for 16 hours, a transitional alumina phase is formed with a surface area of 410 m^2/g and a pore volume of 1.9 cc/g. This silica-alumina had an average pore diameter of 18 nm, similar to the silica-free material discussed previously. Steam treatment of this 18 nm pore diameter silica-alumina at 870°C (1600°F) in 90% H_2O-10% N_2 for 16 hours resulted in a material with surface area of 196 m^2/g. This surface area is much higher than expected for an amorphous gel and is consistent with silica enrichment of the outer surface during the recrystallization step where water was added to the pores of the amorphous gel. Silica stabilization of boehmite alumina by formation of a surface phase complex has been reported in recent work (9). ESCA analysis also indicates silica surface enrichment when compared to the amorphous gel.

If ten times the amount of water is added to the undried 5% SiO_2 content gel than required to fill the pore voids a much more dense silica-alumina phase is formed than in the previous example. After drying at 120°C and calcination at 500°C a material of 283 m^2/g surface area and a pore volume of 0.9 cc/g is formed. The average pore diameter of this material is 8 nm. There is an unmistakable parallel in the recrystallization of the silica-free gel to the 5% silica-content silica described here. Addition of limited water to the amorphous gel produces large pores and a high pore volume. Addition of water so that a slurry is formed with the amorphous gel results in smaller pores and a relatively dense phase upon recrystallization.

Higher silica content systems were also considered. In the case of 15 wt % SiO_2-content SiO_2-Al_2O_3 the "amorphous" co-gel was formed as discussed previously. After drying and calcination at 500°C a material of 523 m^2/g was formed. The gas oil cracking activity of this amorphous gel was relatively low and was not changed significantly by steam treatment at 760°C or 870°C (see Table 1.). This is surprising as the surface area decreased to about 140 m^2/g for these steaming conditions. The high temperature steam stability of the amorphous 15% SiO_2-content SiO_2-Al_2O_3 gel is consistent with an earlier patent reference, however (1).

If water was added to the 15% SiO_2 co-gel to fill the pore voids a partially recrystallized boehmite was formed with a surface area of 464 m^2/g and with a pore volume of 1.8 cc/g. If water was added to the 15% SiO_2 co-gel to form a slurry and then dried and calcined at 500°C a partially recrystallized boehmite was formed with a surface area of 334 m^2/g. Steam treatment at 760°C of this second, small pore, boehmite-like silica-alumina resulted in no change in the surface area. The gas oil cracking activity of the steamed sample was definitely higher than that for the amorphous co-gel, i.e., a Micro Activity Test (MAT) Activity Number of 38 (see Table 1.).

We then proceeded to prepare 20 wt. % SiO_2-content co-gels by the above discussed procedures. The relative surface area stabilities of the amorphous and recrystallized boehmite-like materials are given in Table 2. Both of the recrystallized, boehmite-like materials have about twice the surface area compared to that for of the amorphous gel after 1050°C calcination. The gas oil cracking activities of boehmite-like 20% SiO_2-content SiO_2-Al_2O_3 recrystallized with excess water before and after steam treatment are given in Table 1. Although the cracking activity of the 500°C calcined 20 wt % SiO_2 prepared by the slurry recrystallization procedure (small pores) sample is similar to that for the 15 wt % SiO_2-content amorphous gel sample there is a unique activation by the subsequent steam treatment for the recrystallized sample. Steam treatment at 760°C leads to a doubling of the cracking activity. Steam treatment at 870°C which causes a 50% decrease in surface does not cause a severe loss in the cracking activity. Steam treatment at 760°C of the 20% SiO_2, large pore boehmite-like SiO_2-Al_2O_3, prepared by the pore filling recrystallization procedure, also leads to a material with high cracking activity.

Table 1.

COMPARISON OF GAS OIL CRACKING ACTIVITIES OF AMORPHOUS AND
CRYSTALLINE HIGH-ALUMINA SILICA-ALUMINAS AFTER STEAM TREATMENT

Sample	Sample Treatment	Surface Area (m^2/g)	MAT[1] Activity	Volume % 400^- Liquids
Amorphous 15% SiO_2 content SiO_2-Al_2O_3	500°C	523	24	4.5
Amorphous 15% SiO_2 content SiO_2-Al_2O_3	Steam[2] 760°C	149	31	11.1
Amorphous 15% SiO_2 content SiO_2-Al_2O_3	Steam[2] 870°C	--	28	9.9
Recrystallized 15% SiO_2 content SiO_2-Al_2O_3 (small pore)	500°C	334	--	--
Recrystallized 15% SiO_2 content SiO_2-Al_2O_3 (small pore)	Steam[2] 760°C	334	38	15.6
Recrystallized 20% SiO_2 Content SiO_2-Al_2O_3 (small pore)	500°C	318	28	10.1
Recrystallized 20% SiO_2 content SiO_2-Al_2O_3 (small pore)	Steam[2] 760°C	240	58	25.7
Recrystallized 20% SiO_2 content SiO_2 Al_2O_3 (small pore)[3]	Steam[2] 870°C	169	45	18.9
Recrystallized 20% SiO_2 content SiO_2-Al_2O_3 (wide pore)[4]	Steam[2] 760°C	311	49	19.7

[1]Micro Activity Test Activity Numbers. (See references 4-5)

[2]Steamed for 16 hours in 90% H_2O-10% N_2.

[3]Pore distribution centered at 7 nm diameter, excess water recrystallization conditions.

[4]Pore distribution centered at 1.6 nm diameter, water pore-filling recrystallization conditions.

Table 2. RELATIVE STABILITY OF AMORPHOUS AND CRYSTALLINE
HIGH-ALUMINA SILICA ALUMINAS

	Calcination Temp. (^{O}C, 16 hours)	Surface Area (m^2/g)
Amorphous 20% SiO_2-Content SiO_2-Al_2O_3 Gel	500	405
Amorphous 20% SiO_2-Content SiO_2-Al_2O_3 Gel	1050	75
Recrystallized 20% SiO_2 Content SiO_2-Al_2O_3 (small pore)	500	318
Recrystallized 20% SiO_2 Content SiO_2-Al_2O_3 (small pore)	1050	142
Recrystallized 20% SiO_2 Content SiO_2-Al_2O_3 (large pore)	500	477
Recrystallized 20% SiO_2 Content SiO_2-Al_2O_3 (large pore)	1050	133

In conclusion, a novel class of boehmite-like SiO_2-Al_2O_3 has been discovered which show a unique combination of properties: 1) surface area stability characteristic of aluminas with a terminating outer surface of silica; 2) variations in pore diameter and pore volume depending on recrystallization conditions; 3) unusual steam activation leading to enhanced acidity as measured by gas oil cracking activity; and 4) remarkable stability of high cracking activities after severe steaming conditions compared to high silica content silica-aluminas.

References

1. White, R.J., Hickson, D.A., Rudy, Jr., C.E., U.S. Patent 3 933 621, 1976.
2. Lussien, R.J., Magee, Jr., J.S., Albers, E.W., Surland, G.T., U.S. Patent 3 974 099, 1976.
3. Personal communication with D.E.W. Vaughan at Exxon Research and Engineering Company and confirmed with experimental samples.
4. Soled, S.L., McVicker, G.B., Murrell, L.L., Sherman, L.G., Dispenziere Jr., N.C., Hsu, S.L., and Waldman, D., J. Catal. 1988, 111, 286.
5. Ciapetta, F.G. and Henderson, D.J., Oil Gas J. 1967, 65, 88.
6. Kramer, G.M. and McVicker, G.B., Acc. Chem. Res. 1986, 19, 78.
7. Ono,T., Ohguchi, Y., Togari, O. In Preparation of Catalysts III, Poncelet, G., Grange, P., and Jacobs, P.A., Eds., Elevier Publishers, 1983, p. 631.
8. Trimm, D.L. and Stanislaus, A., Appl. Catal., 1986, 21, 215.
9. Murrell, L.L. and Dispenziere, Jr., Jr. Catal., 1988, 111, 450.

RECEIVED May 9, 1990

LAYERED   STRUCTURES

Chapter 10

Hydrothermal Stability and Catalytic Cracking Performance of Some Pillared Clays

J. Sterte

Department of Engineering Chemistry I, Chalmers University of Technology, Göteberg, S–412 96 Sweden

Several types of pillared clays were prepared and characterized with the objective to make materials suitable as active components in catalysts for heavy oil cracking. Incorporation of Si into the oligomeric Al-cations acting as pillaring precursors resulted in materials with approximately the same basal spacings, somewhat lower surface areas but thermally more stable than samples prepared using no Si. TiO_2-pillared montmorillonites were prepared by reacting the montmorillonite with partially hydrolyzed $TiCl_4$-solutions. The TiO_2-pillared products showed basal spacings of about 28 Å, surface areas of 200-350 m^2/g and were stable up to 700°C. Alumina-montmorillonites were prepared by ion-exchange of the montmorillonite with fibrillar boehmite in colloidal suspension. This resulted in materials with improved stability but with larger and less uniform pores compared with conventional Al_2O_3-pillared montmorillonite. The pillared products were tested as cracking catalysts, after deactivation, using a micro activity test. Initial results of pillaring of regularly and randomly interstratified illite/montmorillonites are reported.

Since smectites pillared with inorganic polycations were first introduced in the late seventies, much work has been undertaken in order to produce materials of this type suitable as active components in catalysts for cracking of heavy oil fractions. Pillared smectites are particularly interesting for this application as they can be prepared with pore openings larger than those in zeolite Y, which is the active component in most of todays commercial cracking catalysts. The larger pore openings would allow entrance and conversion of molecules too large to enter the zeolite structure. The use of most types of pillared smectites for this purpose is, however, limited by their lack of thermal and hydrothermal stability. At the thermal and hydrothermal conditions in the regenerator of a fluid catalytic cracker (FCC), pillared smectites rapidly lose most of their surface area and catalytic activity.

The potential of pillared smectites as catalysts for catalytic cracking

0097–6156/90/0437–0104$06.00/0

was first demonstrated by Vaughan et al. (1) by cracking a gas oil over a sample of aluminum oxide pillared montmorillonite. The sample was exposed to a temperature of 530°C for 3 hr. prior to the test and it showed a micro activity test (MAT) conversion of 77.3%. By hydrothermal treatment (reflux conditions) of the aluminum chlorohydrate solution and by copolymerisation with silica Vaughan et al. (2) succeeded in preparing a pillared smectite with a MAT activity, after steam treatment at 677°C for 8 hr., comparable to that of a commercial REY-catalyst treated in the same manner.

Recently, Jie et al. (3) showed that a hydrothermally stable pillared clay can be prepared by using rectorite instead of smectites. Their product retained most of its original surface area and cracking activity after deactivation in 100% steam for 20 hr. Rectorite is one example of a regularly interstratified clay mineral which is an alteration of pyrophyllite-like and, water swelling, montmorillonite-like layers. As the pillaring only involves the water swelling layers, this results in a pillared material in which the distance between pores in the c-direction is twice that in a conventional pillared smectite, i.e. about 19 Å. This, more rigid, structure of the pillared rectorite, is believed to explain the remarkable hydrothermal stability of this material.

McCauley (4) found that hydrothermally stable pillared smectites can be prepared by using hydrothermally treated pillaring solutions containing mixtures of aluminum chlorohydrate and a cerium salt. Pillared smectites prepared from such solutions differ from conventional alumina-pillared smectites in that the basal spacing is considerably larger, i.e. about 26 Å compared with 18-19 Å for conventional ones. When tested as catalysts for heavy oil cracking, these large-pore pillared smectites showed a high cracking activity even after steam deactivation at 760°C for 5 hrs. When used in combination with zeolite USY, a synergistic effect was observed since the resulting activity was greater than what would be expected from the activities of the components.

At the Department of Engineering Chemistry I, Chalmers University of Technology, most of our research on pillared smectites has been directed towards the preparation of materials with improved thermal and hydrothermal stability. Part of this work has been performed in cooperation with the Department of Fuels Engineering, University of Utah. In the present paper, some of our work on Al_2O_3-, SiO_2/Al_2O_3- and Ti-pillared smectites is discussed mainly from the point of hydrothermal stability and cracking performance. Some initial results of a study on pillaring of regularly and randomly interstratified illite/montmorillonite mixed layer clays are also included.

Experimental

Starting clays. A Wyoming montmorillonite (commercial designation, Volclay SPV 200) was used in the preparation of conventional Al_2O_3-pillared montmorillonite, alumina-montmorillonite complexes and TiO_2-pillared montmorillonite. This clay is further described in reference (5) . In the studies on SiO_2/Al_2O_3-pillared smectites, another Wyoming montmorillonite (Accofloc, 350) was used together with a synthetic Li-fluorhectorite (6) . Samples of interstratified illite/montmorillonites (I/M) from Kinnekulle in southwest Sweden were obtained from the Geological Survey of Sweden. These clays have been extensively characterized by Byström (7) and Brusewitz (8) . All clay samples were fractionated using conventional sedimentation techniques and the <2 micron fractions were used in the preparation of pillared products.

Pillaring solutions. Conventional Al_2O_3-pillared montmorillonite and interstratified illite/montmorillonites were prepared using an aluminum chlorohydrate (ACH) solution having an Al-concentration of 0.25 M and an OH/Al-ratio of 2.5, prepared by dilution of a commercial ACH-solution (Locron L, Hoechst). The solution was aged for 20 hr. after dilution, prior to use for preparation of pillared products.

SiO$_2$/Al$_2$O$_3$-pillared smectites were prepared according to two alternative procedures: treatment of a mixture of orthosilicic acid and AlCl$_3$ with aqueous NaOH, followed by aging of the solution (method A); and preparation and aging of hydroxy-Al oligocations followed by reaction of the latter with orthosilicic acid (method B). The preparation of these solutions is further described in reference (6) .

The preparation of hydrothermally treated ACH-solutions used in the preparation of alumina montmorillonite complexes (AMC) is described in references (9,10) . The ACH-solution (C_{Al} = 0.176 M, Al$_2$O$_3$/Cl = 0.93) was hydrothermally treated in an autoclave for 24 hr. at temperatures in the range 120-160°C. After deionization and reacidification (to pH 3.9) the colloidal suspensions formed were used at a concentration corresponding to 6.0 g Al$_2$O$_3$ per l, for the preparation of AMCs.

The partially hydrolyzed TiCl$_4$-solution used for preparation of TiO$_2$-pillared montmorillonite was prepared by first adding TiCl$_4$ to an amount of 6.0 M HCl corresponding to a final HCl-concentration of 1.0 M. This mixture was then diluted by slow addition of distilled water to obtain a final Ti-concentration of 0.84 M. The resulting solution was aged at room temperature for 3 hr prior to use.

Preparation of pillared products. The conventional pillared montmorillonite and the pillared interstratified I/M were prepared by addition of a volume of the ACH-solution calculated to give an Al/clay ratio of 5 mmoles per g to a dispersion (1 g/l) of the clay.

SiO$_2$/Al$_2$O$_3$-pillared smectites were prepared using a continuous mixing procedure and a clay dispersion having a concentration of 2.45 g/l (see further ref. (6)).

AMCs were prepared by adding calculated volumes of the hydrothermally treated solutions to a montmorillonite dispersion (0.5 g/l) to obtain Al$_2$O$_3$/montmorillonite ratios of 0.3, 0.6 and 1.2 g/g for the sample prepared from the untreated solution (AMC) and from those treated at 120°C (AMC-120) and 140°C (AMC-140), respectively. The AMCs were also used as matrices in catalysts containing 20 wt% REY (sample designations: ZAMC, ZAMC-120 and ZAMC-140) (11) . TiO$_2$-pillared montmorillonite was prepared by adding a calculated volume of the pillaring solution to a clay dispersion (4 g/l) to obtain a Ti/montmorillonite ratio of 20 mmoles per g.

Characterization of products. N$_2$-adsorption-desorption isotherms were recorded at liquid nitrogen temperature after outgassing at 200-250°C for 2hr. Surface areas were calculated using the BET-equation and pore volumes were estimated to be the liquid volume adsorbed at a relative pressure of 0.995. Pore-size distributions were calculated from the desorption branches of the isotherms using parallel plates as a geometrical model. X-ray diffraction analyses were performed on samples oriented in order to amplify the 00l-reflexions.

Catalytic cracking. A modified micro activity test (MAT) was used for testing cracking performance. The feed oils used in this study were a hydroprocessed mixture of Arabian light and North Sea heavy vacuum gas oil (HVGO) and a heavy fraction of Wilmington crude (no.6). Characteristics of these oils are given in reference (12) . Testing was carried out using reactor temperatures of 500 and 560°C for the lighter (HVGO) and heavier (no.6) feed, respectively.

Results and Discussion

Aluminum oxide pillared smectites. The effect of steam deactivation (18 hr, 100% steam) of conventional Al_2O_3-pillared montmorillonite on surface area and cracking performance is shown in Table I.

Table I.   MAT-Cracking results and surface areas of pillared smectites

catalyst designation	t^a (°C)	SA (m^2/g)	conv (wt%)	gasoline (wt%)	gas (wt%)	coke (wt%)	LCO (wt%)
Al-PiM	550	175	81.3	43.5	22.2	15.6	12.5
Al-PiM	650	111	62.7	39.4	15.0	8.4	17.9
Al-PiM	750	42	25.3	14.3	5.3	5.7	10.9
Si/Al-PiM	750	63	16.6	10.5	3.2	2.9	6.6
Ti-PiM	750	108	53.1	25.5	16.8	10.8	16.7

[a]Temperature of steam deactivation. All samples were steam deactivated in 100% steam for 18 hr.

After steaming at 550°C, the MAT conversion is very high and most of the surface area of the fresh catalyst is retained. Steaming at 650°C reduces the surface area as well as the MAT conversion significantly while deactivation at 750°C results in a collapse of the microstructure and, consequently, a low MAT activity. These results demonstrate the potential of pillared smectites as catalysts for catalytic cracking but also point at the major problem with these materials, namely their lack of thermal and hydrothermal stability. Furthermore, they point at the other major problem associated with the use of pillared smectites as cracking catalysts, namely their great selectivity for making coke.

Vaughan et al. (2) found that, due to further polymerization of the pillaring cations, hydrothermal treatment of the aluminum chlorohydrate solution (reflux conditions) used in the preparation of the pillared smectites resulted in a material with improved hydrothermal stability. Tokarz and Shabtai (13) conducted a similar study using base hydrolyzed $AlCl_3$-solutions. They found that refluxing of the solutions for 6-48 hr. was sufficient to produce pillared smectites with improved thermal stability and higher porosity as compared with those prepared from solutions aged at room temperature for two weeks.

At our laboratory we have studied aluminum-oxide montmorillonite complexes prepared from ACH-solutions hydrothermally treated at temperatures up to 160°C (9) . Hydrothermal treatment of ACH at temperatures above about 120°C yields positively charged, fibrillar boehmite in colloidal suspension (14) . The size of the boehmite fibrils increases with increasing temperature and time of hydrothermal treatment. Ion-exchange of montmorillonite with these positively charged fibrils resulted in AMCs with

surface areas of 150-400 m^2/g and pore volumes in the range 0.20-0.45 cm^3/g.
The surface areas decreased with increasing temperature of hydrothermal
treatment of the ACH-solutions while the pore volumes increased with an
increase in this temperature. Pore size distributions calculated from the
desorption branches of N_2-adsorption-desorption isotherms as well as X-ray
powder diffraction analyses indicated an increase in pore size and a broadening
of the pore size distribution with increasing temperature of hydrothermal
treatment of the ACH-solutions (10). The surface area retention of the
AMCs after thermal and hydrothermal treatment increased with increasing
temperature of hydrothermal treatment of the ACH-solution used in the
preparation. Hydrothermal treatment of the AMCs at 750°C for 18 hr.
resulted in increased average layer distances (pore sizes) primarily due to
collapse of micropores. Three types of AMCs were evaluated as cracking
catalysts, alone and as matrices for rare earth exchanged zeolite Y (REY),
using the MAT and three different feed oils (11). Prior to the test, all
catalysts were steam deactivated at 750°C for 18 hr. MAT results are given in
Table II together with surface areas of the catalysts before and after steam
deactivation.

Table II.   Catalytic Cracking over Alumina-Montmorillonite Complexes alone
and as Matrices for REY

catalyst designation	SA (m^2/g)	feed oil	conv. (wt%)	gasoline (wt%)	coke (wt%)	gas (wt%)	LCO (wt%)
ACH-K-REY	115	no.6	52.9	23.3	13.2	16.5	12.4
AMC	61	no.6	33.4	17.9	6.5	9.0	15.0
AMC-120	113	no.6	54.5	24.7	15.0	14.8	18.2
AMC-140	143	no.6	61.3	26.3	19.2	15.8	18.6
ZAMC	176	no.6	62.3	26.6	15.3	20.4	9.9
ZAMC-120	139	no.6	63.1	27.9	15.6	19.6	11.1
ZAMC-140	214	no.6	72.1	30.4	19.6	22.1	12.7
ACH-K-REY	92	HVGO	73.3	47.1	8.6	17.6	14.5
ZAMC	176	HVGO	78.2	46.5	9.9	21.8	10.7
ZAMC-120	139	HVGO	79.7	46.8	10.4	22.5	11.3
ZAMC-140	214	HVGO	83.7	49.5	12.6	21.6	10.4

For cracking of the heavy oil (no.6), the AMCs showed conversions similar to
that of a reference catalyst containing 20 % REY in a kaolin-binder matrix.
The AMCs showed higher coke and lower gas selectivity compared with the
reference catalyst while the gasoline yields were similar on the two types of
catalysts. The selectivity for light cycle oil (LCO) was considerably greater
for the AMCs. When used as matrices for REY, the AMCs resulted in
considerably more active catalysts at the same zeolite content compared with
the reference catalyst while the selectivity pattern differed little between the
two types.

Silicon oxide / Aluminum oxide pillared smectites. SiO_2/Al_2O_3-pillared
smectites were prepared using two different methods (6). The surface areas
of the pillared products decreased with increasing Si/Al-ratio in the pillaring
solution while the basal spacings were essentially independent of this ratio.
The surface area of the SiO_2/Al_2O_3-pillared Li-fluorhectorites remained

almost constant after heat treatment at temperatures up to 600°C. At higher temperatures the areas decreased indicating a collapse of the pillared structure. The SiO_2/Al_2O_3-pillared montmorillonites showed a more gradual decrease in surface area with increasing temperature of heat treatment. An interesting effect of the silica incorporation was a remarkable increase in the amount of acid sites as measured by ammonia adsorption.

A sample of SiO_2/Al_2O_3-pillared montmorillonite was tested as catalyst for catalytic cracking after steam treatment at 750°C for 18 hr. The MAT activity and the surface area, both given in Table I, indicate an almost complete collapse of the material upon steaming.

Titanium-oxide pillared smectites.   Another interesting type of pillared clays studied at our laboratory are the large spacing TiO_2-pillared smectites (5) . These materials were prepared by cation exchange of the smectite with polymeric Ti-cations, formed by partial hydrolysis of $TiCl_4$ in HCl. On further hydrolysis and heating, TiO_2 pillars in the form of anatase were formed between the smectite layers. The resulting TiO_2-pillared smectites possessed surface areas in the range 200-350 m^2/g and pore volumes of about 0.2 cm^3/g. The basal spacing of products heated at temperatures above 200°C was about 28 Å, as determined by X-ray powder diffraction and by N_2-desorption pore-size analysis. The uptake of TiO_2 by the montmorillonite, the surface area, and the pore volume increased with increasing amount of Ti added in the preparation up to about 10 mmoles of Ti per g of montmorillonite. A further increase in the amount of Ti added resulted in a decrease in surface area while the pore volume and the uptake of TiO_2 remained almost constant. Fig.1 shows the surface areas of a sample prepared using 20 mmoles of Ti per g of montmorillonite thermally treated in air or hydrothermally treated in steam. The material was stable after being heat treated to 700°C, at which temperature the surface area started to decrease. The steam treated sample showed a more gradual decrease in surface area with increasing temperature.

A sample of TiO_2-pillared montmorillonite was evaluated as cracking catalyst using the MAT, after steam treatment at 750°C for 18 hr. The results of this test are shown in Table I. Compared with a conventional Al_2O_3-pillared smectite, the activity of the TiO_2-pillared montmorillonite is high which is reasonable as the TiO_2-pillared sample retains a larger fraction of its original surface area. The TiO_2-pillared smectite does, however, show a greater selectivity for coke than the other types of pillared smectites investigated at this laboratory.

Pillared interstratified illite/montmorillonites.   Pillared interstratified illite/montmorillonites were prepared using four different K-bentonite samples from the Kinnekulle region, southwest Sweden. Some chemical and physicochemical properties of these clays and a Wyoming montmorillonite used as reference clay are given in Table III. While sample A2 is regularly interstratified with a maximum I/M ordering, the three B-samples are all randomly interstratified. Fig. 2 shows X-ray diffraction patterns of air dried, oriented samples of the different clays together with the corresponding diffractograms for the Al_2O_3-pillared products. For all clays the pillaring results in a considerable increase in the basal spacing. The fact that the shift of the 001-peak is more pronounced for the reference montmorillonite is believed to be associated with the fact that this clay predominantly contains sodium as charge compensating cation while the major exchangeable cation in the interstratified clays is Ca^{2+}. Surface areas and pore volumes of the different pillared clays are given in Table IV. The surface areas of the pillared

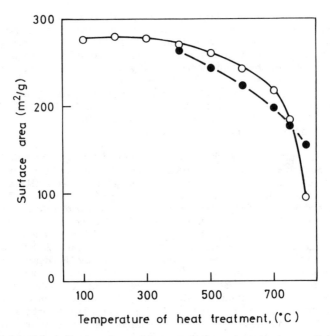

Figure 1. Dependence of surface area upon temperature of heat treatment (open circles) and of steam treatment (solid circles).

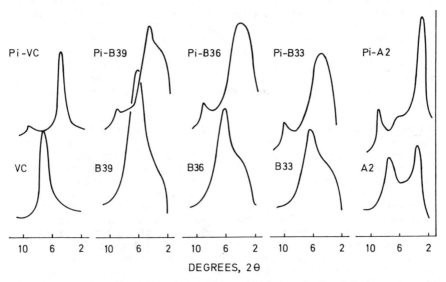

Figure 2. X-ray powder diffraction patterns of air dried, oriented samples of pillared interstratified illite/montmorillonites and of the corresponding starting clays.

Table III.   Properties of Interstratified Illite/Montmorillonites and of Reference Montmorillonite

clay sample	%smectite in I/M	CEC (meq/100g)	SA (m^2/g)	P_v (cm^3/g)
A2	32[a]	46[b]	80.9	0.133
B33	41[c]	78[c]	68.5	0.118
B36	46[c]	78[c]	75.4	0.110
B39	54[c]	91[c]	85.6	0.130
VC	100	89	51.2	0.072

[a]Estimated using the method of Ścrodoń (16) .
[b]From Byström (6) .
[c]From Brusewitz (7) .

Table IV.   Properties of Pillared Interstratified Illite/Montmorillonites and of Pillared Reference Montmorillonite

sample	SA (m^2/g)	P_v (cm^3/g)	Surface area (m^2/g)				steamed[b] 750°C
			thermally treated[a]				
			400°C	600°C	700°C	800°C	
Pi-A2	207	0.183	181	168	165	62	32
Pi-B33	229	0.172	199	175	173	75	27
Pi-B36	254	0.184	201	200	189	77	28
Pi-B39	243	0.188	201	169	158	80	36
Pi-VC	362	0.248	267	174	172	91	44

[a]Samples exposed to given temperatures in air for 3 hr.
[b]Samples exposed to 100% steam for 18 hr.

products increase with increasing fraction of smectite layers in the starting clay with the exception of Pi-B39. The relatively low surface area of Pi-B39 is probably explained by the high ion-exchange capacity of this clay resulting in uptake of more Al inbetween the clay layers thus leaving less of the interlayer space available for nitrogen adsorption. Table IV further shows the surface areas of samples thermally treated in air for 3 hr. at temperatures up to 800°C and of samples treated in 100% steam at 750°C for 18 hr. As seen, the pillared interstratified clays are relatively stable up to about 700°C while treatment at 800°C in air or at 750°C in steam results in an almost complete loss of surface area in micropores. X-ray powder diffraction analysis of thermally and hydrothermally treated samples also indicates an almost complete breakdown of the pillared structure.

The clay minerals used in this study differ from the rectorite used by Jie et al. (3) in that the negative charge of the clay layers is due, primarily, to isomorphous substitutions in the octahedral layers while the corresponding charge in the rectorite is caused by substitutions in the tetrahedral layers. Plee et al. (15) showed that, upon calcining, there is a reaction between the pillars and the clay surface in tetrahedrally substituted smectites (e.g. beidellite). This reaction did not occur in smectites without tetrahedral substitution. The pillared I/M do not show the remarkable hydrothermal stability reported for the pillared rectorite. According to the authors, the enhanced stability of the pillared rectorite is explained by the fact that, in this material, the pillared expanded layers are separated by two 2:1 layers compared to one in pillared smectites. Our investigation indicates that an increase in thickness between the expanded layers does not necessarily result in improved hydrothermal stability. On the contrary, little difference was found between the stability of the pillared interstratified clays and the pillared montmorillonite used in our study. It is, however, possible that the stability of the pillared rectorite is explained by a combination of the increase in thickness between the expandable layers and the beidellitic nature of the clay.

Conclusions

Several attempts were made to prepare pillared smectites with sufficient hydrothermal stability for use as active components in catalysts for catalytic cracking of heavy oil fractions. Although improvements were made, none of the attempts resulted in pillared materials stable enough to withstand the hydrothermal conditions found in the regenerator of a commercial FCC. One type of materials studied, i.e. alumina-montmorillonites, may be attractive alternatives to the active matrices, often alumina, currently used in FCC-catalysts designed for cracking of heavy oils. The alumina-montmorillonites can, perhaps, not be considered to be bona fide pillared smectites as they have considerably larger pores and a wider pore-size distribution than what is characteristic for pillared smectites.

    At this point, the most promising types of pillared clays for use as cracking catalysts are the pillared rectorite developed by Jie et al. (3) and the large pore Ce/Al-pillared smectites developed by McCauley (4). Further studies will have to be performed in order to determine the feasibility of these to types of materials for this application.

Acknowledgments

The author wishes to thank the Swedish Board for Technical Development (STU) for financial support of this project. Helpful advice from Jan-Erik Otterstedt of this Department and from Joseph Shabtai and Frank Massoth of the Department of Fuels Engineering, University of Utah, is greatly appreciated.

Literature Cited

1.   Vaughan, D. E. W.; Lussier, R.; Magee, J. U.S. Patent 4 175 090, 1979.
2.   Vaughan, D. E. W.; Lussier, R.; Magee, J. U.S. Patent 4,248,739, 1980.
3.   Jie, G. J.; Ze, M. E.; Zhiquing, Y. Eur. Pat. Appl. 197 012, 1986.
4.   McCauley, J. R. Int. Pat. Appl. PCT/US88/00567, 1988.
5.   Sterte, J. Clays and Clay Minerals, 1986, 34, 659.
6.   Sterte, J.; and Shabtai, J. Clays and Clay Minerals, 1987, 35, 429.
7.   Byström, A. M. Sver. Geol. Unders. Serie C, No 540, 1956.
8.   Brusewitz, A. M. Clays and Clay Minerals, 1986, 34, 442.
9.   Sterte, J.; Otterstedt, J-E., In Studies in Surface Science and Catalysis, 31, Ed.; B. Delmon, P. Grange, P. A. Jacobs and G. Poncelet, Elsevier Science Publishers, Amsterdam, 1987; p. 631.

10. Sterte, J.; Otterstedt, J-E.; Thulin. H.; Massoth, F. E., <u>Appl. Cat.</u> 1988, <u>38</u>, 119.
11. Sterte, J.; Otterstedt, J-E., <u>Appl. Cat.</u> 1988, <u>38</u>, 131.
12. Otterstedt, J-E.; Gevert, B.; Sterte, J., <u>ACS Symposium Series,</u> 1988, <u>375</u>, 266.
13. Tokarz, M.; Shabtai, J., <u>Clays and Clay Minerals</u>, 1985, <u>33</u>, 89.
14. Sterte, J.; Otterstedt, J-E., <u>Mat. Res. Bull.,</u> 1986, <u>21</u>, 1159.
15. Plee, D.; Borg, F.; Gatineau, L.; Fripiat, J. J., <u>J. Am. Chem. Soc.,</u> 1985, <u>107</u>, 2362.
16. Scrodon, J., <u>Clays and Clay Minerals</u>, 1980, <u>28</u>, 401.

RECEIVED May 9, 1990

Chapter 11

Photodegradation of Dichloromethane with Titanium Catalysts

James F. Tanguay[1], Robert W. Coughlin[2], and Steven L. Suib[1,2]

[1]Department of Chemistry and [2]Department of Chemical Engineering, University of Connecticut, Storrs, CT 06268

Titanium dioxide and titanium incorporated into alumino-silicate photocatalysts have been studied for the photodegradation of dichloromethane into carbon dioxide and HCl. Different forms of titanium dioxide have been produced such as rutile and anatase as well as an amorphous form. Titanium pillared clays have been found to be more active than titanium clays. Carbon felt was also used as a support for titanium species and this material is the most active we have studied. By enriching solutions of dichloromethane with oxygen or by pre-irradiating the titanium catalyst, faster rates of reaction and larger conversions are obtained.

Titania (TiO_2) occurs in nature in three crystal modifications as anatase, rutile and brookite. Rutile is the most common form which has octahedral coordination of titanium ions. Anatase and brookite contain distorted octahedra. Anatase is thermodynamically 8 to 12 kJ/mol more stable than rutile ([1]). Brookite on the other hand is thermodynamically unstable.

Titania is of great interest as a catalyst or photocatalyst. Matsudo and Kato have reviewed the catalytic (thermal) chemistry of TiO_2 ([2]). For photochemical activity, titania has been used to hydrogenate alkynes and alkenes ([3]), to oxidize H_2O_2 ([4]), to oxidize ethylene ([5]), to oxidize 2-propanol ([6]), for amine production ([7]) and in water splitting reactions ([8]).

Another reaction of great concern is the photodechlorination of chlorinated hydrocarbons. Titania has been used as a photocatalyst in the heterogeneous catalytic decomposition of chloroform and dichloromethane to form carbon dioxide and HCl by Ollis and co-workers ([9]). These titania catalysts are particularly useful at low levels (parts per million) of chlorinated hydrocarbon. Titania catalysts require near

0097–6156/90/0437–0114$06.00/0

ultraviolet light to degrade such hydrocarbons. One of the common problems of such systems, however, is the suspension of titania and reactant which is not easily separated.

This paper deals with a comparison of the activity of various titania catalysts in the photodegradation of dichloromethane. In addition, we will report on the possibility of supporting titania on a carbon felt to ease the separation of catalyst and reactant/product. Finally, the importance of oxygen and pre-irradiation of the catalyst will be reported.

EXPERIMENTAL

Clays and pillared clay materials were prepared by adding TiI_4 or $TiCl_4$ to aqueous solutions and ion exchanging with the clay or incorporating titanium ions into the aluminochlorohydrate pillar precursor prior to pillaring. Titanium oxide was deposited onto carbon felt by acidifying titanium isopropoxide and forming a sol into which the carbon felt was dipped. Pure titania of the amorphous, anatase, and rutile phases were prepared by treating titania sols at 70°C, 375°C , and 850°C, respectively.

Solutions of 1 mL dichloromethane and 100 mL distilled deionized water were irradiated in the presence and absence of titania catalysts with a 1000 watt Xe lamp. About 1 g catalyst was used in these studies. A Corning pH meter was used to monitor pH of the reaction. A 300 nm cutoff filter was used to filter out low wavelength ultraviolet light. The conversion of dichloromethane was monitored with gas chromatography methods.

RESULTS

Results of photocatalytic experiments are given in Table I. The percent conversion was calculated from the observed amount of Cl⁻ in solution after photolysis.

Table I. Conversion of Dichloromethane

Catalyst	% Conversion
Amorphous	15
Anatase	37
Rutile	59
Anatase on Felt	29
Amorphous Titania on Felt	6
Pre-irradiated Rutile	77
Titania Pillared Clay	9
Pre-irradiated Pillared Clay	10
Rutile, Oxygen Bubbling	83
Rutile, Nitrogen Bubbling	41

The pre-irradiation lasted for two hours in all cases and all measurements were taken after four hours of photolysis. Oxygen bubbling was done by attaching an oxygen tank, regulator and flow meter to the photolysis apparatus. A plot of the chloride production of pre-irradiated titania pillared clays is given in Figure 1 for samples with no pre-irradiation and for those irradiated for one and two hours.

Data for a fresh anatase catalyst and a used anatase catalyst are given in Figure 2. The amount of chloride produced from dichloromethane as a function of time for both materials is given there. At long periods of time, the used material shows less chloride ion production than the fresh catalyst.

DISCUSSION

The data of Figure 1 clearly show that irradiation of the titania pillared clay systems leads to an enhancement in the moles of Cl⁻ produced during photodegradation of dichloromethane. Prolonged pre-irradiation leads to lower conversion than shorter periods of pre-irradiation. This is a general result and was found for all supported systems studied here.

The significance of pre-irradiation is that defect sites must be created at the titania surface during irradiation that serve as adsorption sites for the dichloromethane. These may be sites for dissociation of Cl⁻ since we have found (10) that poisoning of various substrates by NaCl causes a general decrease in overall conversion. Prolonged irradiation may cause degradation of these activated sites.

Data for several titania systems are given in Table I for the photodegradation of dichloromethane. First of all, data for unsupported titania should be compared. These data suggest that the rutile form of titania is more active than the anatase form which in turn is more active than the amorphous titania. In general this is also true for the supported analogs. Reasons why an ordered high temperature form is more active than a low temperature or amorphous form may be due to the propensity of the former to form defect adsorption sites (See below).

For the amorphous and anatase forms supported on carbon felt, the percent conversion is markedly higher for the anatase form than the amorphous form. It was not possible to produce the rutile form on the carbon felt due to degradation of the felt during thermal activation at the high temperature ($> 600°C$) needed to produce rutile.

Data for pre-irradiated rutile and titania pillared clay clearly show that pre-irradiation causes the % conversion to increase with respect to non-irradiated materials as discussed earlier.

Data in Table I for oxygen bubbling of rutile versus nitrogen bubbling show that oxygen can enhance the overall % conversion of dichloromethane into HCl and CO_2. These data may indicate that the rutile surface provides oxygen for the oxidation of the carbon part of the chlorinated hydrocarbon and that lattice oxygen is then replaced with gas phase oxygen that absorbs at the surface.

Data of Figure 2 clearly show that the used anatase catalyst produces less Cl⁻ than the fresh catalyst. Poisoning studies with titania

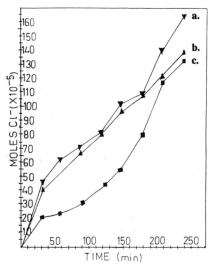

Figure 1. Chloride production of Pre-irradiated Titania Pillared Clays, 750 Watts Power, Wavelengths of 300 nm to 800 nm. (a) 1 Hr Pre-irradiation, (b) 2 Hr Pre-irradiation, (c) No Pre-irradiation.

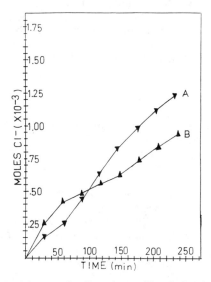

Figure 2. Chloride production from Used Anatase: (A) Fresh catalyst, (B) Catalyst in 2A, washed with distilled deionized H_2O.

catalysts treated with NaCl suggest that Cl⁻ ions adhere to the surface of titania and react with sites that are active in the photodechlorination process. In our systems it appears that Cl⁻ ions are produced during the photodegradation and then poison active sites.

Ollis and co-workers have suggested that dichloromethane absorbs as an ion by reacting with a hole on the titania surface (9). Other studies have shown, however, that both Cl⁻ and H^+ inhibit the degradation of simple hydrocarbons (11). These inhibitors could influence the Brönsted acidity of the titania surface. This could lead to a different mode of binding of dichloromethane and other hydrocarbons to the surface such as by abstraction of H^+ from the hydrocarbon as well as binding of a hydrocarbon anion to the surface of titania (i.e. as $CHCl_2^-$).

In this case, H^+ could be bound to lattice oxygen and the anionic hydrocarbon fragment could be bound to Ti^{4+} in the lattice. This mechanism is supported by the data of Ollis and co-workers (9). Our pre-irradiation data can be explained by the fact that light consists of an electron-hole pair that can form on the titania surface during irradiation. Pre-irradiation of the surface of titania leads to a build-up of such defect sites that are not initially destroyed before reaction with dichloromethane since there is no chloride ion around.

CONCLUSION

We have shown here that different crystalline forms of titania as well as different supports for the titania influence the overall reactivity of TiO_2 in the photodechlorination of dichloromethane. The importance of pre-irradiation to produce defect centers on the surface has also been observed. Oxygen plays a role in these photodegradations since saturation of the solution with oxygen enhances the rate of photodegradation. Chloride ions poison TiO_2 during irradiation. Further studies should focus on removing Cl⁻ ions from the surface.

Literature Cited

1. Cotton, F. A.; Wilkinson, G. Advanced Inorganic Chemistry, Fourth Edition, John Wiley and Sons, NY, 1980.
2. Matsudo, S.; Kato, A. Appl. Catal. 1983, 8, 149-165.
3. Anpo, M.; Aikawa, N.; Kadama, S.; Kubukawa, Y. J. Phys. Chem. 1984, 88, 2569-2572.
4. Rivers-Arnau, V. J. Electroanal. Chem. 1985, 190, 279-281.
5. Gonzalez-Elire, A. R.; Che, M. J. Chim. Phys. 1982, 79, 355-359.
6. Henglein, A. Ber. Buns. Phys. Chem. 1982, 86, 241-246.
7. Miyama, H.; Nosaka, Y.; Fukushima, T.; Toi, H. J. Photochem. 1987, 36, 121-123.
8. Grätzel, M.; Borgarello, E.; Kiwi, J., Pelizetti, E.; Visca, M. J. Am. Chem. Soc. 1981, 103, 6324-6329.
9. Ollis, D. F.; Hsiao, C-Y.; Lee, C-L. J. Catal. 1983, 82, 418-423.
10. Tanguay, J. F.; Coughlin, R. W.; Suib, S. L. J. Catal. 1989, 117, 335-347.
11. West, A. R. Solid State Chemistry, John Wiley and Sons, NY, 1987.

RECEIVED May 9, 1990

Chapter 12

New Tubular Silicate-Layered Silicate Nanocomposite Catalyst

Microporosity and Acidity

Todd A. Werpy, Laurent J. Michot, and Thomas J. Pinnavaia

Department of Chemistry, Center for Fundamental Materials Research, Michigan State University, East Lansing, MI 48824

The microporosity of a new tubular silicate-layered silicate nanocomposite formed by the intercalation of imogolite in Na^+-montmorillonite has been characterized by nitrogen and m-xylene adsorption. The nitrogen adsorption data yielded a liquid micropore volume of ~0.20 cm^3g^{-1}, as determined by both the t-plot and the Dubinin-Radusikevich methods. The t-plot provided evidence for a bimodal pore structure which we attributed to intratube and intertube adsorption environments. The m-xylene adsorption data indicated a much smaller liquid pore volume (~0.11 cm^3g^{-1}), most likely due to incomplete filling of intratubular pores by the planar adsorbate. The FTIR spectrum of pyridine adsorbed on the TSLS complex established the presence of both Bronsted and Lewis acid sites. The TSLS complex was shown to be active for the acid-catalyzed dealkylation of cumene at 350°C, but the complex was less reactive than a conventional alumina pillared montmorillonite.

Imogolite is a naturally occurring aluminosilicate mineral with a unique tunnel-like or tubular structure[1]. The external and internal van der Waals diameters of the tubes are approximately 25Å and 10Å respectively[1,2]. The tube dimensions are larger for the synthetic form than for the naturally occurring derivative owing to a difference in the number of repeat units defining the tube walls. The molecular sieving properties of imogolite already have been demonstrated[3-5]. The adsorption of small molecules occurs readily on the internal surfaces, but larger species with kinetic diameters $\geq$ 10Å, such as perflurotributylamine, are

0097–6156/90/0437–0119$06.00/0

excluded[5]. The potential for shape-selective catalysis by imogolite also has been recognized, but the thermal instability of the tubular structure is a limiting factor[2,5-7].

We have been investigating the use of imogolite as a pillaring agent for smectite clays with layer lattice structures[8-9]. The regular intercalation of the tubes within the layered host results in the formation of a tubular silicate-layered silicate (TSLS) complex. These new nanocomposite materials may be viewed as pillared clays in which the pillars themselves are microporous. Significantly, the TSLS structure is thermally stable up to 450°C when montmorillonite is selected as the layered host[9]. A schematic representation of a TSLS complex is provided in Figure 1. On the basis of preliminary XRD and stochiometric studies, it appears that the imogolite tubes are in van der Waals contact, most likely in a log-jam-like array in the layer silicate galleries. Although the tubes stuff the galleries, two unique adsorption environments are available, namely, the intra-and inter-tube pores designated A and B in Figure 1.

In the present work we examine the microporosity of a TSLS complex formed from synthetic imogolite and natural montmorillonite. Nitrogen adsorption and desorption isotherms are reported and analyzed in terms of microporous volume and surface area. Also, the adsorption isotherm for an organic adsorbate, m-xylene, is reported. Preliminary FTIR results for the chemisorption of pyridine and catalytic studies of the dealkylation of cumene suggest that TSLS complexes are promising microporous acids for shape selective chemical conversions.

Experimental Section:

An imogolite pillaring solution was prepared by the acid hydrolysis of 30mM tetraethylorthosilicate and 60mM aluminum sec-butoxide for 2 days at 98° C according to the method of Farmer[10]. The reaction solution was dialyzed 4 days against distilled water to remove ionic and molecular by-products prior to being used as a pillaring reagent for montmorillonite. Intercalation of imogolite into sodium montmorillonite was achieved by reacting ~1.0 wt% aqueous suspensions of the two reagents according to previously described methods[8,9].

Nitrogen adsorption/desorption isotherms were determined at -196°C on a Quantachrome Autosorb Sorptometer. Ultrahigh purity nitrogen was used as the adsorbate and ultrahigh purity helium was the carrier gas. The adsorption isotherm for m-xylene was obtained on a McBain balance equipped with quartz glass springs and buckets. The samples were outgassed at 325° C under vacuum for about four hours prior to adsorption.

The FTIR spectrum for chemisorbed pyridine was obtained on an IBM IR44 spectrometer. The TSLS sample was prepared by evaporating an aqueous suspension of the complex onto a silicon wafer at room temperature and subsequently outgassing the film at 350°C prior to pyridine adsorption at 25°C. Physically adsorbed pyridine was removed by pumping under vacuum at room temperature.

Catalytic dealkylation of cumene was carried out in a fixed bed reactor operated at 350°C and atmospheric pressure. The contact time was 1.5 sec and the WHSV was 0.4g cumene/g/hr. The conversion of cumene was determined using a Hewlett Packard 5890 gas Chromatograph equipped with a Supelco wide bore capillary column.

Results and Discussion

Isotherms for the adsorption and desorption of nitrogen (kinetic diameter, 3.6Å) were obtained for the imogolite-montmorillonite TSLS complex at -196°C (see Figure 2). The shape of the adsorption isotherm below a partial pressure of 0.5 was similar to the classical type I isotherm for a microporous (<20Å) adsorbent according to the classification of Braunauer et al.[11] Above P/P_o of ~0.50, tailing of the adsorption isotherm was observed due to mesoporosity in the range 20-500Å. The presence of some mesopores was confirmed by the hysteresis loop observed upon desorption.

The nitrogen adsorption data over the partial pressure range $0 < P/P_o < 0.15$ were fitted to the BET equation:[12]

$$\frac{P/P_o}{V(1-P/P_o)} = \frac{1}{V_m C} + \frac{(C-1)(P/P_o)}{V_m C}$$

An equivalent surface area of 460 $m^2 g^{-1}$ was determined from the monolayer volume, V_m. The value obtained for the dimensionless energetic constant, C=260, was characteristic of a microporous material. Although the BET surface area may not be a physically precise quantity due to the fact that the nitrogen molecule does not exhibit the same cross-sectional area in a microporous environment as on a flat surface,[13] the BET value is useful for comparisons of relative porosities among a related class of adsorbents. For instance, smectite clays pillared by metal oxide aggregates typically exhibit BET surface areas in the range 150 - 400 m^2/g. Thus, the TSLS complex is among the more porous intercalated nanocomposites derived from smectite clays.

The total pore volume (V_o) of the TSLS complex was assessed by applying the Dubinin-Radusikevich equation[14] to the nitrogen adsorption data:

$$\text{Log } V = \text{Log } V_o + D[\log(P/P_o)]^2$$

As shown by the dashed line in Figure 3, this equation provided a good fit of the data over the low and

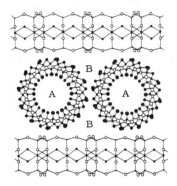

Figure 1.   Schematic illustration of the TSLS complex
             formed from imogolite and Na^+- montmorillonite.
             The larger filled circles represent hydroxyl
             groups, the open circles are oxygens and the
             small filled circles are the positions of metal
             ions such as silicon and aluminum.   A and B
             denote two possible types of micropores.

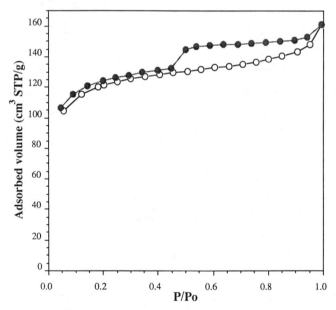

Figure 2.   Isotherms for the adsorption and desorption of
             nitrogen on the TSLS complex at -196^OC.

intermediate pressure ranges. The intercept of the best straight line fit gave a liquid pore volume V_o = 0.20 cm^3g^{-1}.

A t-plot[15] of the N_2 adsorption data, shown Figure 4, exhibited three distinct regions. In the first region, the data were fitted to a straight line passing through the origin. The slope of this line yielded an equivalent surface area of 480 m^2g^{-1}, comparable to the value of 460 m^2g^{-1} obtained from the BET treatment of the data. Another domain of data points were fitted to a second straight line with an intercept (67 cm^3 STP g^{-1}) corresponding to a microporous liquid volume of 0.10 cm^3g^{-1}. A third domain of points yielded an overall liquid volume of 0.16 cm^3g^{-1}. This bimodal microporous behavior is consistent with the two types of micropores expected for an intercalated TSLS complex. As illustrated in Figure 1, there are two types of environments available for nitrogen adsorption, namely, the type A intratubular pores and the type B intertubular pores. We tentatively suggest that the experimentally observed bimodal microporous behavior arises from these two types of environments.

The nitrogen desorption isotherm was treated according to the model developed by Delon and Dellyes for parallel plate pores in phyllosilicates.[16] The mesopore liquid volume obtained using this treatment was relatively small, 0.03 cm^3g^{-1}. Thus, the pore volumes derived from t-plot analysis (0.16 cm^3g^{-1}) and the nitrogen desorption branch (0.03 cm^3g^{-1}) sum to a value (0.19 cm^3g^{-1}) in good agreement with the pore volume obtained from the Dubinin-Radusikevich equation (0.20 cm^3g^{-1}).

In order to examine the adsorption of an organic adsorbate by the TSLS complex, the adsorption isotherm for m-xlene (kinetic diameter, 5.6Å) was obtained at 25°C. As shown in Figure 5, the shape of the isotherm was again type I, as expected for a microporous adsorbent. The Dubinin-Radusikevich equation provided a reasonable fit of the data over the low and medium partial pressure ranges (see Figure 6). The microporous liquid volume obtained by this treatment was 0.11 cm^3g^{-1}. This value was significantly lower than the microporous volume obtained by nitrogen adsorption (0.20 cm^3g^{-1}). Thus, Gurvitsch's rule,[17] which states that the microporous volume should be independent of adsorbate size, does not hold for the TSLS adsorbent. This result suggested to us that the planar m-xylene molecule, for which kinetic diameter is ~2/3 the internal 10Å pore diameter of the intercalated imogolite, was geometrically incapable of filling completely the available pore volume. An analogous geometrical effect was believed to be important in determining the adsorbate-dependent filling of pristine imogolite tubes[5]. Thus, we conclude that a substantial portion of the microporous volume of the TSLS complex arises due to adsorption on the internal surfaces of the intercalated imogolite tubes.

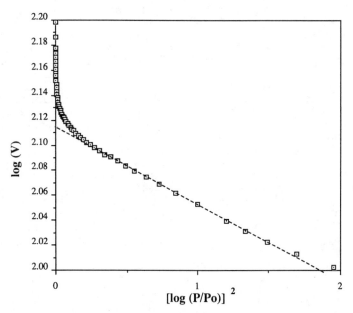

Figure 3.   Nitrogen adsorption data plotted according to the Dubinin-Radusikevich equation.

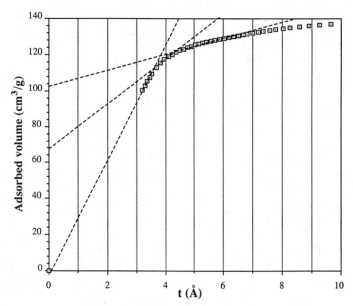

Figure 4.   t-Plot for N_2 adsorption on the TSLS complex.

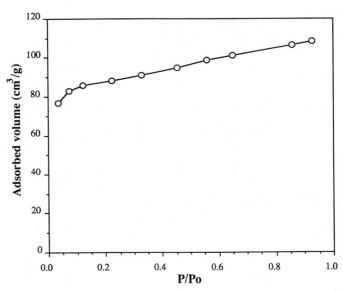

Figure 5.   Adsorption of m̲-xylene by the TSLS complex at 25°C.

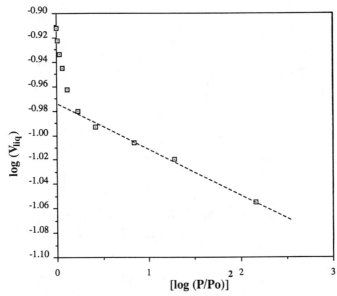

Figure 6.   m̲-Xylene adsorption data plotted according to the Dubinin-Radusikevich equation.

In an effort to identify the nature of acid sites in the TSLS complex and the potential for acid catalysis, we investigated the chemisorption of pyridine by FTIR spectroscopy. This adsorbate is known to exhibit aromatic ring stretching frequencies characteristic of pyridinium ion (Bronsted acidity) and of coordinated pyridine molecules (Lewis acidity)[18,19]. The FTIR spectrum shown in Figure 7 was obtained after outgassing the TSLS complex at 350°C, adsorbing pyridine at 25°C and P/P^o = 1.0, and subsequently outgassing to remove much of the physically adsorbed pyridine. Both Bronsted and Lewis acid sites were clearly evident in the FTIR spectrum. An intense ring stretching mode characteristic of pyridine coordinated to Lewis acid sites was found at 1450 cm^{-1}, whereas a weaker band due to the N-H deformation of pyridinium ion was observed at 1540 cm^{-1}. Also, a ring stretching band of intermediate intensity due to both Bronsted and Lewis acid sites was found at 1490 cm^{-1}.

The acidic functionality of the TSLS complex was demonstrated in the catalytic dealkylation of cumene. For this model reaction at 350°C the conversion of cumene to benzene was monitored as a function of time on stream (see Figure 8). Included in the study for comparison purposes

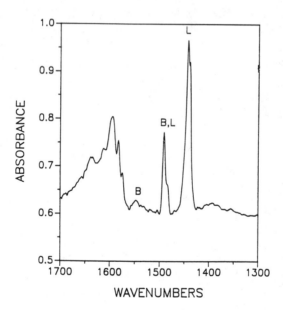

Figure 7.   FTIR spectrum (1700 - 1300 cm^{-1}) for pyridine adsorbed on the TSLS complex.

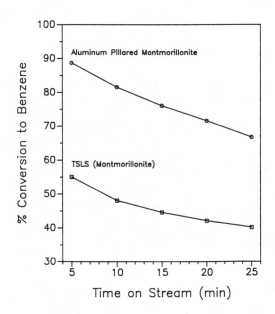

Figure 8.   Catalytic dealkylation of cumene at 350°C over the imogolite-montmorillonite TSLS complex and alumina pillared montmorillonite (APM).

was a typical alumina pillared montmorillonite (APM) prepared by the reaction of aluminum chlorhydrate and the Na^+ exchange form of the clay.[20] For both catalysts the conversion decreased with increasing time on stream, most likely due to the formation of coke and restricted accesses to the acid sites.

Significantly, the TSLS complex is less active than APM. Initial conversions were approximately 55% and 88% for the TSLS and APM, respectively. This difference in reactivity was not especially surprising, because the charge on the montmorillonite component of the TSLS complex is balanced by weakly acidic Na^+ ions, whereas in APM the charge-balancing cations are more highly acidic aluminum cations and/or hydrogen ions. Thus, the acidity observed both by pyridine adsorption and by cumene cracking most likely resulted from the intrinsic acidity of the imogolite component of the TSLS complex. Future catalytic studies will focus on the effect of the exchange cation on TSLS catalytic activity.

Acknowledgments

The financial support of the National Science Foundation through grant DMR-8903579 and the Michigan State University Center for Fundamental Materials Research is gratefully acknowledged.

References

1. Cradwick, P.D. 6.; Farmer, V.C.; Russell, J.D.; Mason, C.R.; Wada, K.; Yoshinaga, N. Nat. Phys. Sci. **1972**, 240, 187-189.

2. Farmer, V.C.; Adams, M.J.; Fraser, A.R.; Palmieri, F.; Clay Minerals **1983**, 18, 459-472.

3. Wada, K.; Henmi, T. Clay Sci. **1972**, 4, 127-136.

4. Egashina, K.; Aomine, S. Clay Sci. **1974**, 4, 231-242.

5. Adams, M.J. J. Chromatog **1980**, 188, 97-106.

6. Wilson, M.A.; Wada, K.: Wada, S.I.; Kakuto, Y. Clay Minerals **1988**, 23, 175-190.

7. MacKenzie, K.J.D.; Bowden, M.E.; Brown, I.W.M.; Meinhold, R.H. Clay Minerals **1989**, 37, 317-329.

8. Johnson, I.J.,; Werpy, T.A.; Pinnavaia, T.J. J. Amer. Chem. Soc. **1988**, 110, 8545-8546.

9. Werpy, T.A.; Pinnavaia, T.J. J. Amer. Chem. Soc., to be submitted.

10. Farmer, V.C. U.S. Patent 4,252,779,1981.

11. Brunauer, L.S.; Deming, L.S.; Deming, W.S.; Teller, E. J. Amer. Chem. Soc. **1940**, 62, 1723.

12. Brunauer, L.S.; Emmett, D.H.; Teller, E. J. Amer. Chem. Soc. **1938**, 60, 309.

13. Sing, K.S.W.; Everett,D.H.; Haril, R.A.W.; Moscou, L.; Pierotti, R.A.; Rouquerol, J.; Siemienewska, T. Pure and Applied Chemistry **1985**, 57, 603-619.

14. Dubinin, M.M. Pure and Applied Chemistry **1965**, 10, 309.

15. De Boer, J.M.; Lippens, B.C.; Linsen, B.G.; Broekhoff, J.C.P.; Van den Heuvel, A.; Osinga, Th.J. J. Coll. and Interface Sci. **1966**, 21, 405-414.

16. Delon, J.F.; Dellyes, R. C.R.Ac.Sci. Paris, Série D **1967**, 265, 1161-1164.

17. Gregg, S.J.; Sing, K.S.W. in "Adsorption, Surface Area and Porosity" Second Edition, **1982**, Academic Press, London.

18. Perry, E.P. J. Catal. **1963**, 2, 371.

19. Verdine, J.C. J. Catal. **1979**, 59, 248.

20. Pinnavaia, T.J.; Tzou, M.-S.; Landua, S.D.; Raythathan, R.H. J. Molec. Catal. **1984**, 27, 195-212.

RECEIVED May 9, 1990

Chapter 13

Cobalt Clays and Double-Layered Hydroxides as Fischer–Tropsch Catalysts

L. Bruce, J. Takos, and T. W. Turney

Division of Materials Science and Technology, Commonwealth Scientific and Industrial Research Organisation, Locked Bag 33, Clayton, Victoria, Australia, 3168

Cobalt-containing clays and double layered hydroxides have been synthesized and examined as Fischer Tropsch catalysts. Hydrothermal treatment of cobalt-aluminum hydroxysilicate gels affords 2:1 phyllosilicate clays with structures intermediate between smectites and chlorites. The clays become more chloritic on increasing Al content. Calcination results in decomposition of the hydroxide interlayer region and formation of small $Co(Co,Al)_2O_4$ crystallites decorating the edges of the clay platelets. Hydrogen reduction converts these crystallites to ca. 100Å metallic Co particles. Catalysts containing metallic Co were also prepared from Co/Al hydroxycarbonates with a hydrotalcite structure. Comparison of the Fischer Tropsch activity of these catalysts showed the hydrotalcite-derived system to have longer lifetime and lower water gas shift activity.

This work represents part of a program designed to identify Fischer-Tropsch (FT) catalysts for the processing of H_2-rich synthesis gas derived from natural gas. Current advanced gas processes are generally a combination of partial oxidation and steam reforming, resulting in H_2:CO ratios of about 1.5-2.3. Cobalt- rather than iron-based FT catalysts have been examined, in order to minimize the competing water-gas shift reaction, which would result in a lowered carbon efficiency. Most cobalt FT catalysts have been prepared by coprecipitation of Co salts with various promoters onto a slurried oxide support to afford mixed phase systems (1). Reduction to the active catalyst was controlled by addition of various promoters (e.g. MgO, ThO_2, Al_2O_3) (2). In part, these promoters are necessary to maintain good metal dispersion in the catalyst and resistance to sintering. Dispersion

0097–6156/90/0437–0129$06.00/0

in multicomponent systems should be more effectively controlled through formation of structurally well-defined precursor chemical compounds. Low dimensional, lamellar solids, such as clays or hydrotalcites, are especially attractive for this purpose, offering both readily varied composition and the potential for high component dispersion and surface areas.

Experimental Section

Details of characterization by x-ray and electron diffraction, transmission electron microscopy and XPS are given elsewhere (3). Temperature programmed reduction (TPR) studies using a $3\%H_2/N_2$ gas mixture, were performed on 40 mg catalyst samples in a tubular furnace (heating rate 10°C/min), using TCD and FID detection. Prior to TPR, the sample was heated in air at a specified temperature for 3h.

Hydrothermal syntheses were performed in an autoclave (Parr, 2 L), using portions of a slurry derived from adding a hot (85-90°C) solution of $Co(NO_3)_2.6H_2O$ (0.5M, 1L) to a rapidly stirred solution of hot NaOH (4M, 0.5L), into which $Al(NO_3)_3.9H_2O$ and fumed silica had previously been dissolved. The atomic ratios of Al and Si employed, at constant ratio Co/Al+Si = 1.5, are shown in Table I.

Table I. Dependence of Cobalt Clay Phases
on Aluminum Content

Al/Al+Si	Major Phases
0.0	Smectite
0.1	Fibres + smectite + chrysotile
0.2	Smectite
0.3	Chlorite + smectite
0.4	Chlorite
0.5	Chlorite
0.6	Chlorite + CoO
0.8	Chlorite + CoO + $Co(Co,Al)_2O_4$
1.0	CoO + $Co(Co,Al)_2O_4$

The autoclave was pressurized with hydrogen (400 kPa) and heated to 250°C for 16 h before cooling and venting. The clay was collected by centrifugation and washed repeatedly with water to remove ionic impurities.

Cobalt hydrotalcites were obtained by co-precipitation and ageing of slurries indicated in Table II, using a technique similar to that of Reichle (4), viz, reaction of $NaOH/Na_2CO_3$ (3/1 ratio) solution (3.5 M in OH⁻) with a mixed Co/Al nitrate solution (1.5 M in Co^{2+}), such that the number of equivalents of base added equalled the number of equivalents of Co+Al ions in solution. For ageing the slurry at 200°C, an autoclave was used. The degree of crystallinity was estimated from the width of $d_{(003)}$ peak in XRD.

Table II. Production of Cobalt/Aluminum
Hydroxide Catalyst Precursors

Metal Ratio Co/Al	Temp. (°C)	Time (h)	Phases present $[d_{(003)}$, Å]	Xtallinity
1.7	100	60	Hydrotalcite [7.54]	Medium
3.0	100	60	Hydrotalcite [7.65]	High
7.0	100	60	Hydrotalcite [7.4] + Co(OH)$_2$	Poor
3.0	65	72	Hydrotalcite [7.70]	High
3.0	200	18	Hydrotalcite [7.62]	V.high

Fischer Tropsch performance was examined at atmospheric
pressure, in a down-flow microreactor with associated on-line gas
phase analysis, which was automated and linked to computerized data
reduction (5). Samples were prepared for testing by initial drying
overnight at 90°C, followed by calcination at 620°C for 2 h and
sizing (60-80#). The reactor temperature was ramped in flowing H$_2$
at 6° min^{-1} to 400°C, and held for at least 3 h, before lowering
the temperature to 180°C and commencing flow of H$_2$/CO (1.5/1). Gas
phase analyses were made at 0.5 h intervals, and the reaction
temperature raised in steps to 220°C. An overnight run at about 10%
conversion level enabled collection of condensate for analysis by
liquid injection.

Results and Discussion

Cobalt Clay Syntheses. Precipitation of cobalt salts by alkali
metal silicate/hydroxide mixtures produces cobalt hydroxysilicate
gels, which afford a wide range of 1:1 and 1:2 cobalt clays on
hydrothermal treatment. It has been shown previously that the type
of clay obtained varies with the alkali metal and initial Co/Si
ratio (3). Addition of aluminium salts to these hydroxysilicate
gels enabled the synthesis of new clay phases (as well as a fibrous
phase) as indicated in Table I. The new mixed Al/Co-containing
clays were pink to lilac in colour and were produced as the major
phase (>95 wt%) in the range of atomic ratios, Al/Si+Al = 0.2-0.5
at a cobalt content of Co/Si+Al = 1.5.

Chloritic Structure of Co/Al Clays. Each of the clay preparations
exhibited an XRD basal spacing in the region, 14.7-15.3Å, as well
as higher order (001) and (hk0) reflections typical of a well-
ordered 2:1 phyllosilicate (Figure 1a). The swelling properties of
these clays varied over the Al/Al+Si composition range. In the
absence of Al, a bright blue clay, with swelling and thermal
properties typical of a smectite, was obtained (3). In the range,
Al/Al+Si = 0.1-0.3, heating of the clay phase resulted in layer
contraction in stages [d(001) = 12Å at 300°C and 9.7Å at 600°C].
This is typical behaviour of a swelling smectite clay, such as
montmorillonite. However, with the Al-containing clays, treatment

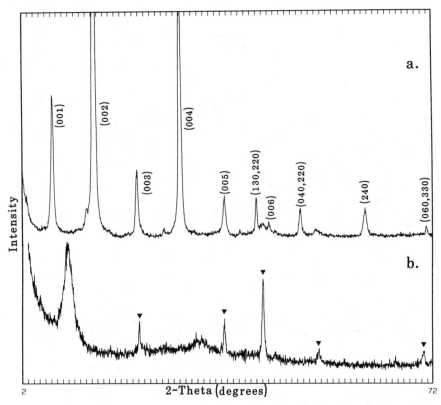

Figure 1. X-ray powder diffraction pattern of a Co/Al chlorite phase, Al/Al+Si = 0.4. a) Calcined at 600°C, 3 h, showing spinel phase, $Co(Co,Al)_2O_4$ (*), b) Air-dried sample.

with ethylene glycol or glycerol did not result in increased
basal spacings, as would also be expected for a smectite. At
Al/Al+Si = 0.33–0.8, layer contraction on mild heating was less
(for Al/Al+Si = 0.4, $d_{(001)}$ = 14.1Å at 450°C).

These Al-containing clay phases are believed to consist of
positively charged, mixed Co/Al hydroxide sheets within
the interlayer region of a trioctahedral Co/Al clay. As Al/Al+Si
increases, the interlayer Co/Al hydroxide sheet becomes
progressively more complete, resulting in the clay structure and
properties grading from those expected for a smectite to those of a
typical chlorite. A transmission electron micrograph of a typical
chloritic phase is shown in Figure 2a.

Calcination and Reduction of Clay Phases. The interlayer region in
chlorites and related hydroxy interlayered clays is reported to
dehydrate between 250°C and 650°C, depending upon the interlayer
cation. This causes collapse of the interlayer region, but
apparently without destruction of the 1:2 layer system (6–7). In
accord with this model, the basal spacing of the present Co/Al
clays decreased to ca 9.7Å, on heating in air to 600°C. This
lattice contraction was accompanied by formation of crystalline
Co_3O_4 (Figure 1b). Electron micrographs (Figure 2b) revealed
that the liberated Co_3O_4 particles decorated the edges of the clay
platelets, whose morphology had otherwise remained unchanged. The
formation of Co_3O_4 microcrystallites was confirmed by selected area
electron diffraction (SAD). As expected, heating caused the clay
samples to turn black, due to oxidation of Co(II) to Co(III).

Temperature programmed reduction of calcined clay samples
revealed that their reducibility was sensitive both to the thermal
pretreatment conditions as well as to Al/Al+Si ratio (Figure 3).
Table III summarizes the degree of reduction to Co, as measured
from the H_2 uptake to 450°C. On samples previously calcined to
600°C, reduction to form metallic Co was confirmed by both XPS and
SAD.

Table III. Dependence of Catalyst Reducibility on Aluminum
Content and Calcination Temperature In Cobalt Clays

Al/Al+Si ratio	Calcination Temperature	
	110°C	620°C
0.0	<0.1[a]	<0.1[a]
0.1	1.5	4.0
0.2	1.0	6.5
0.3	1.6	10.5
0.4	1.7	5.6
0.5	2.4	5.7
0.6	4.7	–

a. Number of moles of hydrogen reacted per Kg of catalyst during
the temperature programmed reduction up to 450°C.

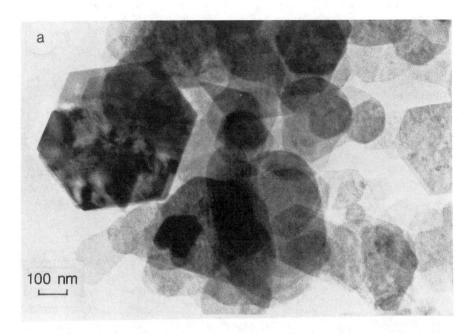

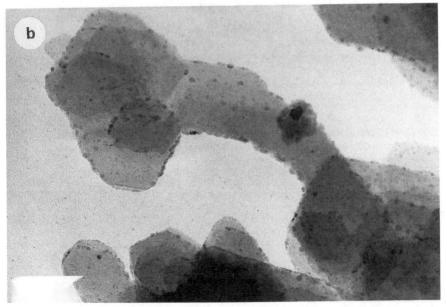

Figure 2. Transmission electron micrograph of a Co/Al chlorite phase, Al/Al+Si = 0.4. a) Air-dried sample. b) Calcined at 600°C, 3h, showing small particles of spinel phase decorating edges of hexagonal clay platelets.

Electron micrographs showed the formation of metallic cobalt particles (ca 100Å), supported on the remnants of the clay after calcination and reduction (Figure 4). In part, the degree of chloritization determined the extent of formation of free Co_3O_4 from the clay and, correspondingly, the amount of Co metal present in the samples was found to reach a maximum at Al/Al+Si of 0.2-0.3. Furthermore, only a fraction of the Co present in these clays was reduced and available for catalytic reaction. The remaining Co was contained within the 2:1 layer clay framework and remained unreduced until >700°C, a temperature typical of that needed to initiate total framework collapse.

Layered Cobalt Hydroxides. The above results demonstrate the importance of the interlayer Co rather than the framework cobalt, in forming the required metallic Co catalyst. A structure with optimal catalytic activity would consist of one where there were only Co/Al hydroxide sheets, a structure closely approximated to by the hydrotalcite series of double layered hydroxides.

Synthesis of Co/Al Hydroxycarbonates. Reaction of a mixture of cobalt and aluminum salts with $NaOH/Na_2CO_3$ afforded a poorly crystalline precipitate, which on aging at elevated temperatures yielded Co/Al hydroxycarbonates, with a lamellar structure, isomorphous with the mineral hydrotalcite (HT) – $[Mg_6Al_2(OH)_{16}][CO_3.xH_2O]$. TEM clearly showed the presence of the layered structure, with a measured basal spacing of about 7Å (Figure 5). XRD indicated the presence of minor and less crystalline hydrotalcite phases in the cases of Co/Al = 3, aged at 60°C, with d(003) = 7.34Å and aged at 200°C with 8.42Å. Such multiple phases imaged in electron micrographs as regions with different degrees of structural order.

Calcination and Reduction of Co/Al Hydroxycarbonates. As found with the Co/Al clays, temperature programmed reduction studies revealed that pre-calcination in air was necessary to produce a metallic cobalt-containing catalyst at reasonable reduction temperatures. TGA and XRD studies showed that dehydration occurred between 200-300°C, without loss of the basal spacing. Further heating was accompanied by partial oxidation and decarboxylation, the latter being complete by about 600°C. This heating yielded a mixed oxide with a spinel structure (XRD). The expected stoichiometry of the calcination reaction for the Co/Al = 3 system should result in a solid solution of Co_3O_4 and $CoAl_2O_4$ spinel phases (ratio 3:5).

$$3[Co_6Al_2(OH)_{16}]CO_3 + 5/2\ O_2$$

$$8Co^{II}(Co^{III},Al)_2O_4 + 24\ H_2O + 3CO_2$$

Hydrogen uptake up to 450°C, measured by TPR, of a catalyst previously calcined at 600°C, yielded 3.3 moles H_2/Kg of Co/Al hydroxide (Co/Al = 3, 60°C aging). This value is somewhat less than the H_2 uptake for the optimal chlorite preparations, and was

Hydrogen Uptake

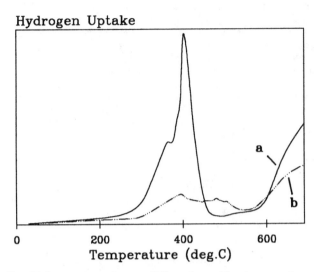

Temperature (deg.C)

Figure 3. Hydrogen uptake profiles from the temperature programmed reduction of a Co/Al chlorite phase, Al/Al+Si = 0.4. Samples previously heated in air to a) 110°C and b) 620°C.

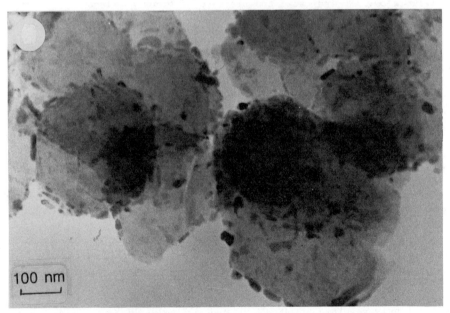

100 nm

Figure 4. Transmission electron micrograph of a used F.T. catalyst, derived from a Co/Al chlorite phase, Al/Al+Si = 0.4, showing small particles of metallic Co on clay platelets.

reflected in a poorer F.T. activity (vide infra). In the case of
the spinel reduction, more Co appeared to be stabilized by a
$CoAl_2O_4$-like phase, which reduced at temperatures well above 500°C.

Fischer-Tropsch Activity

Both chlorites and hydrotalcites performed as Fischer Tropsch
catalysts, with an Anderson-Schulz-Flory value of about 0.8, as
expected from the presence of metallic cobalt. Both yielded
condensates consisting mainly of n-alkanes, peaking at C_{10-11} and
extending to C_{23}, as shown in Figure 6. Within the stability field
for the synthesis of chlorites, catalyst activity was independent
of Al/Al+Si; the activity and selectivity of an average
preparation, with Al/Al+Si = 0.2, is illustrated in Table IV.

Table IV. Selectivity of Cobalt Chlorite and Hydrotalcite

Catalyst	Rate/μ mole/s/g		Hydrocarbon Selectivity/wt%						%ene
	to HCs	to CO_2	C_1	C_2	C_3	C_4	C_5	C_{6+}	C_{2-5}
Chlorite[a]	0.65	0.02	21	3	11	12	13	40	57
Hydrotalcite[b]	0.20	0.2E-3	23	5	14	16	16	26	78

a) GHSV = 1260 h^{-1}; P = 113 kPa; T = 210°C; conversion 22%.
b) GHSV = 700 h^{-1}; P = 113 kPa; T = 220°C; conversion 6%.

The activation energy of the reaction was 106 kjoules/mole. The
product was moderately hydrogenated and predominantly straight
chain, such that 82% of the C_{2-5} alkenes was 1-alkene. This
catalyst deactivated at 1% per h. The hydrotalcite preparations
were generally of lower activity than the chlorites, as illustrated
in Table IV by the results from a preparation aged at 60°C.
Remarkably, however, over a period of 24 h no fall in activity
could be detected. For this catalyst, isomerization of the primary
product was evident, in that while 78% of C_{2-5} was alkene, only 43%
was 1-alkene, thus demonstrating substantial Broensted acidity from
the hydrotalcite residue support.
    Two factors, namely the very low CO_2 production (Table IV) and
its long-term stability, have led us to pursue further activation
of the hydrotalcite preparations, using anion exchange to introduce
promoters into its interlayer region; this has been briefly
reported elsewhere (8), and will form the basis of a more detailed
paper.

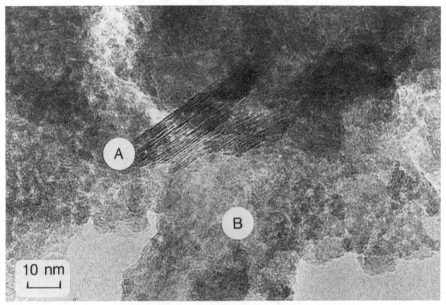

Figure 5. Transmission electron micrograph of Co/Al
hydroxycarbonate (Co/Al = 3, aged 60°C), showing (at A) well-
ordered layers standing on end along a [hk0] projection and (at
B) regions of low crystallinity.

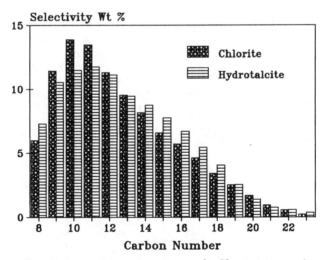

Figure 6. Hydrocarbon selectivity (wt%) within condensate
fraction from chlorite-and hydrotalcite-derived catalysts.
Conditions are shown in Table IV.

Acknowledgments

We wish to thank Dr. D. Hay for X-ray diffraction and the late Dr J.V. Sanders for electron microscopy as well as Messrs S. Hardin and M. Hoang for experimental assistance. This work was supported by the National Energy Research, Development and Demonstration Program, Grant #85/0905.

Literature Cited.

1. Frohning, C.D.; Rottig, W; Schnur, F. In Chemical Feedstocks from Coal; Falbe, J., Ed.; Wiley: New York, 1982; p 351.
2. Sexton, B.A.; Hughes, A.E.; Turney, T.W. J.Catal. 1986, 97, 390.
3. Bruce, L.A.; Sanders, J.V.; Turney, T.W. Clays Clay Miner. 1986, 34, 25.
4. Reichle, W.T. J. Catal. 1985, 94, 547.
5. Bruce, L.A.; Turney, T.W. Catalysts for Synthetic Liquid Fuels; NERDDP Report No. EG88/751; Aust. Department of Primary Industries and Energy: Canberra, 1988.
6. Calliere, S.; Henin, S. Bull. Soc. Fr. Ceram. 1960, 48, 63.
7. Brindley, G.W.; Lemaitre, J. In Chemistry of Clays and Clay Minerals; Newman, A.D.C., Ed.; Mineralogical Soc.: London, 1987; p 334.
8. Bruce, L.A.; Takos, J.; Turney, T.W. Prepr. Amer. Chem. Soc, Div. Petrol. Chem., 1989, 34, 502.

RECEIVED May 9, 1990

Chapter 14

Pillared Hydrotalcites

Synthesis, Characterization, and Catalytic Activity

Mark A. Drezdzon

Amoco Chemical Company, Amoco Research Center, Naperville, IL 60566

Exchange of terephthalate-pillared hydrotalcite with
molybdate or vanadate salts under mildly acidic conditions
affords the corresponding isopolymetalate-pillared
hydrotalcite in high yield. Synthesis of mixed
terephthalate/isopolymetalate-pillared hydrotalcites
followed by calcination yields higher surface area
products. Isopolymetalate-pillared hydrotalcites have
been found to produce both propylene and acetone when
screened for 2-propanol conversion activity, indicating the
presence of both acidic and basic catalytic sites.

Over the past 15-20 years, there has been a renewed and growing
interest in the use of clay minerals as catalysts or catalyst
supports. Most of this interest has focused on the pillaring of
smectite clays, such as montmorillonite, with various types of
cations, such as hydrated metal cations, alkylammonium cations and
polycations, and polynuclear hydroxy metal cations ([1-17]). By
changing the size of the cation used to separate the anionic sheets
in the clay structure, molecular sieve-like materials can be made
with pore sizes much larger than those of conventional zeolites.
    Clays containing hydrated metal cations or organic-based
cations collapse upon moderate heating due to the thermal
instability of these pillaring agents ([10-16]). Using polynuclear
hydroxy metal cations such as $[Al_{13}O_4(OH)_{24}(H_2O)_{12}]^{7+}$, stable porous
clay materials can be made ([1-9]). However, the number of metals
that form suitable polymeric species is limited.
    The anionic polyoxometalates comprise a much larger class of
metal oxide-based pillaring agents ([18]). It was proposed that
hydrotalcites (also referred to as layered double hydroxides, or
LDHs) would be suitable host materials for pillaring with
polyoxometalates. Subsequently, a route for synthesizing
isopolymetalate-pillared hydrotalcites via organic-anion-pillared
precursors was developed ([19-20]).
    This paper summarizes work done on the synthesis and
characterization of terephthalate-, isopolymetalate-, and
terephthalate/isopolymetalate-pillared hydrotalcites. Several test

0097–6156/90/0437–0140$06.00/0
© 1990 American Chemical Society

reactions aimed at understanding the catalytic properties of these materials, including their acid-base properties, will also be reported.

Experimental

Details on the materials and methods used to synthesize and characterize the pillared hydrotalcites discussed in this paper have been previously published (19). For the interested reader's convenience, a brief description of the synthesis of several pillared hydrotalcites follows.

Synthesis of Terephthalate-Pillared Hydrotalcite. A 5-L, three-neck, round-bottom flask equipped with a reflux condenser, thermometer, mechanical stirrer, and electric heating mantle was charged with 1600 mL deionized water, 133.1 g terephthalic acid, and 575 g 50% NaOH solution. A solution containing 1280 mL deionized water, 410.1 g $Mg(NO_3)_2 \cdot 6H_2O$, and 300.0 g $Al(NO_3)_3 \cdot 9H_2O$ was then added dropwise to the terephthalate/NaOH solution at room temperature over a period of 90 min. The resulting gel-like mixture was digested at 73-74°C for 18 h. Upon cooling, the product was isolated by filtration, washed with deionized water, and stored as a 7.39 wt % water slurry (net weight of pillared hydrotalcite = 203 g, 86% yield).

A product d-spacing of 14.4 Å was determined by XRD. The formula $Mg_{4.27}Al_2(OH)_{12.54}(TA) \cdot 4H_2O$ (TA = terephthalate) is consistent with elemental analyses performed on the product. BET surface area of the product = 35 m^2/g (outgassing temperature = 200°C), which increased to 298 m^2/g after calcination at 500°C for 12 h.

Synthesis of Molybdate-Pillared Hydrotalcite. To a 2500-g portion of the preceding terephthalate-pillared hydrotalcite slurry was added a solution consisting of 254 g $Na_2MoO_4 \cdot 2H_2O$ in 450 mL deionized water. (The amount of molybdate added corresponds to a 50% excess of that needed for stoichiometric exchange of terephthalate with $Mo_7O_{24}^{6-}$.) Approximately 350 mL 4N HNO_3 was slowly added to the mixture with vigorous stirring, resulting in a pH drop from 12.0 to 4.4. After 5 min additional stirring (pH = 3.7), the product was filtered, washed, and dried at 125°C overnight. The yield of hard, slightly off-white chunks of product was 215 g (94%).

A product d-spacing of 12.2 Å was determined by XRD. The formula $Mg_{12.26}Al_6(OH)_{36.52}(Mo_7O_{24}) \cdot 6H_2O$ is consistent with elemental analyses performed on the product. BET surface area of the product = 27 m^2/g (outgassing temperature = 200°C, which increased to 32 m^2/g after calcination at 500°C for 12 h.

Synthesis of Vanadate-Pillared Hydrotalcite. To a 1400-g portion of a terephthalate-pillared hydrotalcite was added a solution containing 73.91 g $NaVO_3$ (50% excess) in 500 mL deionized water. Approximately 320 mL 4N HNO_3 was slowly added to the mixture with vigorous stirring, resulting in a pH drop from 9.0 to 4.5. After 5 min additional stirring (pH = 4.9), the product was filtered, washed, and dried at 125°C overnight. The yield of hard, yellowish chunks of product was 105 g (85%).

A product d-spacing of 11.8 Å was determined by XRD. The formula $Mg_{12.83}Al_6(OH)_{37.66}(V_{10}O_{28})\cdot 6H_2O$ is consistent with elemental analyses performed on the product. BET surface area of the product = 30 m^2/g (outgassing temperature = 200°C), which increased to 32 m^2/g after calcination at 500°C for 12 hr.

Synthesis of Terephthalate/Vanadate-Pillared Hydrotalcite. To a 1350-g portion of a terephthalate-pillared hydrotalcite was added a solution consisting of 14.0 g $NaVO_3$ in 100 mL deionized water. (The amount of vanadate added corresponds to 25% of that needed for stoichiometric exchange of terephthalate with $V_{10}O_{28}^{6-}$.) Approximately 100 mL 4N HNO_3 was slowly added to the mixture with vigorous stirring, resulting in a pH drop from 9.3 to 4.5. After 5 min additional stirring (pH = 4.8), the product was filtered, washed, and dried at 125°C overnight. The yield of hard, pale yellowish chunks of product was 97 g (81%).

A product d-spacing of 14.43 Å was determined by XRD. The formula $Mg_{12.90}Al_6(OH)_{37.96}(V_{10}O_{28})_{0.21}(TA)_{2.37}\cdot 6H_2O$ is consistent with elemental analyses performed on the product. BET surface area of the product = 40 m^2/g (outgassing temperature = 200°C), which increased to 166 m^2/g after calcination at 500°C for 12 hr.

Results and Discussion

Previously reported methods for synthesizing polyoxometalate-pillared hydrotalcites involve the exchange of chloride-containing hydrotalcites (21-23). In this work, the novel approach taken in synthesizing pillared hydrotalcites has been to prepare an organic-anion-pillared precursor, which is subsequently exchanged with the appropriate isopolymetalate under mildly acidic conditions (Figure 1).

Synthesis and Characterization of Terephthalate-Pillared Hydrotalcite. Molecular modeling studies indicated that organic anions containing an aromatic ring, such as the terephthalate dianion, would be large enough to preseparate the metal hydroxide layers of hydrotalcite for polyoxometalate exchange. Therefore, a series of hydrotalcite syntheses was performed by published coprecipitation/digestion procedures (24-27) except for the substitution of the desired organic anion, such as terephthalate, for carbonate. In this manner, terephthalate-pillared hydrotalcite was prepared, as confirmed by X-ray diffraction and elemental analysis.

The 14.4 Å d-spacing measured for terephthalate-pillared hydrotalcite agrees well with that calculated from molecular models assuming the plane of the aromatic ring is perpendicular to the brucite layers. In samples dried below 125°C, the remaining gallery space is filled with water. Heating to 300-350°C removes this interlayer water (approximately 4 H_2O per terephthalate anion). Calcination at higher temperatures results in decomposition of the organic anion.

In neutral or basic media, terephthalate-pillared hydrotalcite is stable indefinitely. Acidification with dilute HCl or HNO_3 results in the formation of the corresponding chloride- or nitrate-

containing hydrotalcite, as indicated by a collapse of the d-spacing
to 7.8 Å. Treatment of terephthalate-pillared hydrotalcite with
monometalate salts (Na_2MoO_4 or $NaVO_3$) at room temperature without
acidification results in no exchange.

Synthesis and Characterization of Isopolymetalate-Pillared
Hydrotalcites. Exchange of terephthalate-pillared hydrotalcite with
the appropriate metalate under mildly acidic conditions (pH = 4.5)
proceeds smoothly to yield the corresponding isopolymetalate-
pillared hydrotalcite, as indicated by X-ray diffraction and
elemental analysis. The d-spacings of the products confirm that the
shortest dimension of the isopolymetalate is perpendicular to the
metal hydroxide layers.

Figure 2 illustrates the thermal behavior of a heptamolybdate-
pillared hydrotalcite; the thermal behavior of a decavanadate-
pillared hydrotalcite is similar, with gallery water loss below 300-
350°C and brucite layer dehydroxylation up to 500°C. High-
resolution TEM micrographs of materials calcined up to 500°C show
that the layered hydrotalcite structure is still intact; however,
the exact fate of the polyoxometalate pillars with respect to
possible rearrangement is still under investigation. Total
destruction of the layered structure occurs above 600°C, where the
corresponding magnesium metalate is produced.

Variable Pillar Spacing in Pillared Hydrotalcites. As illustrated
in Figure 3, the pillars in a decavanadate-pillared hydrotalcite are
very closely spaced, with just enough room for gallery water
molecules. Pillared hydrotalcites with larger pores may be
synthesized by the partial exchange of terephthalate-pillared
hydrotalcite, followed by calcination to remove the remaining
organic material (20). Table I summarizes data obtained on a series
of pillared hydrotalcites with varying terephthalate/decavanadate
pillar ratio. As expected, the surface area of the calcined
pillared hydrotalcites increases as the number of decavanadate
pillars in the gallery decreases.

Table I. Exchange Reactions of Terephthalate-Pillared
Hydrotalcite Clays with Decavanadate

	d-Spacing (Å)		Surface Area (m²/g)	
% Exchange	TA	Exchanged	Dry	Calcined
91	14.39	11.76	38	38
82	14.39	11.89	33	45
60	14.40	11.76,14.27	38	64
39	14.38	14.32	40	123
21	14.40	14.43	40	166
0	14.40	---	36	298

Clays which have a large majority (>80% exchange) of
polyoxometalate pillars exhibit a d-spacing consistent with a
polyoxometalate-pillared material. When less than half of the
terephthalate has been replaced by polyoxometalate, the d-spacing
does not change, reflecting the larger effective pillaring size of
the terephthalate anion. At intermediate exchange levels (50-75%)

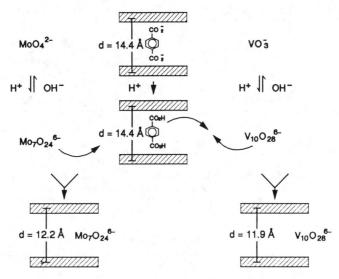

Figure 1. Schematic representation of terephthalate exchange
with heptamolybdate and decavanadate in hydrotalcite. Pillared
hydrotalcite d-spacings were calculated from models.
(Reproduced from Ref. 19. Copyright 1988 American Chemical Society.)

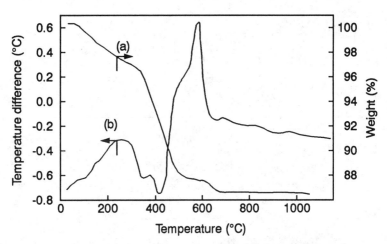

Figure 2. Thermal analysis for heptamolybdate-pillared
hydrotalcite: (a) TGA; (b) DTA. (Reproduced
from Ref. 19. Copyright 1988 American Chemical Society.)

d-spacings corresponding to both the terephthalate- and polyoxometalate-pillared hydrotalcites are detected. Since analytical electron microscopy shows a relatively homogeneous elemental distribution throughout the sample, exchange on a particle-by-particle basis is not occurring. One cause of the dual d-spacings might be sagging of the brucite layers in areas where rafts of polyoxometalate pillars may have been incorporated.

Figure 4 shows the ^{51}V MAS-NMR spectra of several terephthalate/decavanadate-pillared hydrotalcites. These spectra are consistent with that previously published for $(NH_4)_6V_{10}O_{28}$ (23). The changes which occur to the decavanadate pillars upon calcination of the pillared hydrotalcite are currently being investigated. It is known, however, that in the case of pillared materials containing as little as 25-30% polyoxometalate pillars the overall layered structure is maintained up to 500°C (Figure 5).

Catalytic Activity of Pillared Hydrotalcites

It has previously been reported that hydrotalcite catalyzes the aldol condensation of acetone (25). Polyoxometalates are known to dehydrate alcohols due to their acidic nature (18). In order to compare the relative basicity of polyoxometalate-pillared hydrotalcites to that of hydrotalcite itself, a variety of hydrotalcites were screened for 2-propanol conversion (Table II). This reaction is known to give propylene when the catalyst contains acidic sites (such as alumina) and acetone when the catalyst contains basic sites (such as magnesium oxide).

Table II.   2-Propanol Conversion Over Various Catalysts (a)

Catalyst	% Conv	% Selectivity (b)	
		Propylene	Acetone
Alumina	100	100	0
V_{10}-Hydrotalcite	85	87	12
Mo_7-Hydrotalcite	35	47	51
Hydrotalcite	24	25	70
Magnesium Oxide	8	6	90

(a) Temperature = 350°C; neat 2-propanol feed rate = 1 g/min (no diluent); catalyst bed length = 3".
(b) The balance of the products are higher-molecular-weight oligomers formed from the aldol condensation of acetone.

The results in Table II suggest that molybdate- and vanadate-pillared hydrotalcites contain both acidic and basic sites, the basic sites located on the metal hydroxide sheets, and the acidic sites located on the polyoxometalate pillars.

Polyoxometalate-pillared hydrotalcites were also screened for activity for the dehydrogenation of t-butylethylbenzene to t-butylstyrene. Table III shows that molybdate-pillared hydrotalcites outperformed other pillared hydrotalcites.

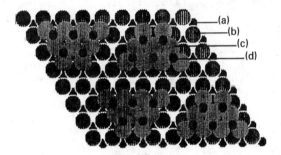

Figure 3. Model of decavanadate-pillared hydrotalcite viewed
normal to the metal hydroxide layers. Atom types are as
follows: (a) Mg or Al; (b) OH group in the metal hydroxide
layer; (c) O from the decavanadate pillar in contact with the
metal hydroxide layer; (d) V. (Model was constructed using
Chem-X, developed and distributed by Chemical Design Ltd.,
Oxford, England.) (Reproduced from Ref. 19.
Copyright 1988 American Chemical Society.)

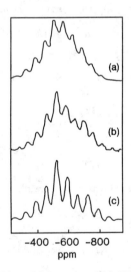

Figure 4. Vanadium-51 solid state MAS-NMR of terephthalate-
pillared hydrotalcites partially exchanged with decavanadate:
(a) 39% exch.; (b) 60% exch.; (c) 91% exch.

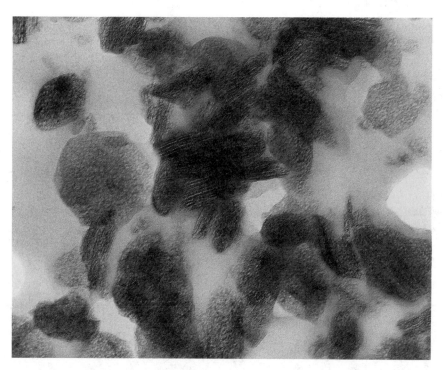

Figure 5. TEM micrograph of a terephthalate/decavanadate-pillared hydrotalcite (39% decavanadate exchanged) after calcination at 500°.

Table III. Dehydrogenation of $\underline{t}$-butylethylbenzene Over Various Isopolymetalate-Pillared Hydrotalcites ($\underline{a}$)

Catalyst	Time (h)	Temp (°C)	% Conv	% Sel
Mo$_7$-Hydrotalcite	8	500	54	80
	12	550	57	68
V$_{10}$-Hydrotalcite	3	515	38	55
	6	515	31	55
W$_7$-Hydrotalcite	6	520	21	57
	9	520	23	83

(a) Reactions were run using 8% oxygen in nitrogen as a diluent. Reaction conditions were not optimized.

It was initially hoped that running the reaction under oxidative dehydrogenation conditions would lead to higher conversions. Studies with molybdate-pillared hydrotalcites under more optimized conditions showed that the presence of oxygen in the feed simply kept the catalyst from coking up as fast as if oxygen were not included. Unfortunately, even with the presence of oxygen in the feed, molybdate-pillared hydrotalcites were found to lose approximately 40% of their activity after 100-150 hours on stream.

Acknowledgments

Technical assistance by W. J. Dangles, S. M. Oswald, and P. A. Shope
has been greatly appreciated. The New Catalytic Materials Group at
Amoco is gratefully acknowledged for useful discussions. Marc
Kullberg and Eric Moore ran t-butylethylbenzene dehydrogenation
reactions. Claire Grey collected and aided in the interpretation of
NMR data. Consultation with A. K. Cheetham and A. Clearfield has
also been helpful during the course of this project. Finally, I
wish to thank Amoco Chemical Company for permission to publish and
present this work.

Literature Cited

1. Pinnavaia, T. J. Science 1983, 220, 365.
2. Barrer, R. M. Zeolites and Clay Minerals as Sorbents and
   Molecular Sieves; Academic: New York, 1978.
3. Carradom, K. A.; Suib, S. L.; Skoularikis, N. D.; Coughlin,
   R. W. Inorg. Chem. 1966, 25, 4217.
4. Pinnavaia, T. J.; Tzou, M.-S.; Landau, S. D. J. Am. Chem. Soc.
   1985, 107, 4783.
5. Pinnavaia, T. J.; Tzou, M.-S.; Landau, S. D.; Rasik, H. R.
   J. Mol. Catal. 1984, 27, 195.
6. Yamanaka, S.; Brindley, G. W. Clays Clay Miner. 1979, 27, 119.
7. Yamanaka, S.; Brindley, G. W. Clays Clay Miner. 1978, 26, 21.
8. Lahav, N.; Shani, U.; Shabtai, J. Clays Clay Miner. 1978,
   26, 107.
9. Brindley, G. W.; Samples, R. E. Clay Miner. 1977, 12, 229.
10. Mortland, M. M.; Berkheiser, V. E. Clays Clay Miner. 1976,
    24, 60.
12. Pinnavaia, T. J. ACS Symp. Ser. 1982, 192, 241.
13. Pinnavaia, T. J.; Raythatha, R.; Lee, J. G.-S.; Halloran,
    C. J.; Hoffman, J. F. J. Am. Chem. Soc. 1979, 101 6891.
14. Traynor, M. F.; Mortland, M. M.; Pinnavaia, T. J. Clays Clay
    Miner. 1978, 26 319.
15. Thomas, J. M.; Adams, J. M.; Graham, S. H.; Tennakoon, T. B.
    Adv. Chem. Ser. 1977, 163, 298.
16. Knudson, J. I.; McAtee, J. L. Clays Clay Miner. 1973, 21, 19.
17. Maes, A.; Cremers, A. ACS Symp. Ser. 1986, 323, 254.
18. Pope, M. T. Heteropoly and Isopoly Oxometalates; Springer-
    Verlag: New York, 1983.
19. Drezdzon, M. A. Inorg. Chem. 1988, 27, 4628.
20. Drezdzon, M. A. U.S. Patent 4 774 212, 1988.
21. Iyagba, E. T. Ph.D. Thesis, University of Pittsburgh,
    Pittsburgh, PA, 1986.
22. Woltermann, G. M. U.S. Patent 4 454 244, 1984.
23. Kwon, T.; Tsigdinos, G. A.; Pinnavaia, T. J. J. Am. Chem. Soc.
    1988, 110, 3653.
24. Reichle, W. T. CHEMTECH 1986, 58.
25. Reichle, W. T.; Kang, S. Y.; Everhardt, D. S. J. Catal. 1986.
    101, 352.
26. Reichle, W. T. J. Catal. 1985, 94, 547.
27. Nakatsuka, T.; Kawasaki, H.; Yamashita, S.; Kohjiya, S.
    Bull. Chem. Soc. Jpn. 1979, 52 2449.

RECEIVED May 9, 1990

CLUSTERS

Chapter 15

Low-Nuclearity Platinum Clusters Supported on Graphite

Structural Flexibility

P. Gallezot, D. Richard, and G. Bergeret

Institut de Recherches sur la Catalyse, Centre National de la Recherche Scientifique, 2 Avenue Albert Einstein, 69626 Villeurbanne, Cedex, France

Platinum clusters prepared by ion–exchange of a functionalized, high surface area graphite are selectively located on graphite steps. The interatomic distances obtained by radial electron distribution (RED) are elongated with respect to Pt–bulk distances because of an electron transfer from graphite to metal. Platinum clusters prepared by decomposition of platinum–dibenzylidene acetone complex are selectively located on the basal planes. The combined use of RED and HRTEM shows that they have a raft–like morphology and that their lattice is distorted because there is a strong epitaxy between Pt(110) and graphite (00.1) planes. The different interaction of the two types of clusters with graphite leads to differences in morphology, electronic structure and chemoselectivity in cinnamaldehyde hydrogenation.

Heterogeneous catalysts used in organic synthesis are still largely based on Raney–nickel and charcoal–supported metal catalysts. Innovative preparations of new catalytic materials are needed for the production of fine chemicals. Thus the selective synthesis of biologically active isomers used in drugs will require the design of new stereoselective and enantioselective catalysts. We have shown recently (1–3) that platinum catalysts supported on high surface area graphite are much more selective for the hydrogenation of cinnamaldehyde into cinnamyl alcohol than charcoal–supported catalysts. The aim of the present work was to describe in more details the structure of platinum clusters supported on graphite with two specific locations : clusters on graphite steps anchored at the edges of basal planes and clusters on top of basal planes.

Experimental

Graphite-supported platinum catalysts were prepared from a high surface area graphite (HSAG 300 m^2g^{-1} from Lonza). It was oxidized by stirring for 24 h a graphite suspension in a solution of sodium hypochlorite (15 % active chlorine), the suspension was filtered and washed with hydrochloric acid (1.25N) to eliminate the excess of NaClO. The graphite was thoroughly washed by water until the wash–water was Cl–free, then it was dried overnight at 373 K in a vacuum oven. The acidic groups created by the

0097–6156/90/0437–0150$06.00/0

oxidizing treatment were dosed by NaOH titration (4). The HOPG graphite used for STM studies was obtained from Union Carbide and Carbone Lorraine. It was cleaved and then submitted to the same oxidizing treatment as the HSAG graphite.

Catalyst Pt/Gex was prepared by ion-exchange of the oxidized HSAG graphite in suspension in an ammonia solution (1N). A solution of $Pt(NH_3)_4(OH)_2$ (obtained by passage of a solution of $Pt(NH_3)_4Cl_2$ on an anionic exchange resin) was added dropwise. After overnight stirring at room temperature, the suspension was filtered and washed to eliminate the excess of metallic salt and dried overnight. Chemical analysis indicated a 3.6 wt % Pt-loading. The exchanged graphite was reduced under flowing hydrogen at 573 K. Higher reduction temperatures were also used to test the stability of platinum toward sintering.

Catalyst Pt/Gc was obtained by carbon monoxide decomposition of the zerovalent $Pt(dpo)_2$ complex (dpo = 1,5-diphenyl-1,4-pentadiene-3-one). A suspension of graphite in a CH_2Cl_2 solution of the complex was stirred while carbon monoxide was bubbling in the solution. After 30 min the slurry was filtered, washed and dried. The platinum loading was 1,1 wt %.

The high resolution TEM study was carried out with a JEOL 200 CX microscope equipped with high resolution polar pieces. The STM used for the study of HOPG graphite was of the same type as that described by Hansma and Tersoff (5). Imaging was performed in air, in the constant current mode. Further details were given previously (6).

Radial electron distribution (RED) was calculated from Fourier analysis of X-ray scattering data as described previously (7). The RED of a Pt-free graphite was subtracted from the RED of graphite-supported platinum to obtain the distribution of distances between platinum atoms. The experimental distribution was compared to distributions calculated from model f.c.c. aggregates of different morphology.

Results and Discussions

Structure of the Oxidized Graphite. The oxidizing treatment with sodium hypochlorite produces two effects on the structure of graphite namely (i) etching of the microcrystals resulting in an increase of the number of steps (ii) functionalization of the steps i.e. creation of carboxylic and other acidic groups at the extremities of basal planes.

TEM observations before and after NaClO treatment of HSAG graphite do not allow a good evaluation of the etching effect because steps are present initially and new steps can be evidenced only if there is enough contrast i.e. if step heights are not too small with respect to the total thickness of graphite platelet. Scanning tunneling microscopy (STM) is the best probe to map surface heterogeneities. However, since the technique cannot be applied to the high surface graphite powder, measurements have been done on a macroscopic piece of HOPG graphite. STM images of freshly cleaved lamellas show that the surface is smooth over areas extending over several hundreds of nanometers with only occasional steps. In contrast after NaClO treatment, STM images present a high density of etching figures clearly evidenced on Figure 1.

There are two types of heterogeneities (i) parallel or wedge-shaped steps with heights in the range from one nanometer or less to a few tens of nanometers (ii) domains of high surface roughness with bumps as high as 5 nm. It can be concluded that graphite layers have been etched by the oxidizing treatment, steps and bumps are left as a result of incomplete layer attack.

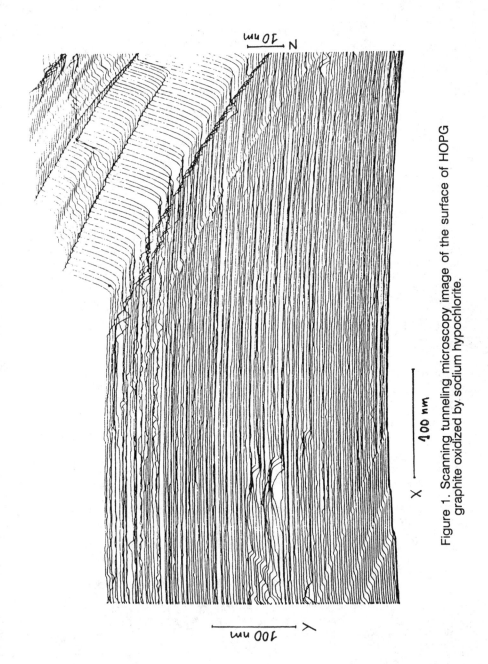

Figure 1. Scanning tunneling microscopy image of the surface of HOPG graphite oxidized by sodium hypochlorite.

As the surface area of frontal planes (rise of steps and bumps) increases upon the oxidizing treatment, the number of functional groups on these planes, i.e. at the extremities of basal planes, also increases. Thus, NaOH titration of acidic groups shows that the acidity increases from 1.7 mmol m^{-2} (total graphite surface) to 6.1 mmol m^{-2}. Therefore, steps and associated acidic groups exist initially but their number is greatly increased after NaClO treatment.

Structure of Clusters on Graphite Steps. The TEM study of Pt/Gex catalyst,obtained by ion-exchange of the oxidized HSAG graphite and hydrogen reduction at 573 K, shows that platinum clusters are mainly located along the steps of graphite (Figure 2). A few of them do not seem to be on steps but rather on zones with a "granular" contrast which correspond probably to the bumps imaged by STM. The size distribution (Figure 3) shows that most of the particles are smaller than 2 nm, the mean size is 1.3 nm. We failed to detect these clusters by STM probably because of their small size and of their particular location. Indeed because they are nested at the foot or on the rise of steps, they escape detection by the tip ascending the high heterogeneities of the graphite surface.

The structure of the clusters has been studied by radial electron distribution (RED) from X-ray scattering data recorded under hydrogen atmosphere. Figure 4a gives the RED of Pt/Gex obtained after subtraction of the RED of a Pt-free graphite. All the interatomic distances correspond to the f.c.c. structure but they are significantly elongated with respect to the normal bulk platinum distances. The lattice expansion could be attributed to an electron transfer from graphite to metal which would increase the population of the antibonding levels and thus decrease the cohesive energy of the lattice. Evidence for an electron transfer was given by the decrease of the ratio $K_{T/B}$ of the adsorption coefficients of toluene and benzene with respect to that observed on platinum catalysts supported on silica or on active charcoal (1). Interestingly the reverse was observed on electron-accepting zeolite supports, namely a contraction of the lattice and an increase of $K_{T/B}$ ratios (8).

The electron transfer could happen as a result of an equalization of the Fermi levels of graphite and platinum the process being favored by the coupling of platinum atoms with the carbon atoms terminating the basal planes. However, functional groups are still present as shown by TPD experiments where CO_2 and CO outgassed at temperatures much higher than the reduction temperature (573 K). Electron-donating O$^-$ or COO$^-$ groups terminating the basal planes could act as ligands increasing the electron density on low-nuclearity clusters.

The experimental RED (Figure 4a) has been compared to various RED calculated from model clusters of different size and shape. Spherical models give always the best agreement, thus Figure 4b is a distribution calculated for a 50:50 mixture of 6-atoms (7 Å) octahedral clusters and of 135-atoms (17.4 Å) spherical clusters (96 % of the atoms are in the large clusters).

Structure of Clusters on Basal Planes. The particles in Pt/Gc catalyst prepared by decomposition of the Pt(dpo)$_2$ complex are randomly distributed on graphite basal planes. They have irregular rounded shapes with sizes in the range 1 to 10 nm (Figure 5a). Their contrast on TEM images is low whatever the particle size. This suggests a flat morphology since one would expect quite different contrasts for spherical 1 and 10 nm particles. It was not possible to record nanodiffraction pattern with a FEG-STEM because the particles recrystallize under the beam to give more contrasted and smaller particles which indicates a change from a raft-like to a more spherical morphology.

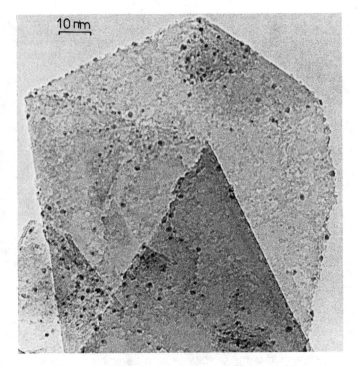

Figure 2. TEM views taken on Pt/Gex catalyst.

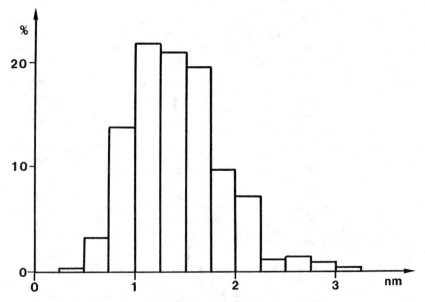

Figure 3. Particle size distribution on Pt/Gex.

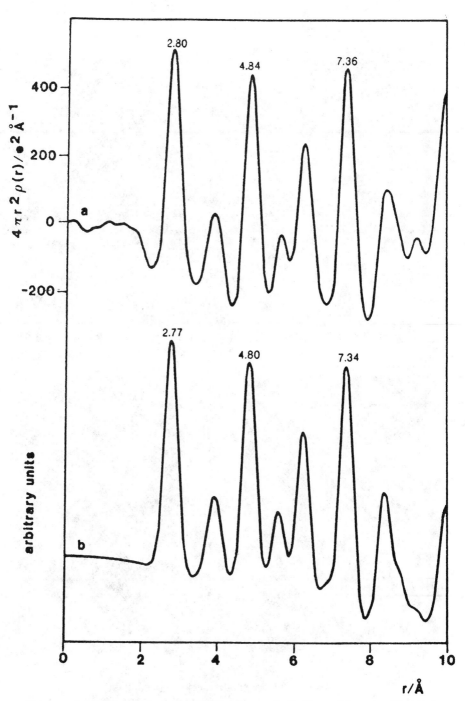

Figure 4. Radial electron distribution of Pt/Gex (a) experimental distribution obtained by Fourier transform of X-ray scattering data (b) model distribution calculated for spherical clusters.

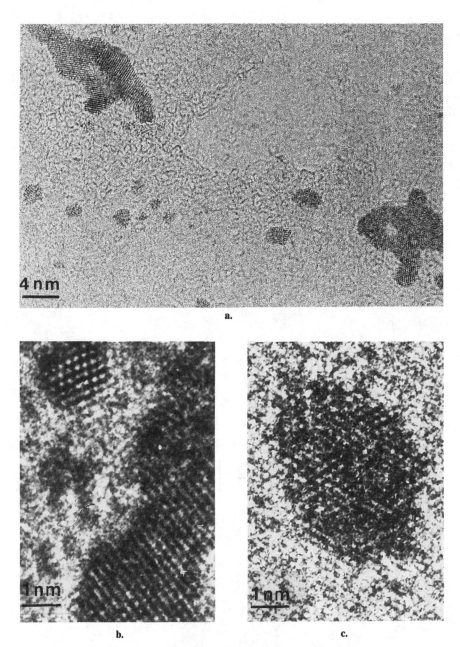

Figure 5. TEM views taken on Pt/Gc catalyst.

The raft–like morphology was confirmed by combined HRTEM and RED studies. The TEM views of Pt/Gc taken at high magnification with the JEOL 200 CX microscope show the lattice images of the (111) planes of platinum (d = 2.26 Å) (Figures 5b, 5c) and, in some zones, the image of the (10.0) planes of graphite (d = 2.13 Å). It is noteworthy that the larger particles are polycrystalline (Figure 5a). In many monocrystalline particles two sets of (111) planes can be imaged (Figures 5b, 5c). This implies that the particles are oriented with the (1$\bar{1}$0) plane perpendicular to the electron beam i.e. parallel to the basal plane (00.1) of graphite. Measurements of the angles between (111) planes indicate that in some aggregates they are close to the theoretical value of 70° 53' (Figure 5b). In contrast, in a number of particles, the angle between the (111) planes is smaller, thus in Figure 5c, it is smaller than 65°. This lattice distortion is due to the strong epitaxial interaction between Pt(110) and graphite (00.1). The epitaxy between these planes is very favorable because the two lattices match closely. Therefore when there is a large fraction of surface atoms, the interaction between the platinum and carbon atoms is sufficient to distort the whole lattice, especially the angle between Pt(111) planes tends to bend toward 60°, the angle between carbon atom rows in the graphite basal plane.

The RED of Pt/Gc given in Figure 6a shows that the relative magnitudes of the Pt–Pt peaks do not correspond to spherical particles. Thus, the 4th and 5th peaks have about the same magnitude and the 6th peak corresponding to a low multiplicity distance is observed. We have tried to match the RED of Pt/Gc with a distribution calculated from various models. The best fits are obtained for plate–like fcc aggregates oriented parallel to the (110) plane. Thus Figure 6b gives the RED calculated for a model aggregate containing 23 atoms arranged in two (110) atomic layers. The interatomic distances in Pt/Gc are not systematically elongated like in Pt/Gex. Some distances are elongated, other are contracted and the peaks are broader than in Pt/Gc. This means that the fcc lattice of the raft–like particles is distorted in agreement with the TEM study.

Since there is no significant lattice expansion, the electron transfer from graphite to metal should be smaller. Indeed the ratio $K_{T/B}$ measured on Pt/Gc (7.2) is intermediate between those observed on Pt/Gex (5.5) and Pt/SiO$_2$ (8.0) where there is no electron transfer. This could be due to a less favorable coupling between platinum and graphite atoms when the particle are top–on rather than edge–on. However this could also be due to the fact that the electron–donating species are the functional groups at the edges of basal planes. There is a good correlation between the extent of charge transfer from support to metal and the selectivity to cinnamyl alcohol in the hydrogenation of cinnamaldehyde. Thus the initial selectivities are 72, 54 and 0 % on Pt/Gex, Pt/Gc and Pt/active charcoal (3). This is because the higher electron density on the cluster, the lower the probability for the activation of the C = C bond which involves as a first step an electron transfer of π–electrons to the metal.

Conclusion

Previous studies on zeolite–encaged metal clusters have shown that the structure of low–nuclearity metal clusters is flexible in presence of adsorbates acting as surface ligands. The flexibility i.e. the ability to change morphology and the extent of atomic displacement with respect to the normal f.c.c. packing depends upon the cohesive energy of the metal and of the enthalpies of metal–adsorbate bondings (9). Thus it was shown (7, 8) that the most probable structure of platinum clusters in Y–zeolite cages is a 40–atoms, truncated tetrahedron which has the same symmetry as the

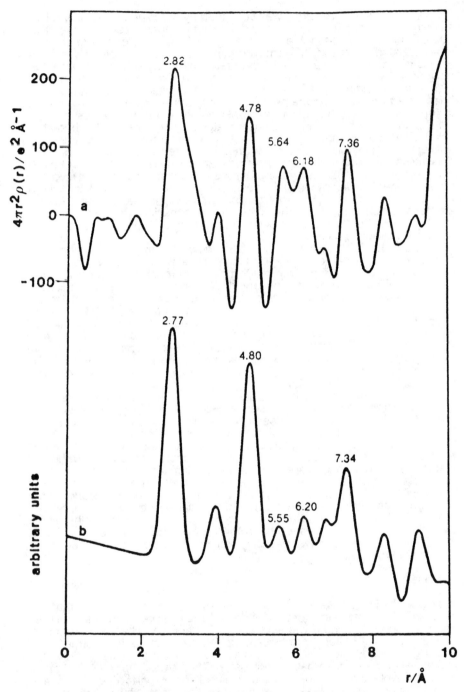

Figure 6. Radial electron distribution of Pt/Gc (a) experimental distribution
(b) model distribution calculated for two (110) layers (23 atoms).

supercage and which fits perfectly within its inner space. However the f.c.c. packing of the cluster was deeply perturbed by adsorption of O_2, CO or H_2S. The present investigation shows also that the graphite support commands the particle morphology ; thus clusters located on steps tend to have a spherical morphology whereas clusters on basal planes tend to have a raft-like morphology. The most interesting conclusion of this paper is that the arrangement of atoms in the clusters is flexible in response to the support acting as a macroligand. Two kinds of deformations with respect to the normal f.c.c. packing were observed (i) a lattice expansion of the clusters interacting with the graphite steps (ii) a lattice distortion of the raft-like cluster interacting with the basal plane. These modifications of the atomic structure are connected with modifications of the electronic structure especially for the clusters anchored on steps. Ultimately, these different atomic and electronic structures lead to different catalytic properties for reactions very sensitive to the structure of catalyst such as the chemioselective hydrogenation of cinnamaldehyde.

Literature Cited

1. Richard, D. ; Fouilloux, P. ; Gallezot, P. ; In Proc. 9th Int. Cong. on Catalysis ; Phillips, M.J. ; Ternan, M., Eds., The Chemical Institute of Canada : Ottawa, 1988, p. 1074.
2. Giroir-Fendler, A. ; Richard, D. ; Gallezot, P. ; In Heterogeneous Catalysis and Fine Chemicals, Guisnet, M., Ed. ; Elsevier : Amsterdam, 1988, p. 171.
3. Richard, D. ; Gallezot, P. ; Neibecker, D. ; Tkatchenko, I. ; Catalysis Today, 1989, 6, 171.
4. Tomita, A. ; Tamai, Y. ; J. Phys. Chem., 1971, 75, 649.
5. Hansma, P.K. ; Tersoff, J. ; J. Appl. Phys., 1987, 61, 1.
6. Porte, L. ; Richard, D. ; Gallezot, P. ; J. Microsc., 1988, 152, 515.
7. Bergeret, G. ; Gallezot, P. ; In Proc. 8th Int. Cong. on Catalysis, Verlag Chemie : Weinheim 1984, p. 659.
8. Gallezot, P. ; Bergeret, G. ; In Catalyst Deactivation, Bell, A. ; Petersen, E. Eds. ; Marcel Dekker : New York, 1987, p. 263-296.
9   Bergeret, G. ; Gallezot, P. ; Z. Phys. D-Atoms, Molecules and Clusters, 1989, 12, 591.

RECEIVED May 9, 1990

Chapter 16

Adsorbed Carbon Monoxide and Ethylene on Colloidal Metals

Infrared and High-Resolution ^{13}C NMR Spectroscopic Characterization

John S. Bradley, John Millar, Ernestine W. Hill, and Michael Melchior

Exxon Research and Engineering Company, Route 22 East, Clinton Township, Annandale, NJ 08801

Colloidal solutions of highly dispersed platinum (<10Å particles) and palladium (18Å particles), prepared by the condensation of platinum or palladium vapor into cold solutions of isobutylaluminoxane oligomers in methylcyclohexane, absorb carbon monoxide. Infrared spectroscopy on these solutions shows bands at 2035(s) cm-1(Pt), and 2062(s) and 1941(m) cm-1 (Pd), corresponding to the absorptions for linear and bridging carbonyls adsorbed on the surface of the metal particles. Solution 13C n.m.r. spectra (75MHz.) of the carbonylated platinum and palladium colloids exhibit broad lines ($w_{1/2}$ = 50 p.p.m. and 20 p.p.m. respectively) centered near 190 p.p.m. At low temperature the Pd/CO resonance shows a low field shoulder suggestive of bridging carbonyls. The carbonylated Pt colloid is easily transformed into the molecular cluster $[Pt_{12}(CO)_{24}]^{2-}$ by reaction with water. The reaction of ethylene with aluminoxane stabilized colloidal solutions of palladium has been investigated in a similar fashion. Exchange between free and adsorbed ethylene are observed by ^{13}C n.m.r.

The preparation of metal colloids and their use in catalysis is almost as old as the study of catalysis itself, and their efficiency as catalysts is well established. In fact the high activity of colloidal metals in a number of catalytic processes has sometimes been a complicating factor in the investigation of homogeneously catalysed reactions, where the presence of even a minute fraction of the catalytic metal in colloidal form can be active enough to compete with a less active homogeneous catalyst (1). A recent example of this was reported by Lewis *et al.* (2) in a careful analysis of the homogeneous hydrosilylation of olefins catalyzed by (COD)PtCl$_2$ which revealed that under reaction conditions colloidal platinum was produced, and that the colloid catalyzed the reaction extremely efficiently.

As part of our continuing interest in the preparation and chemistry of small metal particles in organic media (3) we have begun to investigate the surface chemistry of colloidal metals in terms of the concepts of molecular cluster chemistry, such as the coordination of ligands to the surface of the colloids, spectroscopic analysis of ligand reactions in the "coordination sphere" of the colloids, and the development of the surface organometallic chemistry of colloidal metals. This work was undertaken with a view to making comparisons with both heterogeneous supported metals and molecular organometallic clusters. In this paper we report the application of the two most commonly encountered spectroscopic techniques of molecular organometallic chemistry (high resolution liquid phase n.m.r. and infrared spectroscopy) to the characterization of small molecules adsorbed on colloidal palladium and platinum in non-polar solvents. It seemed possible that such adsorbed molecules could be reasonably compared with their counterparts in molecular clusters since the metal particle sizes in the colloidal systems we have prepared approach those found for some of the larger molecular metal carbonyl clusters, for example [Pt$_{19}$(CO)$_{22}$]$^{4-}$(4) and [Ni$_{38}$Pt$_6$(CO)$_{48}$H]$^{5-}$ (5) to which high resolution molecular spectroscopy is routinely applied. We have chosen to focus on carbon monoxide, the most common ligand in organometallic cluster chemistry, adsorbed onto colloidal palladium and platinum, and of ethylene on colloidal palladium. These molecules were selected because there is a considerable body of spectroscopic data in the literature reflecting the coordinated state (for molecular carbonyls and ethylene complexes) and adsorbed state (of ethylene and carbon monoxide on single crystals or supported metals) with which to compare our results, and to help guide us in our assignments.

METAL COLLOID PREPARATION

The metal colloid solutions were prepared by the condensation of the metal vapors into methylcyclohexane solutions of *iso*butylaluminoxane at -120°C, a general method we have developed for the preparation of stable colloidal transition metals in non-polar organic solvents (3). The aluminoxane, (^{i}C$_4$H$_9$AlO)$_n$, which is a low

molecular weight oligomer (6), fulfills the typical role
of a polymer stabilizer in colloid chemistry, associating
with the surface of the metal particles and preventing
aggregation. We chose this stabilizing agent with a view
to synthesizing colloids with a reactive organometallic
"coating", to allow for subsequent chemical manipulation
of the colloidal metal in a manner compatible with
applications in catalysis and the synthesis of novel
oxide based materials (7).

Platinum vapor, (generated by electron beam
evaporation of the molten metal at $ca.$ $2x10^{-5}$ torr), or
palladium vapor (by evaporation from a resistively heated
hearth at $ca.10^{-4}$ torr)) was allowed to dissolve in a
solution of poly(isobutyl-aluminoxane), $[^1C_4H_9AlO]_n$,
(prepared by partial hydrolysis of 50 mmol
triisobutylaluminum) in methylcyclohexane (500 mL.) at -
120°C in the 12 L. flask of a rotary metal vapor
synthesis reactor. In a typical preparation under these
conditions $ca.$ 200 mg/h of platinum or 1.0 g/h of
palladium could be evaporated into a liquid
methylcyclohexane solution of the aluminoxane in our
apparatus. The deep brown liquid thus produced was
transferred under helium to a Schlenk tube. Any bulk
metal suspended in the liquid product was removed by
passage through a 0.2m teflon filter and the colloidal
metal solution was stored under helium. All subsequent
manipulations were performed using standard inert
handling techniques.

CHARACTERIZATION of ALUMINOXANE STABILIZED METAL COLLOIDS

The colloidal metals prepared in the manner described
above were characterized by transmission electron
microscopy. The solutions, as prepared, were diluted to
a concentration adequate to allow for the formation of a
film of aluminoxane on a sample grid of sufficient
thinness for adequate imaging of the metal particles.
The film of aluminoxane is, of course, sensitive to
atmospheric moisture, and the sample was handled
anaerobically to the point of insertion into the
microscope. The colloidal particles of palladium are
extremely highly dispersed, ca. 18± 3Å (3). Attempts to
image the platinum particles by transmission electron
microscopy on a film cast from the solution onto a carbon
film failed to reveal any recognizable metal clusters
until beam damage had occured and the platinum clusters
had grown to >8Å in diameter (M. Disco and S. Behal,
personal communication). Although this is not a
satisfactory determination of the dispersion of the
platinum clusters as prepared, it sets an upper limit of
8Å for the particle size, placing these metal particles
in a size range comparable to that of some of the larger
molecular carbonyl clusters, and raising the prospect
that the data provided by the application of molecular

spectroscopic techniques to molecules adsorbed on the
colloids might be interpretable in terms of molecular
phenomena.

It is not clear just how the aluminoxane is
associated with the metal particles. There is a
possibility that during the preparation of the colloids,
the *iso*butyl groups on the oligomer are attacked by the
clustering metal atoms, with the formation of metal
carbon bonds. This was deemed unlikely from the results
of deuterolysis of a concentrated palladium colloid with
D_2SO_4 in D_2O, in which only isobutane-d_1, the product of
deuterolysis of the aluminum-carbon bond, was detected by
mass spectroscopy. Attack by the dissolved metal atoms
or clusters on the *iso*butyl groups would presumably lead
to C_4 fragments with more than one metal-carbon bond, and
deuterolysis would therefore lead to isobutane-$d_{2,3}$....
It is intuitively more probable that the polar inorganic
backbone of the oligomer interacts in some fashion with
the surface of the metal. This speculation is supported
by the fact that poly*iso*butylene, a purely paraffinic
polymer, is ineffective in stabilizing the colloids in
methylcyclohexane. Although we have not made a thorough
investigation of this aspect of the chemistry of the
colloid preparation, we have observed that a minimum
ratio of aluminoxane (as monomer equivalents)to metal of
ca. 5 is necessary to obtain a stable solution. Given
this condition, the colloidal metal solutions are stable
for months at room temperature, and can be heated to
moderate temperatures without precipitation of bulk
metal.

SPECTROSCOPIC STUDIES of SMALL MOLECULE ADSORPTION

INFRARED STUDIES of ADSORBED CARBON MONOXIDE   In surface
chemistry, carbon monoxide in its adsorbed state is
probably the molecule most widely studied by vibrational
spectroscopy, and in its coordinated state it holds a
similar position in molecular cluster chemistry . It was
thus a natural candidate in our investigation of the
surface chemistry of colloidal metals. Carbon monoxide
was passed through the colloid solutions for several
minutes. The rapid adsorption of CO was revealed by
infrared spectroscopy as shown in Figure 1. No further
change in intensity of the carbonyl stretching bands was
observed after the initial spectrum was recorded one
minute after exposure of the solution to CO. The
infrared bands at 2062(s) and 1941(m) cm^{-1} for palladium
(Figure 1a) and 2035(s) for platinum (figure 1b)
correspond to the expected absorptions for linear and
bridging carbonyls. The broad profile of the bands
resembles that found for CO on metal surfaces and
crystallites (8), and contrasts with the relatively sharp
bands characteristic of molecular cluster carbonyls. The
frequencies, especially that of the linear CO, are lower

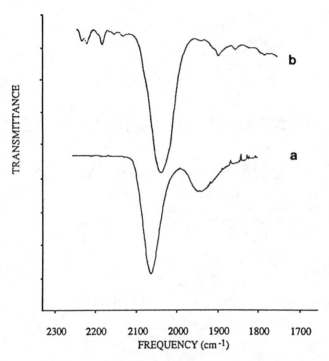

Figure 1.   Infrared absorptions of carbon monoxide on
(a) colloidal palladium (18Å) and (b)colloidal
platinum (<10Å).   1 wt% solutions in methylcyclohexane, stabilized with *iso*butylaluminoxane (see text)

than the range often reported for CO on the respective
metals, but it is not possible to draw any precise
conclusions on the nature of the carbonylated colloidal
metal particles from simple comparisons with the infrared
spectrum of CO on platinum or palladium since the
stretching frequency for surface adsorbed CO is markedly
affected by coverage, particle size and the presence of
impurities. There are few other examples of CO stretching
frequencies for carbon monoxide on colloidal metals
available for comparison. Lewis has reported (2)
frequencies of $2050cm^{-1}$ and $1880cm^{-1}$ for CO on colloidal
platinum (size range 6-60Å, mean 23Å) in methylene
chloride. The infrared absorption of carbon monoxide on
aqueous colloidal platinum was also recently reported
(9).

The linear CO stretching frequency for the
carbonylated platinum colloid, while lower than that
found for surface bound CO, is in the range reported for
the platinum carbonyl clusters $[Pt_3(CO)_6]_n^{2-}$, and we find
that the carbonylated colloid is easily transformed into
the molecular cluster $[Pt_{12}(CO)_{24}]^{2-}$ (10) by reaction
with water. The cluster was isolated in ~ 50% yield based
on platinum content of the precipitate by extraction with
tetraethylammonium bromide in methanol from the aluminum
hydroxide precipitated when water is added to the
aluminoxane solution. The isolation of the platinum
carbonyl cluster reveals nothing about the size or
structure of the colloidal platinum particles, but merely
emphasizes the high reactivity of metals in this highly
dispersed state. The cluster isolated is presumably more
a reflection of the stability of the $[Pt_3(CO)_6]_n^{2-}$ family
of clusters than a clue to the nuclearity of the
colloidal metal particles – in a similar series of
experiments with colloidal cobalt with a mean particle
size of 20Å carbonylation results in the direct formation
of $Co_2(CO)_8$.

NMR STUDIES of ADSORBED CARBON MONOXIDE. In the n.m.r.
investigation of molecules adsorbed on solid supported
metals, such as palladium on silica or platinum on
alumina, both the chemical shift and line width of the
observed nucleus in the adsorbate are drastically
affected by the physical properties of the sample. The
chemical shift anisotropy of a ^{13}C nucleus in a molecule
such as CO adsorbed on a supported metal particle in a
solid sample results in a broadening of its resonance of
as much as several hundred ppm (11). The shift is also
susceptible to the effect of the metallic properties of
the metal particle, and Knight shifts of several hundred
ppm have been reported. The small size of the metal
clusters in the aluminoxane stabilized colloids raised
the prospect that the adsorbed CO might be ammenable to
^{13}C n.m.r. in solution, where tumbling of the colloid

particles might be sufficient to average the chemical
shift anisotropy.

All our ^{13}C n.m.r. experiments were performed on a
commercial instrument (Bruker MSL-300) operating at
75.47MHz. The probe was a standard Bruker 10mm high
resolution liquids probe equipped for proton decoupling.
All chemical shifts are referenced to TMS, and the 26.7
ppm resonance of the methylcyclohexane solvent was used
as an internal secondary reference.

Samples of a palladium colloid were prepared by
concentrating the original organosols to ~1-2% metal
concentration (a factor of 5-10), filtering as before,
and exposing the concentrated solution to 99% ^{13}CO at 1at.
for several minutes. From the infrared sudy it was clear
that after this time the colloid surface was saturated
with CO. ^{13}C n.m.r. spectra of CO on colloidal palladium
at room temperature show a broad ($w_{1/2}$=19 ppm)
inhomogeneously broadened line (T_2 = 12 ms) centered at
190 ppm (Figure 2b). This lineshape is essentially
unchanged at 333°K (figure 2a); however cooling the
sample to 220°K results in a broadened line with a
shoulder to lower field (Figure 2c). A saturated solution
of ^{13}CO in methylcyclohexane containing
poly(*iso*butylaluminoxane) shows a low intensity sharp
singlet at 184 ppm due to the saturation concentration
CO, which is absent in the palladium containing sample.
^{13}C nmr spectra of CO adsorbed on platinum colloids gave
similar results, although the linewidth is about 50ppm.
No variable temperature experiments have yet been
performed on the platinum sample, and any explanation of
the linewidth in this system would be premature.

The resonance at 190 ppm for CO on colloidal
palladium is within the range reported both for terminal
carbonyls in transition metal cluster carbonyls (12,13)
and for CO adsorbed on supported rhodium and ruthenium
(14-16 and Thayer, A. M. and Duncan, T. M., J. Phys.
Chem., in press). The room temperature lineshape in our
system extends to ~ 230 ppm, and bridging carbonyls in
supported metals and carbonyl clusters typically have
chemical shifts to low field of terminal carbonyl
resonances. As noted above, cooling the sample results
in the appearance of a low field shoulder, and we
conclude that the spectrum reflects the presence both
types of carbonyls, consistent with the infrared
spectrum, Figure 1a. Neither the observed lineshapes nor
room temperature relaxation measurements allow an
unambiguous assignment of the relative populations in
terminal and bridge sites, and it is probable that in the
inhomogeneous surface environment provided by the polymer
stabilized colloid particles the terminal and bridging
carbonyls have overlapping chemical shifts. We note that
the resonance is quite far from the isotropic terminal
carbonyl shifts obtained by solid state n.m.r. for CO on
palladium/ η-alumina (675 ppm) (15) or palladium/γ-

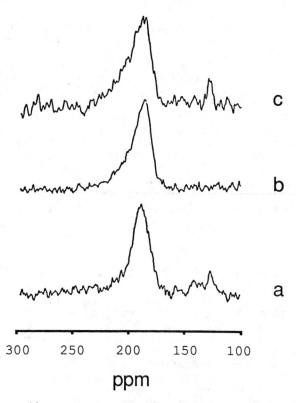

Figure 2. ^{13}C n.m.r. (70 MHz) of ^{13}CO on colloidal palladium, at (a)333°K, (b) 293°K, (c) 222°K. 1Wt% metal in methylcyclohexane, stabilized with *iso*butyl-aluminoxane (see text).

alumina (750 ppm) (Zilm, K. W. , Bonneviot, L., Simonsen,
D. M., Webb, G. G. and Haller, G.L., personal
communication). The metal particles in those studies,
however, were larger and presumably more metallic than
those studied here, and the large shifts have been
attributed to a Knight shift interaction with the
conduction electrons of the metal particles.
     The relatively large linewidth of the CO resonances
could be attributed to either (or both) of two factors –
(a) a distribution of both linear and bridging sites on
the surface of the colloid particles, and (b) incomplete
averaging of the CO chemical shift tensor by the non-
isotropic tumbling of the colloid particles. We feel
that (a) is the most plausible, and rule out (b) on the
basis of the following observations. Suppose tumbling
about one molecular frame axis is much slower than about
the others, leading to a residual shift tensor with
appreciable anisotropy. In a time equal to the
correlation time τ for reorientation about this axis, the
resonance should sweep out all values characterizing the
residual tensor. Experimentally we have found that a
hole (i.e. a sharp saturation pulse) burned in the CO
resonance (Duncan, T. M., Root, T. W. and Thayer, A.
M.,Phys. Rev. Lett., in press.) does not broaden with
time and recovers with T_1 ($T_1(298°K) = 160ms$). For (b)
to be operative would require $\tau \approx 160$ ms which seems
unreasonably long in a non-viscous liquid. Details of
these and other NMR experiments will be published
separately.

N.M.R. STUDIES of ADSORBED ETHYLENE   We have also
investigated the reaction of ^{13}C ethylene with colloidal
palladium. Our initial intent was to attempt to observe
the formation of ethylidyne from ethylene on the surface
of the colloidal palladium particles, a reaction which is
known to occur readily on the surface of supported
palladium and on palladium single crystals (17). Such a
reaction has been identified for ethylene on supported
platinum by magnetic resonance experiments in which spin
echo double resonance techniques were used to
characterize the organic species (18,19), but direct
observation of resonances for adsorbed ethylene or
ethylidyne was not possible in the highly inhomogeneous
solid samples used. The ^{13}C chemical shift differences
which would result from the transformation of ethylene
=CH2 groups to ethylidyne C-CH3 and C-CH3 groups would
make the presence of ethylidyne readily detectable in our
liquid phase system, even given the broad lines we assume
will be typical for colloid adsorbed small molecules
based on our observations in the case of carbon monoxide
described above.
     The colloid solutions used for this part of the
study were essentially the same as those used in the CO
adsorption study. A methylcyclohexane solution of

colloidal palladium (18Å±3Å) was prepared as above and concentrated to 1-2% metal. The solution, as before, contained about 5:1 aluminoxane (monomer equivalents) to palladium. The sample was exposed to 100% $^{13}C_2H_4$ and sealed in a standard 10 mm tube under *ca.* 1 atm. of the enriched ethylene. ^{13}C n.m.r. spectra were obtained as above.

The $^{13}C\{^1H\}$ resonance of dissolved $^{13}C_2$-ethylene in methylcyclohexane containing poly*iso*butylaluminoxane is a typically sharp singlet at 123ppm, with a T_1 of 5.4sec. In the presence of colloidal palladium, the singlet is broadened ($w_{1/2}$= *ca.* 100Hz) and has a T_1 of 0.7sec. Recalling the shortened relaxation time of adsorbed CO on colloidal palladium (see above), we interpret this as due to exchange between the slowly relaxing free ethylene and a more rapidly relaxing adsorbed form.

A resonance for adsorbed ethylene was not observed. At this stage we may only speculate on the reasons for this, and two obvious factors come immediately to mind. If, as we propose from the observation of its short T_1, the free ethylene is exchanging with the adsorbed form, the resonance for the adsorbed ethylene will be expected to be broadened by this exchange process. Although we cannot say at this time what are the relative proportions of free and coordinated ethylene in the sample, there is undoubtedly a considerable excess of free ethylene, and thus the adsorbed ethylene resonance will be exchange broadened to a degree proportionately larger than the free ethylene. The effect of coordination to a colloidal metal particle in these systems is also apparently to broaden the resonance of the adsorbed molecule, as seen in the case of adsorbed CO described above, and this broadening would be expected to be dependent on the size of the metal particle (which determines its electronic properties). It should be remembered that the colloid particle size distribution in our various sample preparations are not all identical, as would be the case in a solution of a molecular cluster, and so some irreproducibility, for example in particle size distribution is unavoidable. This variation occurs in a crucial range when one recalls that the transition from molecular to metallic properties would be occuring in the 20±5Å range, and this could have severe effects on the n.m.r. experiment. For example in some cases the ^{13}C we have observed that the resonance of adsorbed CO is severely broadened, in others it is relatively sharp. In the case of the ethylene experiments, we have obtained, although infrequently, colloids which do enable us to derive more detailed information, including the observation of a resonance at 119.0 ppm, which we tentatively asssign to adsorbed ethylene, and a measurement of its T_1 of .034 sec. The controlled preparation of such colloid solutions is under investigation.

CONCLUSIONS

Colloidal metal particles in non-polar solvents are an instructive model system for investigating the adsorbed state of small molecules. Whereas in heterogeneous supported metal systems the exact molecular constitution of the adsorbate is often a matter of conjecture or controversy, owing to the inhomogeneity of the support and of the metal particles, the use of a liquid medium to study surface chemistry introduces sufficient homogeneity into the system that molecular spectroscopic techniques can be used to advantage to diagnose the molecular identity of adsorbed molecules. These advantages are of course found ostensibly to an even greater extent in the use of molecular clusters to model heterogeneous surface chemistry, but as has so often been observed, an inert molecular cluster makes a poor model for a highly reactive catalyst particle. The colloidal metal particles used in this study are, of course, active catalysts, and as such can be seen as a bridge between heterogeneous catalysis and its homogeneous counterpart.

LITERATURE CITED

1   Bradley, J. S. In Fundamental Research in Homogeneous   Catalysis vol.3, Tsutsui, M. ed., Plenum: New York,   1979.
2   Lewis, L. N.; Lewis, N. J. Amer. Chem. Soc., 1986, 108, 7228.
3   Bradley, J. S.; Hill, E. W.; Leonowicz, M. E.; Witzke,H., J. Mol. Catal., 1987, 41, 59
4   Dahl, L. F. et al.,J. Am. Chem. Soc., 1979, 101, 6110;
5   Cerriotti, A. et al. Angew. Chem., Int. Ed. Engl., 1985, 24, 696
6   See for example Manyik, R. M., Walker, W. E. and Wilson, T. P., J. Catal., 1977, 47, 197
7   Bradley, J. S., Hill, E. W., US 4,857,492, 1989
8   Sheppard, N.; Nguyen, T. T. in 'Advances in Infrared and Raman Spectroscopy', R. J. H. Clark , R. E. Hester eds. 1978, 5, 106 and references therein.
9   Mucalo, M. R.; Cooney, R. P., J. Chem. Soc., Chem. Comm. 1989, 94.
10   Longoni, P., Chini, P. J. Am. Chem. Soc., 1976, 98, 7225
11   Slichter, C. P., Ann. Rev. Phys. Chem., 1986,37, 25-51
12   Todd, L. J. and Wilkinson, J. R., J. Organomet. Chem., 1974, 77, 125
13   Duncan, T. M., J. Phys. Chem. Ref. Data, 1987, 16, 125-51

14   Duncan, T. M. and Root, T. W., J. Phys Chem., 1988, 92, 4426-32

15   Duncan, T.M., Zilm, K.W., Hamilton, D.M., and Root, T.W., J. Phys Chem., in press

16   Shoemaker, R.K. and Apple, T.M., J. Phys. Chem., 1985, 89 3185-8

17   See for example Beebe, T. P.,Jr., Albert, M. R. and Yates, J. T.,Jr., J. Catal, 1985, 96, 1, and references therein.

18   Wang, P.-K., Slichter, C. P. and Sinfelt, J. H., J. Phys. Chem., 1985, 89, 3606.

19   Shore, S. E., Ansermet, J. P., Slichter, C. P., and Sinfelt, J. H., Phys. Rev. Lett., 1987, 58, 953-6

RECEIVED May 9, 1990

Chapter 17

Effect of Cluster Size on Chemical and Electronic Properties

D. M. Cox, A. Kaldor, P. Fayet[1], W. Eberhardt, R. Brickman, R. Sherwood, Z. Fu, and D. Sondericher

Exxon Research and Engineering Company, Route 22 East, Clinton Township, Annandale, NJ 08801

Dependent on the exact number of metal atoms in a transition metal cluster, small clusters (n<30) are shown to exhibit pronounced variation in their chemical and electronic properties. For instance, under kinetically controlled conditions orders of magnitude change in the rate constant for dissociative chemisorption of di-hydrogen and alkanes is found even though the cluster size changes by only a few atoms. Interestingly, the largest variation of a key electronic property, the IP, also occurs over this same size range leading us to propose a simple model based upon partial charge transfer which qualitatively explains the general trends. At the other extreme at near saturation uptake, we find that small transition metal clusters are "hydrogen rich" and can bond an abnormally large number of hydrogen (deuterium) atoms per metal atom in the cluster (D/M as high as 8 for $Rh_n{}^+$, 5 for $Pt_n{}^+$ and $Ni_n{}^+$ and 3 for Pd_n) and that the smaller the cluster the greater the D/M ratio. It now appears that $(H(D)/M)_{max}$ greater than one is more the rule rather than the exception for small transition metal clusters, an effect which has important implications in chemical and catalytic processes involving hydrogen. Lastly, we will describe the results of the first measurements probing the evolution of core and valence band electronic structure as a function of cluster size for mass selected, monodispersed, platinum clusters.

[1]Current address: Ecole Polytechnique Federale Delausanne, Lausannne, Switzerland

0097–6156/90/0437–0172$06.00/0

What are transition metal clusters and why are they interesting? For example what are the properties of a 5-atom platinum cluster? Is it similar to bulk platinum or does it behave more like the atom? What we are discovering is that below a certain size each small n-atom cluster has unique chemical and electronic properties, it behaves neither like the atom, the bulk, nor even like other clusters of the same metal. Thus there are opportunities to exploit their novel properties by making it possible to create new materials with potential applications in such diverse areas as solid state physics, electronics, chemistry and catalysis.

In this paper we will be dealing either with "naked" gas phase metal clusters, i.e. metal clusters which contain no ligands other than the one(s) which may be attached during our experiments, or with deposits of "naked" clusters on some support material such as silica, carbon or alumina. Thus the typical cluster is coordinatively highly unsaturated, and as such might be expected to very reactive. In many instances this is indeed the case, but surprisingly, depending on the particular metal and reaction, certain size "naked" clusters are actually highly unreactive! For example, the 10, 12 and one isomer of the 19 atom niobium cluster cations Nb_{10}^+, Nb_{12}^+ and Nb_{19}^+ are found to be several orders of magnitude less reactive towards di-hydrogen than other niobium cluster cations, even those differing by only one more or one less niobium atom.[1] Similarly Fe_8 is more than an order of magnitude less reactive towards di-hydrogen than Fe_{10}.[2,3] In both instances di-hydrogen readily undergoes dissociative chemisorption onto the bulk metal surface. Several different examples showing strong size selective chemical behavior of small clusters will be presented and discussed in this paper. The main point to remember is that for a given chemical reaction even a small change in cluster size, e.g. by only a single metal atom, can result in dramatic variations in reactivity.

In addition to being coordinatively highly unsaturated, small clusters also are dominated by their surface properties. For example, a 19 atom cluster may have at most one or two interior atoms. Even a 100 atom cluster has only about 28 interior (bulklike?) atoms. Thus the surface sensitive properties become increasingly important as the size becomes smaller. Clusters do not possess the long range periodicity one would have with a bulk crystal. Thus we expect that each small cluster will be a unique entity, and that its ground state structure (or structures if different isomorphic structures are nearly isoenergetic) will depend sensitively on size and likely will not be simply a subunit of the bulk lattice.

Similarly the electronic properties of small clusters are observed to rapidly change as a function of cluster size in an as yet unpredictable, non-monotonic fashion.[4-7] In general terms this may not be unexpected. The atom possesses a sparse set of electronic states whereas the density of states for a bulk metal is quite high. Similarly for the bulk material the ionization potential, IP, is equal to the electron affinity, EA. This does not hold for a metal cluster where generally the IP>EA. Also the cluster IPs (EAs) are significantly greater(less) than the bulk work function. As will be discussed below, recent experiments probing the electronic structure of mercury (8) and platinum clusters(9) show that the small (n<20-40) clusters behave more like non-metals than like metals.

The next section will describe the experimental techniques we use to synthesize and study clusters and present typical data. The following section will discuss the cluster size sensitive behavior observed in kinetic studies of H-H and C-H bond activation reactions as well as the size sensitive behavior of hydrogen uptake, and discusses the potential implications of these experiments in catalysis and chemisorption. The last section gives the highlights of recent studies of the electronic properties of mass selected, monodispersed, platinum clusters containing up to 6 Pt atoms supported on SiO_2.

Experimental Techniques

Gas Phase Clusters. The experimental techniques used to generate, react and detect transition metal clusters will only be briefly described here since most details are available from previous publications. (10-12) Figure 1 shows the schematic layout of the experimental apparatus. The clusters are synthesized by using the combination of laser vaporization and gas aggregation. Briefly, focussing an intense laser onto the surface of a metal rod vaporizes metal from the surface. This vaporized plume of atoms, ions and electrons is entrained in a relatively high pressure of helium gas which is passing over the rod at the instant of vaporization. The helium both serves as a heat reservoir which removes the heat of condensation arising from the cluster formation and serves to direct the flow into a fast flow reactor into which a reagent may be injected. The geometry is such that the cluster growth is terminated prior to the clusters entering the fast flow reactor and the clusters have been cooled to near room temperature (the temperature of the helium carrier gas). The seeded gas expands into vacuum upon exiting the reactor. Under most operating conditions this latter expansion effectively freezes the cluster distribution which contains neutral clusters and cluster-adducts, and well as cationic and anionic clusters and cluster ion-adducts. The beam of neutral clusters and cluster-adducts are then detected and analyzed using UV photoionization followed by time-of-flight mass spectrometry, and the ionic species are detected using pulsed extraction followed by time-of-flight mass analysis. Pulsed laser vaporization of solid substrates is a general technique which allows us to generate and study even the most refractory metals, i.e. we are no longer restricted to refractory ovens using low melting point metals.

A variety of different experiments can now be carried out which probe the properties of the gas phase clusters. One of the first investigations centered upon how an electronic property, the cluster ionization threshold, varied as a function of the number of

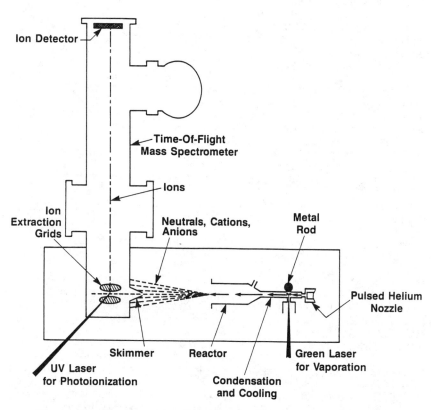

Figure 1. A schematic diagram of the pulsed cluster beam apparatus.

metal atoms in the cluster. Ionization threshold energies, obtained by measurements of ion yields as a function of ionizing laser photon energy, have been carried out in our laboratory for iron, vanadium and niobium clusters and are reported in a series of papers.(4-7) Similarly by allowing the neutral clusters to pass through a inhomogeneous magnetic field (Stern-Gerlach experiment) we have measured how magnetic properties of iron and aluminum clusters vary with size.(13-14) Other experimental probes such as photodetachment and photoelectron spectroscopy have been applied by others to further probe the electronic structure of metal clusters.(15-17) In addition, both photo- and collisional-dissociation spectroscopy now allows one to obtain the binding energies and activation barriers for metal atoms and adducts.(18-20)

Much of our effort involves studies of the chemical behavior of clusters not only as a function of size, but also as a function of metal type, charge state (neutral, cationic or anionic), and reagent molecule. There are two different operating conditions for which we probe the chemisorption of molecules onto clusters as a function of cluster size. The first is such that the rate of reaction is kinetically controlled. Here we obtain information about the rate at which the first reagent molecule chemisorbs onto the otherwise bare cluster. In the second case, chemisorption studies are carried out under near steady-state conditions. In this instance we attempt to determine how many molecules a particular size cluster can bind, i.e. the degree of saturation. The relative concentration of reagent molecules injected into the reactor is the controlling factor. As will become clear later cluster kinetics shows the strongest size sensitive behavior.

The chemisorption studies are performed as follows: First the clusters are formed as discussed earlier and then the cluster/helium pulse enters the reactor. In the reactor we either inject a pure helium pulse or a reagent/helium pulse at the same total pressure but containing a small concentration of some reagent molecule. The helium only pulse allows us to measure the physical effects of the injection (mixing, scattering, etc.) and serves to set the reference signal levels. The second pulse containing the reagent/helium mix then allows the effect of the reagent on the cluster distribution to be probed. Evidence that a reaction has occurred is indicated by (a) loss of bare cluster ion signal and (b) appearance of new mass peaks M_nR in the mass spectrum, where M_n is a metal cluster of n metal atoms and R is the reagent molecule.

Figure 2 shows the effect of increasing di-deuterium concentration upon the iron cluster mass spectrum. The reaction applicable to the loss of the bare cluster spectrum is

$$Fe_n + D_2 \longrightarrow Fe_nD_2 \qquad (1)$$

Note that the although we expressed the product formed as Fe_nD_2, the experiments are carried out at or slightly above room temperature 295K, and all indications are that only dissociatively chemisorbed hydrogen is present.

Now let us examine Fe_{10} more closely. Note that as the hydrogen concentration is increased the bare Fe_{10} ion signal begins to decrease and is almost entirely depleted at the highest concentration. Fe_8 on the other hand shows little evidence of depletion, and other clusters show varying levels of reactivity towards D_2. This is but one instance of strong size selective behavior observed for transition metal clusters. Next we note that for Fe_{10} at the highest concentration a new product peak $Fe_{10}D_{10}$ is detected. The Fe_{10} cluster saturates at this level of deuterium (hydrogen) uptake and does not chemisorb additional deuterium.

As shown elsewhere(21-22), the natural logarithm of the ratio of the bare cluster ion signals I/I_0 is then a measure of the relative rate constant for reaction of the cluster with reagent R (hydrogen in this instance), where I is the bare cluster ion signal with reagent added and I_0 is the bare cluster ion signal with helium only added. From data such as that shown in Figure 2, the relative rate constants for the reaction of metal clusters with a reagent are calculated. Such data for the iron cluster reactivity towards di-hydrogen is plotted in Figure 4, and that for the reaction of methane with platinum and palladium clusters is shown in figure 5.

Deposition of Monosized Clusters. In order to study the properties of deposited monosized clusters we have the capability to generate nanoamps (nA) of mass selected metal cluster ions. The experimental apparatus is shown schematically in figure 3.(23) The metal cluster ions are produced by sputtering of a metal substrate with high energy rare gas (Ar^+, Kr^+ or Xe^+) ions. The cluster ions are energy and mass selected and then directed onto an appropriate substrate. Typical operating conditions are as follows: Xe^+ at 25KeV, 5ma; cluster beam size 0.25 cm^2; cluster beam currents several nA down to a few tenths of nA depending on the cluster size; surface coverage by clusters is typically limited to less than 15% area coverage, i.e. cluster flux on surface is between 5×10^{13} and 1×10^{14} cm^{-2}; substrates are kept at room temperature; and vacuum during deposition is maintained at about 10^{-6} torr. The advantage of this technique is that continuous beams are produced but one major disadvantage is that the intensity of larger cluster ions drops off rapidly. So although the sputtering techinque is general (most solid substrates can be sputtered), it is difficult to obtain sufficient intensity to make sample deposits of larger (n>10) clusters in a reasonable (say 2 hours) time. We have studied the mass selected deposited clusters using a variety of techniques including scanning transmission electron microscopy, scanning tunnelling microscopy, and UV photoemission. Some of this work will be described in the last section.

Studies of Gas Phase Clusters

H-H Bond Activation. The reactivity of iron clusters toward di-hydrogen is shown in figure 4. Also plotted in figure 4 are electron binding energies (IPs) of iron clusters. Note that the

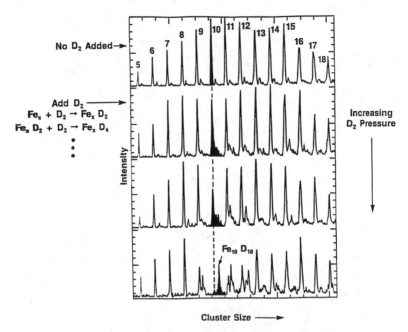

Figure 2. Time-of-flight mass spectra of iron clusters as a function of increasing deuterium partial pressure in the reactor. The upper panel shows the reference mass spectrum obtained when helium only is pulsed into the reactor. The lower curves show the effect of increasing deuterium pressure.

(Reprinted with permission from Ref. 2. Copyright 1985 American Institute of Physics.)

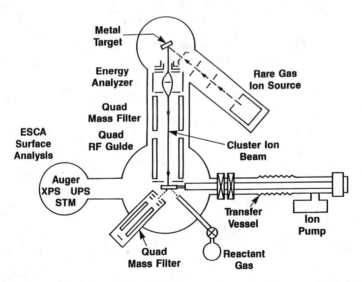

Figure 3. A schematic of the apparatus used to produce mass selected clusters.

reactivity is plotted on a log scale and the electron binding energies on a linear scale. Thus we see that the reactivity can vary by several orders of magnitude for only a few atom size change. Note also that the highest reactivity is observed for the larger clusters n>23.(2,24) In fact nearly constant and high reactivity has been measured by the Argonne group from n=25 out to n=270.(3)

For clusters larger than about 8 atoms we find a strong correspondence between the measured relative reactivity and that which can be inferred from electron binding energies.(2,24) For the smaller clusters, just the opposite is observed.(24) In its simplest terms the size selective behavior may be rationalized as resulting from the competition between Pauli repulsion which creates a barrier to di-hydrogen chemisorption and attractive partial charge transfer interactions. For the larger metal clusters with lower IPs, charge transfer from the metal to hydrogen is the dominant attractive interaction; whereas for the smaller, higher IP, clusters the dominant attractive interaction is hydrogen to metal charge donation.(25) A similar global correspondence has been observed for di-hydrogen chemisorption on niobium and vanadium, the only other transition metal clusters for which both IP and reactivity measurements are presently available.(5-7) As noted above (with the exception of the coinage metals) most transition metal clusters containing more than about 25 metal atoms exhibit a high, nearly size independent reactivity towards di-hydrogen. This suggests that for clusters larger than about 25 atoms non-activated dissociative hydrogen chemisorption occurs (or at least the activation barrier is sufficiently reduced that facile chemisorption occurs at near room temperature), quite analogous to what occurs on on many transition metal surfaces.

It should be pointed out that the correspondence noted above is simply that, a correspondence. There are certain size clusters, e.g. Nb_{16}, Fe_{17}, in the different metal systems for which this correspondence does not seem to apply. This suggests that it is unlikely that the strong size dependence observed for activation of di-hydrogen shown in figure 4 can be explained by a model based solely on charge donation. We expect that a complete explanation will not be shortly forthcoming and likely will require significantly more experimental investigation of structure and stability of the clusters as a function of size. With additional structural data we hope that a detailed theoretical understanding can be obtained.

C-H Bond Activation. We have also examined the chemisorption of various hydrocarbons on different transition metal clusters. In this section we describe results obtained for methane activation on neutral clusters. First, we note that under our experimental conditions (low pressure, near room temperature, short contact time) methane activation readily occurs only on specific type and size metal clusters. For instance, we detect no evidence that methane reacts with iron(12), rhodium(26) or aluminum(27) clusters, whereas as shown in figure 5 strong size selective chemisorption is

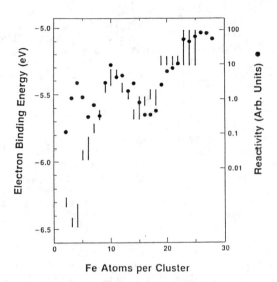

Figure 4. Comparison of the relative reactivity of iron clusters towards hydrogen (right hand scale and circles) with the electron binding energy (left hand scale and vertical lines) as a function of number of iron atoms in the cluster.

(Reprinted from Reference 24. Copyright 1988 American Chemical Society.)

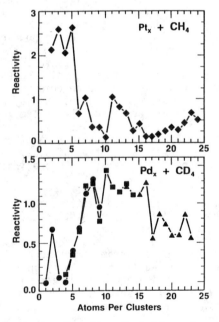

Figure 5. Relative reactivity of platinum (upper panel) and palladium (lower panel) clusters towards methane.

found for small platinum($\underline{28}$) and palladium($\underline{29}$) clusters, but the size selective behavior is different for the two metals. As shown small (n=2-5) Pt clusters are the most reactive, whereas the mid-sized (n=7-16) Pd clusters are the most reactive. In both instances the larger clusters become less reactive. Again a simple charge transfer model can crudely rationalize the gross behavior as follows: Both platinum and palladium clusters have high IPs and the small clusters appear to have the highest IPs of any metal system we have investigated. In this instance the methane to metal charge transfer may dominate for all sizes. This is not unexpected since (i) the larger cluster IPs may never becomes sufficiently small for donation of charge from the metal to occur, and (ii) the lowest unoccupied molecular orbital of methane is at significantly higher energy than that of hydrogen, i.e. as Saillard and Hoffman($\underline{25}$) have pointed out methane should be significantly poorer charge acceptor than hydrogen but is as good or better charge donator. Such arguments also qualitatively explain the non-reactivity of iron and rhodium (lower IP systems) clusters.

Hydrogen (Deuterium) Saturation. In addition to kinetic studies, we also measure hydrogen saturation, i.e. the maximum number of hydrogens which can be bound to a metal cluster as a function of the number of metal atoms in the cluster. Our results show that a very large number of hydrogens can be bound to small transition metal clusters.($\underline{29}$-$\underline{30}$) Figure 6 shows the results of deuterium chemisorption on cationic clusters of three metals, rhodium, platinum and nickel. Let us examine the 10 atom cluster for each of the metals. The deuterium saturation levels are 32, 22 and 16 deuteriums for Rh_{10}^+, Pt_{10}^+ and Ni_{10}^+, respectively. By comparison Fe_{10} saturated at 10 hydrogen atoms (see figure 2). As can be seen from figure 6, the dimer cations of Rh, Pt and Ni have the highest $(D/M)_{max}$ ratios of 8, 5 and 5, respectively. For neutral iron clusters $(H/M)_{max}$ was only 1.1($\underline{31}$), but for neutral clusters of vanadium,($\underline{32}$) niobium($\underline{32}$) and tantalum($\underline{33}$) $(H/M)_{max}$ of 1.4 has been measured. Thus hydrogen saturation displays not only a strong dependence on cluster size, but also a pronounced dependence on the metal type.

These results suggest that small (<10A dia.) highly dispersed, deposited clusters of many different transition metals may take up substantially more hydrogen than one per metal surface atom. Thus the commonly used technique of hydrogen (and CO) chemisorption to titrate the number of exposed surface atoms (where it is typically assumed that H/M=1) may not be accurate for highly dispersed catalysts.

Note however that in terms of electron counting rules Rh_2D_{16} may be considered the hydrogen ligated analogue of $Rh_2(CO)_8$, a stable well characterized organometallic molecule.($\underline{34}$) Since similar analogies may be made for other cluster sizes as well as for different metals, the observation of small "hydrogen rich" clusters may not be as unexpected as first thought. It is simply that the gas phase molecular beam techniques allow us to synthesize and detect such seemingly exotic molecules for the first time.

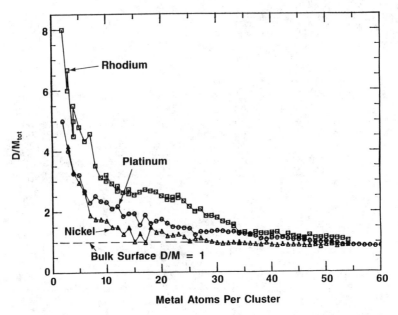

Figure 6.   Hydrogen uptake for rhodium, platinum and nickel
cations as a function of cluster size.   H/M is the measured
hydrogen to metal stoichiometry of the cluster.

Implications for Catalysis. We next discuss possible implications of the above results on the interpretation of certain catalytic processes. For example, ethylene hydrogenation is known to be structure (metal particle size) insensitive for metal particles larger that about 20 A in diameter. Recent studies of this reaction by Masson et al.(35) using differing coverages of evaporated platinum on silica and alumina surfaces showed that the turn over number (TON) was insensitive to particle size for particles larger than about 20 Å in diameter. As the particle size was reduced below 20 Å, the TON increased significantly before peaking and dropping to near zero for the lowest platinum coverage (possibly atomically dispersed platinum). The particle size dependence of TON in this experiment correlates nicely with the cluster size dependence of the hydrogen uptake. In particular, note that not only does H/Pt = 1 for larger particles where the TON is constant, but the increase in TON with decreasing cluster size occurs in the same cluster size range where D/Pt becomes larger than 1, i.e. not surprisingly "hydrogen rich" clusters appear to be more efficient hydrogenators. The drop off towards no reactivity for the lowest coverage suggests that a minimum size cluster is required to accomodate both ethylene and hydrogen. Just the availability of more hydrogen is not likely to be the whole answer. We speculate that in addition to the H/M increasing as the cluster size decreases the activation energy for hydrogen desorption (at least the most weakly bound hydrogen on highly covered clusters) may be lower than that for larger clusters. Thus highly covered small clusters not only have more hydrogen available but the hydrogen is also easier to remove, and so hydrogenation by small "hydrogen rich" clusters is enhanced relative to larger clusters.

Such considerations may also help in understanding results of CH_4/D_2 isotope exchange experiments,(36) in which small Ni, Ir or Pt clusters ($\leq$10 Å) are found to have a higher propensity to form single metal to carbon bonds, whereas larger clusters have a higher propensity to form multiple metal to carbon bonds. Being "hydrogen rich" suggests that not only may the small clusters hinder C-H bond splitting, i.e. dehydrogenation of methane all the way to carbon, but also that small clusters may have lower hydrogen desorption energies, which tends to make hydrogen more easily available and thus hinder complete dehydrogenation of methane.

Size Selective Properties of Mono-sized, Deposited Clusters. As discussed earlier, it is now possible to make and study deposits of monosized, highly dispersed, transition metal clusters.(9) In this section we summarize results from the first measurements of the valence and core level photoemission spectra of mass selected, monodispersed platinum clusters. The samples are prepared by depositing single size clusters either on amorphous carbon or upon the natural silica layer of a silicon wafer. We allow the deposition to proceed until about 10 per cent of the surface in a 0.25 cm^2 area is covered. For samples consisting of the platinum atom through the six atom cluster, we have measured the evolution of the individual valence band electronic structure and the Pt 4f

VALENCE BAND PHOTOEMISSION

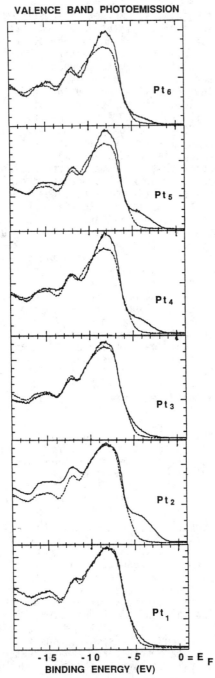

Figure 7. The solid curve in each panel is the UV photoemission signal obtained from the silica substrate without platinum clusters and the dashed curves represent the UV signal obtained when clusters are present.

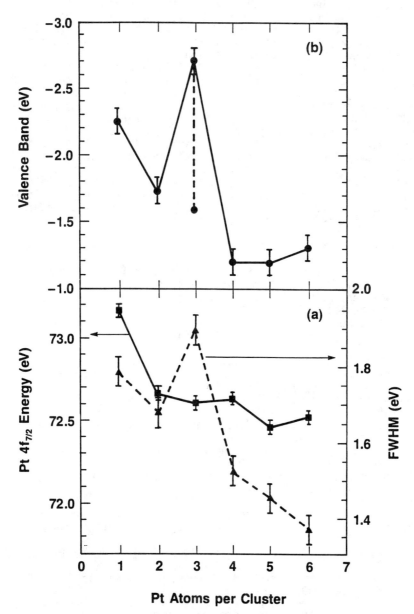

Figure 8. The upper panel plots the measured valence band offset of the Pt$_{1-6}$ samples as a function of cluster size obtained at a photon energy of 40 eV. The lower panel is a plot of the Pt 4f$_{7/2}$ core level shift and bandwidth as a function of cluster size obtained for a photon energy of 280 eV.

core electron binding energy as a function of cluster size. Figure
7 shows the valence band spectra obtained for Pt_1-Pt_6. Note that
the onset of the valence band is offset from the bulk Pt Fermi
energy and that the photoemission intensity rises slowly in the
threshold region, i.e. there is no sharp onset at the Fermi energy.
These effects show that the small platinum clusters do not yet
exhibit characteristic electronic behavior observed for bulk
metals. Figure 8 summarizes some of the key information obtained
from these studies. As can be seen the Pt_1 - Pt_6 samples exhibit
individual, discrete electronic structure features as
characterized by size dependent valence band onsets and Pt $4f_{7/2}$
core level binding energies and bandwidths. This is taken as proof
that samples containing single sized clusters deposited onto
supports can be prepared. Since each size cluster may have unique
materials and chemical properties, further studies will probe not
only how the electronic structure evolves with size, but how it is
perturbed by different substrates, and how the cluster chemical
activity is affected by size, substrate, reagent and temperature.

Literature Cited:

1. Elkind, J. L.; Weiss, F. D.; Alford, J. M.; Laaksonen, R. T.;
   Smalley, R. E. J. Chem. Phys. 1988, 88, 5215.
2. Whetten, R. L.; Cox, D. M.; Trevor, D. J.; Kaldor, A. Phys.
   Rev. Lett. 1985, 54, 1494.
3. Richtsmeier, S. C.; Parks, E. K.; Liu, K.; Pobo, L. G.; Riley,
   S. J. J. Chem. Phys. 1985, 82, 5431.
4. Rohlfing, E. A.; Cox, D. M.; Kaldor, A.; Johnson, K. H. J.
   Chem. Phys. 1984, 81, 3846.
5. Whetten, R. L.; Zakin, M. R.; Cox, D. M.; Trevor, D. J.;
   Kaldor, A. J. Chem. Phys. 1986, 85, 1697.
6. Kaldor, A.; Cox, D. M.; Trevor, D. J.; Zakin, M. R. Z. Phys.
   1986, D 3, 195
7. Cox, D. M.; Whetten, R. L.; Zakin, M. R.; Trevor, D. J.;
   Reichmann, K. C.; Kaldor, A. AIP Conference Proceedings #146,
   Adv. in Laser Science-I, edited by W. C. Stwalley and M.
   Lapp., AIP, New York, 1986.
8. Rademann, K.; Kaiser, B.; Even, U.; Hensel, F. Phys. Rev.
   Lett. 1987, 59, 2319,
9. Eberhardt, W.; Fayet, P.; Cox, D. M; Fu, Z.; Kaldor, A.;
   Sherwood, R.; Sondericher, D. Phys. Rev. Lett. Submitted.
10. Zakin, M. R.; Brickman, R. O.; Cox, D. M.; Kaldor, A. J. Chem.
    Phys. 1988, 88, 3555
11. Rohlfing, E. A.; Cox, D. M.; Kaldor, A. J. Phys. Chem. 1984,
    88, 4497.
12. Whetten, R. L.; Cox, D. M.; Trevor, D. J.; Kaldor, A. J. Phys.
    Chem. 1985. 89, 566.
13. Cox, D. M.; Trevor, D. J.; Whetten, R. L.; Rohlfing, E. A.;
    Kaldor, A. Phys. Rev. 1985, B32, 7290.
14. Cox, D. M.; Trevor, D. J.; Whetten, R. L.; Rohlfing, E. A.;
    Kaldor, A. J. Chem. Phys. 1986 84, 4651.
15. Ervin, K. M.; Ho, J.; Lineberger, W. C.; J. Chem. Phys. 89,
    1988, 4514.

16. Pettiette, C. L.; Yang, S. H.; Craycraft, M. J.; Conceicao, J.; Laaksonen, R. T.; Cheshnovsky, O.; Smalley, R. E. J. Chem. Phys. 1988, 88, 5377.
17. Gantefor, G.; Meiwes-Broer, K. H.; Lutz, H. O. Phys. Rev. 1988, 37, 276.
18. Brucat, P. J.; Zheng, L. S.; Pettiette, C. L.; Yang, S.; Smalley, R. E. J. Chem. Phys. 1986, 84, 3078.
19. Jarrold, M. F.; Bower, J. E.; Kraus, J. S. J. Chem. Phys. 1987, 86 3876.
20. Loh, S. K.; Lian, L.; Armentrout, P. B. J. Am. Chem. Soc. 1989. 111 3167.
21. Cox, D. M.; Reichmann, K. C.; Trevor, D. J.; Kaldor, A. J. Chem. Phys. 1988, 88, 111.
22. Geusic, M. E.; Morse, M. D.; O'Brien, S. C.; Smalley, R. E. Rev. Sci. Instrum. 1985, 56 2123.
23. Our apparatus is quite similar to that developed by P. Fayet and L. Woste. e.g. see Fayet, P.; Granzer, F.; Hegenbart, G.; Moisar, E.; Pischel, B.; Woste, L. Phys. Rev. Lett. 1985 55, 3002.
24. Zakin, M. R.; Brickman, R. O.; Cox, D. M.; Kaldor, A. J. Chem. Phys. 1988, 88, 6605.
25. Saillard, J. Y.; Hoffman, R. J. Am. Chem. Soc. 1984, 106, 2006.
26. Zakin, M. R.; Cox, D. M.; Kaldor, A. J. Chem. Phys. 1988, 89, 1201.
27. Cox, D. M.; Trevor, D. J.; Whetten, R. L.; Kaldor, A. J. Phys. Chem. 1988, 92, 421.
28. Trevor, D. J.; Cox, D. M.; Kaldor, A. J. Am. Chem. Soc. to be published
29. Fayet, P.; Kaldor, A.; Cox, D. M. J. Chem. Phys. in press.
30. Cox, D. M.; Brickman, R. O.; Hahn, M. Y.; Kaldor, A. to be published.
31. Parks, E. K.; Liu, K.; Richtsmeier, S. C.; Pobo, L. G.; Riley, S. J. J. Chem. Phys. 1985, 82, 5470.
32. Kaldor, A.; Cox, D. M.; Zakin, M. R. Evolution of Size Effects in Chemical Dynamics, Part 2, edited by I. Prigogine and S. A. Rice, Wiley, New York, 1988; Adv. Chem. Phys. Vol. 70.
33. Zakin, M. R.; Cox, D. M.; Kaldor, A. unpublished results
34. see for example, Collman, J. P.; Hegedus, L. S. Principles and Applications fo Organotransition Metal Chemistry, University Science Books, Mill Valley, CA, 1980, pg 83.
35. Masson, A.; Bellamy, B.; Romdhane, Y. H.; Che, M.; Roulet, H,; Dufour, G. Surf. Sci. 1986, 173, 479.
36. Van Broekhoven, E. H.; Ponec, V. Surf. Sci. 1985, 162, 731.

RECEIVED May 9, 1990

Chapter 18

Cluster Intermediates in the Molecular Synthesis of Solid-State Compounds

M. L. Steigerwald

AT&T Bell Laboratories, Murray Hill, NJ 07974

We describe the synthesis of several simple binary solid state compounds using organometallic reagents and reactions. Hoping to be able to understand some of the pathways by which the solids assemble we have interrupted the molecules-to-solids transformations using a number of related techniques. We have been able to isolate several intermediates in these reactions and shown that they are organometallic cluster compounds. In some cases these clusters are directly related structurally to the bulk, being simply small fragments of the lattice. In other cases the relationship is less obvious.

Inorganic solid state compounds are most often prepared by the direct combination of the elements in the appropriate stoichiometries(1). This synthesis method is extraordinarily powerful and is a cornerstone of solid state chemistry. A distinct alternative is the molecular precursor method(2) in which the starting materials for the solid state synthesis are molecular compounds. (As noteworthy examples, this general approach has been used to prepare tetrahedral semiconductor materials (Si,(3) GaAs,(4) HgTe,(5) etc.), refractory ceramics(6) and intermetallics(7). In the synthesis from the elements solid state interdiffusion is often the highest energy process, and therefore the distribution products from these syntheses are often determined either by the very complex kinetics of interdiffusion or by the relative thermodynamic stability of the products. In the molecular precursor method the high energy interdiffusion of the elements is avoided and therefore solid state products and distributions can be governed by comparatively simple reaction kinetics. If the molecular precursor approach is to be exploited to its fullest it is important to learn as much as possible about the pathways by which the molecular reagents assemble and reorganize to give the solid state products. In this manuscript we will describe our investigations of two such molecule-to-solid conversions, paying particular attention to the identification of reaction intermediates. We find that these are intermediates not only in the sense of reaction coordinate diagrams but also in the sense of isolated chemical species which are intermediate in size and properties between molecules and solids.

Nanoscale Clusters of CdSe

We found recently(8) that organic ditellurides (RTeTeR, R = some organic moiety) are potentially useful tellurium source compounds for the preparation of Te-containing II-VI compounds by organometallic vapor phase epitaxy (OMVPE). In the course of those investigations we found (9) that di(aryltellurato) mercury compounds [(RTe)$_2$Hg], formed from the ditelluride and Hg,

0097–6156/90/0437–0188$06.00/0

thermally react to form polycrystalline HgTe and diaryl tellurium. We have since shown(10) that essentially all of the II-VI compounds (stoichiometric MX; M=Zn, Cd, Hg; X = S, Se, Te) can be prepared from the corresponding di(arylchalcogenide) metal compound (Equation 1),

$$M(XR)_2 \longrightarrow MX + XR_2 \qquad (1)$$

thus extending the reaction which had been previously known for at least CdS to the entire range of solid state products.

This reaction as written is quite simple but in fact it is quite complex since the solid state product which is formed may be thought of as a very high molecular weight, highly structured "polymer" of the MX unit cell. This implies that rather than a single, simple reaction (Equation 1) occurring a large number of times, a large number of reactions, all rather similar but none exactly the same, occur in a given preparative sequence. Two extreme possible pathways are shown in Figure 1. We believed that by understanding the details of this process as thoroughly as possible we would be able to use it more effectively, and for that reason attempted to characterize some of the chemical intermediates implicit in Equation 1.

Reaction 1 is most readily conducted in the solid state, the precursor being heated in a glass tube sealed under vacuum. This yields the polycrystalline solid state compound as the organic by-product distils to the cooler end of the sealed tube. This reaction environment is not conducive to the interception and characterization of reaction intermediates for the simple reason that initially isolated "strands" of the growing inorganic "polymer" are in close contact with one another and can fuse. It is fortunate that reaction 1 can also be carried out in solution, thereby at least in part isolating different nucleation sites.

When Cd(SePh)$_2$ is heated in dilute 4-ethylpyridine (4EP) solution(10a) a series of changes take place in the UV-visible absorption spectrum of the solution (Figure 2). The initially transparent (400 nm to 800 nm) solution develops a strong, well-defined absorption (λ_{max} = 410 nm) which subsequently fades as a broad absorption at approx 550 nm appears. This behavior is not peculiar to CdSe and its precursors. When Cd(TePh)$_2$ is similarly treated(10b) an analogous set of changes occur in the absorption spectrum of the 4EP solution (Figure 3). In each case we have been able to characterize the intermediates between Cd(XPh)$_2$ and CdX. Transmission electron microscopy (TEM) has shown(10a–b) that the strongly absorbing solutions represented by the data in Figures 1 and 2 contain nanometer-sized fragments of the CdX lattice. This is shown both by electron diffraction and by direct imaging of the solid state lattice. These data establish that the structure of the bulk solid state compound is established very early on in the thermolysis.

In addition to characterization of these reaction intermediates by TEM we have also been able to synthesize the same materials by an independent route.(11) In this procedure Cd^{2+} is dissolved in a surfactant/heptane/water reverse micelle reaction medium and subsequently treated with Se(SiMe$_3$)$_2$, an organometallic equivalent of Se^{2-}. The ensuing reaction is quite fast, yielding intensely colored, apparently homogeneous solutions whose absorption spectra are quite similar to those shown in Figure 2. Sequential addition of alternating aliquots of Cd^{2+} and Se(SiMe$_3$)$_2$ give solutions which absorb further and further to the red. When PhSeSiMe$_3$ is used in place of Se(SiMe$_3$)$_2$ the color leaves the solution and a solid precipitates which has the color of the erstwhile solution. (This process is summarized in Figure 4). This solid dissolves in Lewis basic organic solvents such as pyridine. We have analyzed this material by TEM, X-ray scattering,(12) ^{77}Se–NMR,(13) vibrational spectroscopy,(14) electronic spectroscopy,(15) Auger and energy-dispersive x-ray spectroscopy (16) and have shown that it is made up of nanometer-sized fragments of the CdSe lattice, the surfaces of each fragment being passivated with phenyl moieties from the PhSeSiMe$_3$. The surface passivation is critical since it protects the sample from interparticle fusion.

The source of the absorptions in Figures 2 and 3 is quantum mechanical.(17) The bulk solids, CdSe and CdTe, are semiconductors and therefore absorb all light having energy in excess of their bandgaps (1.8 eV and 1.4 eV respectively). For each photon absorbed an electron-hole pair (exciton) is formed. When this exciton is confined in a volume smaller than that which is characteristic of the crystalline solid the energy required to create the exciton is greater. We have

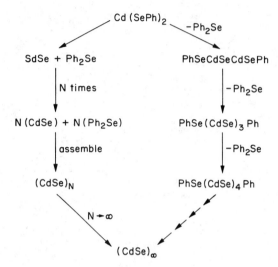

Figure 1. Two possible pathways from Cd(SePh)$_2$ to CdSe.

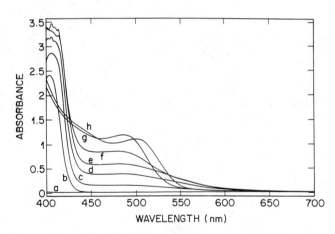

Figure 2. Time evolution of the UV-visible absorption spectrum of the thermolysis (180°C) of Cd(SePh)$_2$ (5×10^{-2} M in 4-ethyl pyridine). Trace (a) shows absorption due to starting material; (b) 4 min; (c) 25 min; (d) 1 hr; (e) 2 hr; (f) 5.5 hr; (g) 9 hr; (h) 84 hr.

(Reprinted from ref. 10a. Copyright 1989 American Chemical Society.)

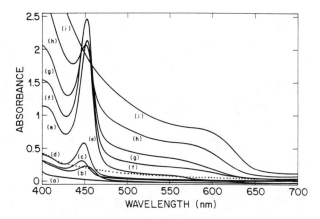

Figure 3. Time evolution of the UV-visible absorption spectrum of the thermolysis (180°C) of Cd(TePh)$_2$.   (Reprinted from ref. 10b.   Copyright 1989 American Chemical Society.)

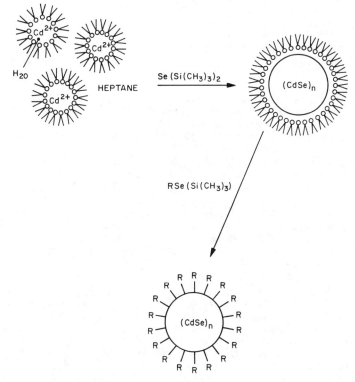

Figure 4.   Schematic description of the preparation of capped CdSe particles by arrested precipitation in reverse micelles.
(Reprinted from ref. 11.   Copyright 1989 American Chemical Society.)

interpreted the electronic spectral absorptions shown in Figures 2 and 3 to the formation of "confined" excitons. They therefore represent the band structure of nanometer-sized semiconductor materials.

Our conclusion from this work has been that the reaction shown in Equation 1 leads initially to the formation of very small fragments of the bulk solid which subsequently either grow by accretion or fusion to give bulk solids.

Molecular Clusters of NiTe

In attempting to extend this conclusion to other reactions and other materials we were led to the reaction shown in Equation 2.(18)

$$Ni(COD)_2 + Et_3PTe \longrightarrow NiTe \qquad (2)$$

Bis(1,5-cyclooctadiene) nickel ($Ni(COD)_2$) is a very convenient and labile source of zerovalent Ni, and Et_3PTe is a similarly convenient source of zerovalent Te. Reaction 2 is essentially a combination of neutral atoms and we were curious whether or not it would proceed as shown and if nanocluster intermediates could be isolated from it as from reaction 1.

When $Ni(COD)_2$ and Et_3PTe are combined in refluxing toluene polycrystalline NiTe is formed quickly, thus reaction 2 does occur and $Ni(COD)_2$ and Et_3PTe are candidates for use in solid state synthesis. It is noteworthy that when the two are combined instead at room temperature there is an immediate reaction to give a black solution, but the bulk NiTe solid is clearly not formed. Treatment of this solution with the appropriate co-solvent led to the crystallization of the cluster compound $Ni_{20}Te_{18}(PEt_3)_{12}$ (1). The structure of this cluster was determined crystallographically, and a representation of that structure is shown in Figure 5. The structure is rather complicated, and at first glance seems to bear little resemblance to bulk NiTe, nevertheless the central Te atom in 1 has a coordination environment similar to Te in the NiTe crystal.(19) Importantly the same reaction mixture which leads to NiTe at 115°C deposits 1 at room temperature. Although this certainly does not establish that 1 is a necessary intermediate in reaction 2 it does assure that 1 is an admissible candidate.

In addition to temperature, reaction 2 has another controllable variable, that being reagent relative stoichiometries. In a large body of outstanding work Fenske & co-workers(20) have shown that a variety of metal-chalcogenide cluster compounds can be isolated by manipulating reagent ratios. Since phosphines stabilize both free Ni and free Te we believed that addition of excess phosphine to reaction 2 (in addition to the use of a high Ni/Te ratio) would yield smaller NiTe building blocks and thus shed some light on the mode of formation of 1. When $Ni(COD)_2$, Et_3PTe and Et_3P are combined in heptane in the ratio 2:1:22 crystals of the compound $Ni_9Te_6(PEt_3)_8$ (2) form. The structure of this cluster was also determined crystallographically (Figure 6). It is much more symmetrical than 1 and is also more clearly related to bulk NiTe.(19) (In bulk NiTe each Ni is surrounded by an octahedron of Te atoms just as is the central Ni in 2). It is important to note that the structure of 2 is closely related to a number of the solid state lattices, viz., Ni_3Te_2,(21) pentlandite(22) (Co_9S_8)and wurtzite,(23) as well as being the inverse of a Chevrel cluster.(24) As mentioned above one of our reasons for studying the Ni/Te system was to test the hypothesis that molecular intermediates in the synthesis of solid state compounds have the same structure as that of the crystalline solid. Clearly the clusters 1 and 2 do not have the structure of NiTe unit cell. Their structures do raise some interesting questions. At what point in size is the bulk structure unavoidable? What distortions are required to make 1 and 2 recognizable fragments of the NiTe structure, and what are the energetic costs of those distortions? Why are structures 1 & 2 selected? While our studies of the Ni/Te and related systems are not yet complete it is certain that a great deal of structural & reaction chemistry remains to be unravelled in this area.

In this paper we have described two of a growing number of molecular reactions which lead to solid state inorganic materials as products. We have been able to follow each of these reactions to a limited degree and have isolated and characterized reaction intermediates in each case. In each case these clusters bear structural resemblance to the intended solid state material. In the case of the II-VI materials CdSe and CdTe this resemblance is direct and obvious while in the case of the NiTe clusters the structural connection is not quite as specific. Further detection of intermediates

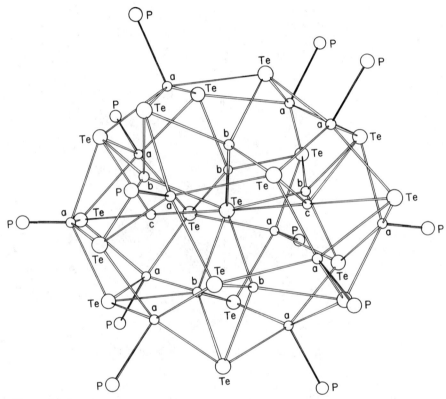

Figure 5. Structure of $Ni_{20}Te_{18}(PEt_3)_{12}$. The labels a, b & c denote Ni atoms in three different coordination environments. Further structural information is included in reference 18.

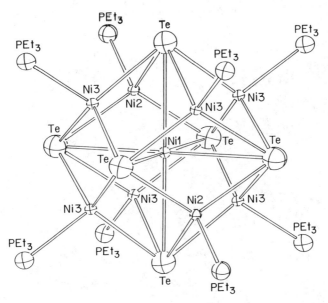

Figure 6. Structure of $Ni_9Te_6(PEt_3)_8$. Detailed structural information is contained in reference 18.

in such molecules-to-solids reactions will help determine how solids are formed and therefore how best to use the molecular precursor method. The independent application of these cluster materials to problems in materials science and catalysis will be interesting and valuable.

Literature Cited

1. West, A. R. Solid State Chemistry and its Applications; Wiley & Sons: Chichester, 1986; Chapter 2.
2. a. Schäfer, H. Chemical Transport Reactions; Academic Press: New York, 1964.
    b. Rao, C. N. R.; Gopalakrishnan, J. New Directions in Solid State Chemistry; Cambridge University Press: Cambridge, 1986; p. 118 et seq.
3. See, for example, Jasinski, J. M.; Meyerson, B. S.; Scott, B. A. Ann. Rev. Phys. Chem. 1987, 38, 109-40.
4. a. Ludowise, M. J. J. Appl. Phys. 1985, 58, R31-55.
    b. Proc. Third Int'l. Conf. on Metalorganic Vapor Phase Epitaxy, Stringfellow, G.B., Ed.; J. Cryst. Growth, 1986, 77, (entire volume).
5. Reference 4b, section VIII.
6. a. Jeffries, P. M.; Girolami, G. S. Chem. Mater. 1989, 1, 8-10.
    b. Boyd, D. C.; Haasch, R. T.; Mantell, D. R.; Schulze, R. K.; Evans, J. F.; Gladfelter, W. L. Chem. Mater. 1989, 1 119-24.
    c. Seyferth, D.; Rees, W. S., Jr.; Haggerty, J. S.; Lightfoot, A. Chem. Mater. 1989, 1, 45-52.
    d. Wu, H-J; Interrante, L.V. Chem. Mater. 1989, 1, 564-8.
    e. Beck, J. S.; Albani, C. R.; McGhie, A. R.; Rothman, J. B.; Sneddon, L. G. Chem. Mat. 1989, 1, 433-8.
7. a. Steigerwald, M. L. Chem. Mater. 1989, 1, 52-7.
    b. Steigerwald, M. L.; Rice, C. E. Amer. Soc. 1988, 110, 4228-31.
8. Kisker, D. W.; Steigerwald, M. L.; Kometani, T. Y.; Jeffers, K. S. Appl. Phys. Lett. 1987, 50, 1681-3.
9. Steigerwald, M. L.; Sprinkle, C. R. J. Amer. Chem. Soc. 1987, 109, 7200.
10. a. Brennan, J. G.; Siegrist, T.; Carroll, P. J.; Stuczynski, S. M.; Brus, L. E.; Steigerwald, M. L. J. Amer. Chem. Soc. 1989, 111, 4141-3.
    b. Brennan, J. G.; Siegrist, T.; Carroll, P. J.; Stuczynski, S. M.; Reynders, P.; Brus. L. E.; Steigerwald, M. L. Chem. Mater. submitted.
11. Steigerwald, M. L.; Alivisatos, A. P.; Gibson, J. M.; Harris, T. D.; Kortan, A. R.; Muller, A. J.; Thayer, A. M.; Duncan, T. M.; Douglass, D. C.; Brus, L. E. J. Amer. Chem. Soc. 1989, 110, 3046-50.
12. Bawendi, M. G.; Kortan, A. R.; Steigerwald, M. L.; Brus, L. E. J. Chem. Phys. 1989, 91, 7282-90.
13. Thayer, A. M.; Steigerwald, M. L.; Duncan, T. M.; Douglass, D. C. Phys. Rev. Lett. 1988, 60, 2673-6.
14. Alivisatos, A. P.; Harris, T. D.; Carroll, P. J.; Steigerwald, M. L.; Brus, L. E. J. Chem. Phys. 1989, 90, 3463-8.
15. Alivisatos, A. P.; Harris, T. D.; Levinos, N. J.; Steigerwald, M. L.; Brus, L.E. J. Chem. Phys. 1988, 89, 4001-11.
16. Kortan, A. R.; Hull, R.; Opila, R. L.; Bawendi, M. G.; Steigerwald, M. L.; Carroll, P. J.; Brus, L. E. J. Amer. Chem. Soc. 1990, 112, p. 1327.
17. Steigerwald, M. L.; Brus, L. E. Ann. Rev. Mat. Sci. 1989, 19, 471-95 and references therein.
18. Brennan, J. G.; Siegrist, T.; Stuczynski, S. M.; Steigerwald, M. L. J. Amer. Chem. Soc. 1989, 111, 9240-1.
19. Barstad, J.; Gronvold, F.; Rost, E.; Vestersjo, E. Acta Chem. Scand. 1966, 20, 2865-79.
20. Wells, A. F. Structural Inorganic Chemistry; Clarendon Press: Oxford, 1975; Chapter 17.

21. Fenske, D.; Ohmer, J.; Hachgenei, J.; Merzweiler, K. Angew. Chem. Int. Ed. Engl. 1988, 27, 1277-96.
22. Kok, R. B.; Wiegers, G. A.; Jellinek, F. Rec. Trav. Chim. Pays Bas 1965, 84, 1585-8.
23. a. Geller, S. Acta Cryst. 1962, 15, 1195-8.
    b. Christou, G.; Hagen, K. S.; Holm, R. H. Inorg. Chem. 1982, 104, 1744-5.
24. a. Reference 1, p. 242.
    b. Reference 2b, p. 23.
25. Yvon, K. In Current Topics in Materials Science; Vol. 3, Kaldis, E., Ed.; North Holland Publishing, 1979.

RECEIVED May 9, 1990

CERAMIC MEMBRANES

Chapter 19

Catalytic Ceramic Membranes and Membrane Reactors

M. A. Anderson[1], F. Tiscareño-Lechuga[2], Q. Xu[1], and C. G. Hill, Jr.[2]

[1]Water Chemistry Program and [2]Department of Chemical Engineering,
University of Wisconsin, Madison, WI 53706

Ceramic membranes represent a comparatively new class of
materials which can be prepared from a variety of
organometallic or inorganic precursors using sol-gel
synthesis routes. The physico-chemical properties of
these membranes depend on both the specific compounds
used as precursors and the experimental protocol used in
their preparation. In this paper, we discuss the key
variables controlling the preparation of these membranes
as well as the physico-chemical properties which make
them good candidates for use as in catalytic membrane
reactors. Finally, by way of a kinetic model for a
specific reaction (the dehydrogenation of diethylbenzene
to styrene), we illustrate the limitations as well as
advantages of employing these membranes in catalytic
reactors.

While ceramic membranes have been heralded as a new type of material
in recent years, their history dates back to the 1940's. Membranes
fabricated from metal oxides were originally created to separate
gases in uranium isotope enrichment processes. The recent rebirth of
interest in ceramic membranes can be attributed primarily to renewed
interest in sol-gel processing of ceramic precursors and to improved
analytical techniques which provide insight into the molecular
events which control hydrolysis and polymerization reactions in
these systems. The improved understanding of these reactions has led
to the use of an ever broadening array of organometallic and
inorganic precursors in the preparation of these materials. The
resultant materials possess different physico-chemical properties.
Hence there is a concomitant increase in the range of possible
applications for these materials. While the original applications of
ceramic membranes involved gas phase separations, we are now using
these membranes for ultrafiltration and as catalysts,
photocatalysts, solar cells, and sensors. This increasing range of
applications is fueling further fundamental studies which increase
the array of precursors, and increase our understanding of sol-gel
reactions.

A major new application of ceramic membranes is in the area of
ultrafiltration. Ceramic membranes can outperform organic polymer

membranes in numerous respects. Some advantages of ceramic membranes relative to their organic polymer counterparts are:

1) Chemical stability - Ceramic membranes are not degraded by organic solvents and can withstand exposure to chlorine. Many crystalline oxides are relatively insoluble in acidic and alkaline media; hence ceramic membranes composed of such oxides should be relatively inert under extreme pH conditions.

2) Stability at high temperature - Once a ceramic membrane is fired at a given temperature, its properties will not change on exposure to lower temperatures. While the appropriate firing temperature varies with the type of oxide or mixed oxide comprising the membrane, the temperature levels at which these membranes can be employed are generally much higher (above 400°C) than those at which organic membranes can be used (typically below 100°C).

3) Stability to microbial degradation - Certain organic membranes are quite susceptible to microbial degradation. Ceramic membranes are not expected to undergo such degradation.

4) Mechanical stability - Organic membranes compact and can undergo inelastic deformations under high pressures, leading to lower permeabilities. Ceramic membranes supported on robust materials such as stainless steel or structural ceramics can be expected to withstand very high pressures.

5) Cleaning conditions - Membrane fouling is a serious problem when using organic membranes. Even if ceramic membranes suffer from similar fouling problems, items 1 and 2 above indicate that harsher, more effective cleaning treatments can be used with ceramic membranes than with organic membranes.

Although an extremely large number of variations in formulation conditions and starting materials can be employed to form organic membranes, these membranes always have a carbon-based backbone. Ceramic membranes composed of inorganic precursors can be fashioned from many of the elements of the periodic chart. Hence there is a wide variety of interesting materials with an equally large variety of physico-chemical properties from which ceramic membranes can be fabricated. The aforementioned advantages led us to consider the use of ceramic membranes in catalytic applications where temperature ranges often exceed several hundreds of degrees and where high pressures and harsh chemical environments prevent the use of organic membranes.

Preparation of Ceramic Membranes

In the fabrication of ceramic membrane modules, several processing steps must be accomplished: sol preparation, gelation, coating of supports, and firing at elevated temperatures. Within each step, several independent variables can be used to tailor the properties of the ultimate product for the specific application of interest. Although discussion of the influence of these variables is beyond the scope of the present paper, a cursory treatment of each step and the most important variables will be given here.

Sol Preparation. The synthesis of oxide sols via the sol-gel process involves the controlled hydrolysis of metal alkoxides and/or metal salts. However, the properties of the resulting sol, including the size of the sol particles, are determined by the interactions of several variables which must be carefully controlled. Such factors as the solvents used, the ratios of alkoxide or salt to water and/or alcohol, the amount of added acid or base, the reaction temperature, the stirring speed, and the rate at which reactants are added to the

solvent must all be studied in order to develop a synthesis protocol
which can produce a sol with the desired properties (1,2). The size
of the primary particles is determined at this stage, although
particle sizes can change as aging occurs (3). NMR studies have
provided useful information about the reactions which occur during
hydrolysis (4).

In general, hydrolysis and condensation reactions can be
described by equations 1 and 2 below:

$$
\begin{array}{ccccc}
\text{OR} & & & \text{OR} & \\
| & & & | & \\
\text{RO-M-OR} & + \ H_2O & \rightleftharpoons & \text{RO-M-OH} & + \quad \text{ROH} \\
| & & & | & \\
\text{OR} & & & \text{OR} &
\end{array}
\qquad (1)
$$

$$
\begin{array}{cccc}
\text{OR} & & \text{OR} \quad \text{OR} & \\
| & & | \qquad | & \\
2 \ \text{RO-M-OH} & \rightleftharpoons & \text{RO-M-O-M-OR} & + \quad H_2O \\
| & & | \qquad | & \\
\text{OR} & & \text{OR} \quad \text{OR} &
\end{array}
\qquad (2)
$$

where M = central metal and R = organic moiety

The relative rates of the hydrolysis and condensation reactions
largely determine whether polymeric or particulate membranes are
obtained. If hydrolysis rates are rapid in comparison to those of
the condensation reactions, larger particulate species are obtained.
On the other hand, if one can reduce the rate of the hydrolysis
reaction to the point where it is slower than that of the
condensation reaction (e.g., by lowering the pH, reducing the amount
to H_2O present, or by changing the chemical nature of the
substituent groups on the precursor, one can produce polymeric or
very small particulate species.

For example, we have used several methods to prepare stable
titania sols by the sol-gel technique. One general technique is
peptization, in which hydrolysis occurs without addition of any acid
or base. Aggregates of primary particles form under these
conditions. These aggregates can then be electrostatically dispersed
by adding acid or base to increase the surface charge on the solid
particles. However, acid peptization of our titania sols appears to
produce smaller aggregates of primary particles rather than a
dispersed sol of the primary particles themselves. We have
successfully prepared suspensions of finer titania particles by
hydrolyzing titanium tetra-isopropoxide under strongly acidic
conditions. Kormann et al. have even synthesized so-called "Q
particles" (quantum size) of titania (<30 Å diameter) by carefully
hydrolyzing titanium tetrachloride (5).

Particle aggregation during and after the synthesis of the sol
can have a significant effect on the properties of the resulting
ceramic membrane. Figure 1 illustrates the effect of aging on
particle size for titania sols prepared with different H_2O/Ti mole
ratios, using 2-methyl-2-butanol (t-amyl alcohol) as the solvent.
For sol 3, it appears that the system does not contain enough water
for hydrolysis to continue after formation of the initial particles.
Under these conditions there is no significant increase in the size
of the primary particles. Particle aggregation is minimized because
the suspension is electrostatically stabilized at pH 2. However, as
water is added to the system (sol 2), slow growth of primary
particles appears to occur until, after 70 days, a sudden rise in
particle size occurs. The observed increase probably indicates the
onset of particle aggregation. The induction period is greatly

affected by the H_2O/Ti ratio. A sol with a H_2O/Ti mole ratio of 3.75 displayed similar behavior, but with an induction period of 105 days. Further addition of water (sol 1) causes general particle growth and aggregation phenomena as in the case of sol 2. In this case, however, rapid aggregation was not observed. It is not clear why this difference in behavior occurs, although it may be related to the structure of the aggregates which form in these systems. Questions about the structure of particle aggregates, as well as those about the structure of gels and membranes, can be studied in some detail using scattering techniques and fractal analysis (6,7).

Gelation. Once a relatively stable sol is prepared, it can be placed on a support. This process involves both coating (or slip-casting) and gelation, although a clear separation of these two steps may not be possible. For instance, one can coat a glass plate by dipping it into a dilute sol and then heating the coated plate. In this case, coating has preceded gelation. In some cases it may be desirable to increase the viscosity of the sol or approach gelation conditions before coating. In this manner a thicker coat can be achieved in a single step, rather than employing the repetitive procedure which would be required to achieve the same thickness using a less concentrated sol.

Stable sols are usually concentrated or gelled by solvent evaporation. This step requires careful control of both system temperature and relative humidity. These variables, in turn, affect the rate of evaporation, although the use of a rotary evaporator under a partial vacuum may limit one's ability to control humidity. As evaporation proceeds, the concentrations of the nonvolatile inorganic salts (introduced either as acids or bases to peptize the system or as precursor salts) in the suspension will increase. For sols in which particles are electrostatically stabilized, the increase in ionic strength accompanying evaporation of the solvent will neutralize some of the charge on the particles, eventually destabilizing the sol. The composition of the sol determines whether a stable sol will gel or flocculate as evaporation proceeds. If a gel is desired, direct addition of inorganic salts will often cause gelation without the need for evaporation of solvent (8,9).

If the presence of inorganic salts will adversely affect the performance of the final membrane, then acids or bases which can be volatilized in the firing stage should be used (e.g., nitric acid or ammonium hydroxide). One should also recognize that dialysis can replace evaporation as a technique for concentrating suspensions, especially when increases in ionic strength cannot be tolerated. In addition, dialysis can be used to achieve controlled addition of a solute to the sol, thereby affecting the surface chemistry of the substrate. However, dialysis tubing must be scrupulously cleaned before use.

Coating of Supports with Sols. Once the solids concentration and viscosity of the sol have been adjusted to desired levels, the sol can be used to coat a support. The procedure to be employed in the coating process, however, depends on whether the support being coated is porous or nonporous. If a dry, porous support is being coated, then slip-casting is utilized. In this case, the thickness of the resulting coat depends on the contact time between the sol and the support, with longer contact times giving thicker coatings (10). Other factors which affect the layer being cast are the ability of the sol to wet the support and the average size of the

sol particles relative to the pore size of the support. However,
when a nonporous support is coated, the primary factor which
controls the thickness of the resulting coating is the rate at which
the support is withdrawn from the sol. In this case, faster
withdrawal rates give thicker layers (11). The viscosity of the sol
also affects the thickness of the coating.

Firing the Supported Membrane. After the support has been coated,
the coating is fixed to the support by firing. Factors governing the
efficacy of this process include the firing temperature, the ramp
rate, the dwell time and the firing atmosphere. The combination of
firing temperature and dwell time at that temperature (essentially a
measure of the thermal energy available to the system) controls both
the crystal structure and the pore size of the resulting ceramic
membrane. The firing atmosphere can also be controlled to either
oxidize or reduce the membrane. Hence the chemical nature of the
local environment is an important factor in preparing photoactive
membranes. The data in Figure 2 indicate the changes which occur in
an unsupported TiO_2 membrane when prepared at different firing
temperatures. At the onset of firing, the membrane is amorphous. As
the temperature reaches 200°C, an anatase form appears. Around
500°C, the anatase is converted to a stabile rutile form. Crystal
size increases with increasing temperature.

The reader should also note that the support itself will often
influence particle growth rates, particle-particle sintering rates,
phase changes and the porosity of the membranes resulting from the
firing process. Some combination of solid state migration and
chemical pinning phenomena changes the crystalline and physical
structure of the supported membrane.

Cracking. Very often a ceramic membrane will develop cracks at some
point in the preparation protocol. Often this occurs when it is
dried, either by evaporation or firing. The unfired membrane still
contains considerable solvent (and possibly other additives) after
the coating and gelation steps. Crack formation may occur as a
consequence of the hydrodynamic stresses which develop in the
membrane as the solvent evaporates (3,12). An obvious approach to
alleviating this problem is to dry the membrane slowly and/or fire
the membrane at a low ramp rate. If evaporation of solvent during
drying is rapid enough to cause cracking, it may be possible to
minimize cracking by drying the module under high humidity
conditions to lower the evaporation rate. It may also be necessary
to reformulate the preparation of the sol so that a less volatile
solvent is used (e.g., by substitution of water for alcohol) or
drying control agents are added to alter surface energies. Such
reformulation often requires trial and error experiments and much
research effort before a successful process is developed.

Pore Size Limitations. Although there are many potential commercial
applications for ultrafiltration using currently available ceramic
membranes, the pore sizes in these membranes are seldom less than 40
Å in diameter, thereby limiting their applications in gas
separations and in ceramic catalytic reactors.

Two approaches can be taken to utilize other oxide ceramic
membranes for separations involving small molecules or ions. Either
the pores of the membrane can be partially plugged by depositing
insoluble materials therein (13,14) or membranes can be prepared
using precursor sols which contain extremely small particles. The

PARTICLE GROWTH KINETICS
In Alcoholic Solution

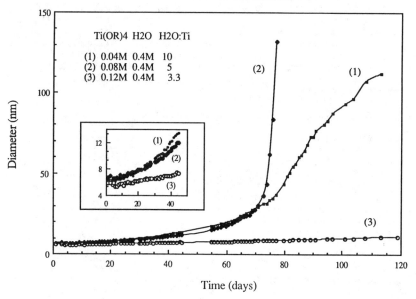

Figure 1. Growth of titania particles as a function of time and the mole ratio of H_2O to Ti.

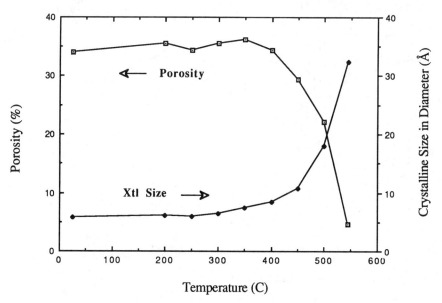

Figure 2. Changes in phase and porosity of unsupported TiO_2 membranes as a function of temperature.

latter situation requires synthesis methods which incorporate techniques used to prepare "Q particles", such as low temperature hydrolysis under pH conditions which lead to highly charged primary particles and dialysis of the resulting sol to reduce the ionic strength of the final sol. By using these techniques (strong acid hydrolysis at room temperature followed by dialysis), we have prepared titania sols for which the hydrodynamic diameters of the primary particles are about 3 nm. Xerogels prepared from this sol have pore diameters of less than 2 nm, while membranes fired at 250°C have pore diameters of less than 3 nm. Further study of these membranes is continuing.

<u>Catalytic Ceramic Membranes and Membrane Reactors</u>

Ceramic membranes can be used in catalytic systems in three modes:
1. those in which the membrane functions only as a catalyst.
2. those in which membrane operates as a permselective barrier.
3. those in which the membrane acts as both a catalyst and a permselective barrier.

The latter two modes circumscribe the role of ceramic membranes in membrane reactors. In the case of membrane reactor applications, the ceramic membranes must be crack-free. In addition, it is desirable for the membrane to exhibit good permselectivity characteristics.

Preliminary results obtained in an effort to model the dehydrogenation of ethylbenzene to styrene in a "membrane reactor" are described below. The unique feature of this reactor is that the walls of the reactor are comprised of permselective membranes through which the various reactant and product species diffuse at different rates. This reaction is endothermic and the ultimate extent of conversion is limited by thermodynamic equilibrium constraints. In industrial practice steam is used not only to shift the equilibrium extent of reaction towards the products but also to reduce the magnitude of the temperature decrease which accompanies the reaction when it is carried our adiabatically.

<u>Reactions of Ethylbenzene</u>. Clough and Ramirez (15) have reported that for the dehydrogenation of ethylbenzene over a potassium promoted iron oxide catalyst in the presence of steam the important reactions are:

$$\text{(3)}$$

$$\text{(4)}$$

$$\text{(5)}$$

$$H_2O \ + \ \tfrac{1}{2} \ CH_2{=}CH_2 \ \longrightarrow \ CO \ + \ 2 \ H_2 \tag{6}$$

$$H_2O \ + \ CH_4 \ \longrightarrow \ CO \ + \ 3 \ H_2 \tag{7}$$

$$H_2O \ + \ CO \ \longrightarrow \ CO_2 \ + \ H_2 \tag{8}$$

Lee (16) has reported that for high conversions some losses may occur as a result of decomposition of styrene. However these effects are not considered in the present study. For the same catalyst, Sheel (17) reported that the reactions which must be considered for this system are the same main reaction (reaction 3), plus side reactions between ethylbenzene and steam resulting in the formation of benzene, toluene, carbon dioxide, and hydrogen.

Wang et al. (18) have studied binary and ternary mixtures of titanium oxide with zirconium oxide, vanadium oxide and iron oxide, among others. They reported that the by-products are produced in small amounts. Some mixtures are as active or more active than the iron oxide catalyst but are less selective. These researchers noted that in addition to the aforementioned reactions, reactions 3 to 5, the following reactions are significant:

$$\text{(styrene)} \ + \ H_2 \ \longrightarrow \ \text{(benzene)} \ + \ CH_2{=}CH_2 \tag{9}$$

$$\text{(ethylbenzene)} \ + \ H_2 \ \longrightarrow \ \text{(benzene)} \ + \ CH_3{-}CH_3 \tag{10}$$

Other reaction networks have been considered by other authors. However, since the technology on which our effort focuses involves the use of membranes made of both iron oxide and titanium oxide, our efforts to analyze the performance of a permselective catalytic membrane reactor center on reactions 3 to 5.

The modelling effort is being carried out in stages. The first stage involves an assessment of the effects of process variables on the equilibrium extent of reaction that can be achieved under conditions such that the membrane is highly permselective for hydrogen, i.e., that only hydrogen permeates the membrane. The second stage introduces kinetic limitations on the reaction and provides for the fact that the membrane is not a perfectly semipermeable membrane. Results from the first stage of our efforts and preliminary results from our second stage efforts are presented below.

Equilibrium Model

Thermodynamic equilibrium limitations exist for the first reaction, but the other reactions can be considered as irreversible, at least in the region where the conversion to styrene is attractive. Only reaction 3 is considered in the first stage calculations because there is considerable experimental evidence that the other reactions occur to only a very small extent.

A study considering all the reactions as being at equilibrium would not be useful because equilibrium would be achieved only in a region which would result in a high proportion of by-products and low yields of styrene. The reaction rate constants for reaction 3 are considerably greater than those for reactions 4 and 5. However, an increase in the temperature of operation would reduce the disparity between the rate constants. This situation implies that as, a first approximation, only reaction 3 occurs to significant degree. This approximation permits one to determine the "equilibrium" conversion attainable when design parameters such as temperature, pressure, steam to oil ratio (r_{H2O}) and the fraction of hydrogen removed (δ_{H2}) are varied. The fraction hydrogen removed is a variable which permits one to alter the equilibrium conversion as a consequence of product removal through a permselective membrane.

If one assumes ideal gas behavior and considers only reaction 3, the following material balance results (basis = 1 mole of ethylbenzene and r_{H2O} moles of water):

Component	Species	Moles remaining in the reactor at extent of reaction ξ
1	Ethylbenzene	$n_1 = 1 - \xi$
2	Styrene	$n_2 = \xi$
3	Hydrogen	$n_3 = (1 - \delta_{H2})\xi$
4	Inert	$n_4 = r_{H2O}$

The equilibrium relation associated with reaction 3 is:

$$K_3 = \frac{(1 - \delta_{H2})\ \xi^2\ P_{tot}}{(1 - \xi)\ [1 + (1 - \delta_{H2})\ \xi + r_{H2O}]} \tag{11}$$

Since the basis is 1 mole of ethylbenzene fed, the concepts of fraction conversion and extent of reaction are interchangeable ($X = \xi$).

The following values of the equilibrium constant are obtained using the thermodynamic data given in Reid et al. (19):

Temperature °C	K_3 (atm)
400	0.0013
600	0.232
800	6.15

Inspection of these values indicates that in order to obtain an attractive conversion the operating temperature will need to be high.

Clough and Ramirez (15) have also reported that for an iron catalyst the optimal steam to oil molar ratio is ca. 10. The plots in Figure 3 indicate that as this ratio is increased, higher conversions to styrene are attainable. Since only one reaction is considered in the analysis, the conversion criterion is used instead of a yield criterion. It should be noted that this analysis does not take into account cost considerations.

The plots in Figure 4 are based on calculations for a steam to oil ratio of 10. Inspection indicates that the conversion to styrene is favored by lowering the pressure. However, the effect of pressure on the conversion is smaller than either the effect of the steam to oil ratio or as will be seen, the effect of the removal of hydrogen.

The shift in conversion resulting from the removal of the hydrogen produced is of particular interest when a permselective membrane reactor is employed. The curves in Figure 5 indicate that in order to gain an appreciable increase in the conversion, it is

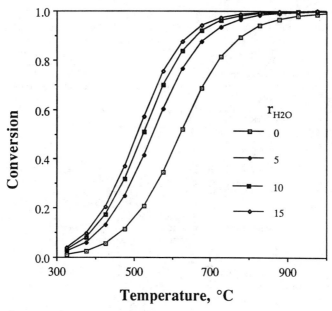

Figure 3.   Combined effects of temperature and steam to oil ratio on the conversion. Total pressure = 1 atm; no hydrogen removal.

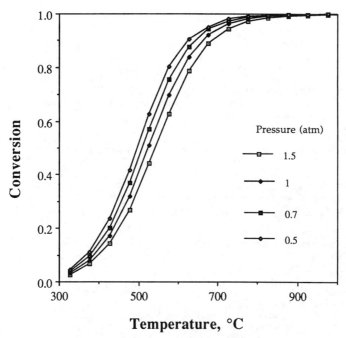

Figure 4.   Combined effects of temperature and pressure on the conversion. No hydrogen removal; steam to oil molar ratio = 10.

necessary to remove over twenty percent of the hydrogen produced. The calculations summarized by the plots in Figures 5 and 6 indicate that when no steam is present initially, consistently lower conversions are achieved at the same degree of hydrogen removal.

The results obtained from the thermodynamic limitation model thus indicate the focus of efforts to develop a process based on a membrane reactor should center on producing a membrane which will
1. be highly permselective
2. be stable at high temperatures and in the presence of water
3. have high activity and selectivity for the desired reaction.

Kinetic Model

Several researchers have modelled other dehydrogenation reactions under similar conditions; for example, Raymond (20) , Ito et. al. (21), and Shinji et al. (22) have studied the dehydrogenation of sulfur anhydride, the decomposition of HI and the dehydrogenation of cyclohexane, respectively. Mohan and Govind (23) have reported results of a generic study of permselective reactors. Although some of their conclusions can be extended to the dehydrogenation of ethylbenzene, an important difference between this system and those cited is the fact that in the case of the dehydrogenation of ethylbenzene, other reactions occur together with the dehydrogenation reaction. This situation introduces the necessity for considering reaction selectivity. This factor can play a very important role in the global optimization problem. The use of a permselective membrane can not only lead to a substantial increase in conversion over the equilibrium conversion, but it can also have a big impact on the selectivity of the process.

Permselective Reactor Model. This model was developed to size an experimental reactor system. Even though, some of our initial assumptions are being refined in our continuing efforts, some important conclusions can be drawn from our early work. The first approach used was to idealize the permselectivity, and assume that only hydrogen diffuses through the membrane. This assumption cannot be justified for the characteristics of the membranes studied experimentally. It does, however, permit one to place an upper limit on the expected performance of the system.

The model uses several assumptions:
1. isothermal operation
2. negligible pressure drop across the reactor
3. plug flow
4. no axial or radial diffusion
5. mass transfer across the membrane occurs only for hydrogen and the transport occurs by Knudsen diffusion
6. the partial pressure of hydrogen on the sweep side of the membrane is constant
7. ideal gas behavior
8. no reaction occurs within the membrane or on the sweep side.

The model equations are solved by an initial boundary value algorithm, D02BBF integrator from the NAG Library (24).

Figure 7 presents the parameters of the reactor model and the operating conditions simulated. A packed bed of catalyst with a loading of catalyst equivalent to 0.11 gr of Shell 105 catalyst/void cc of reactor is assumed. A membrane 5 microns thick with 20 Å pore radii is placed on inner side of a 3/8" I.D. porous support. The pore size of this support is considered to be large enough that it does not exhibit any resistance to mass transfer. A tortuosity

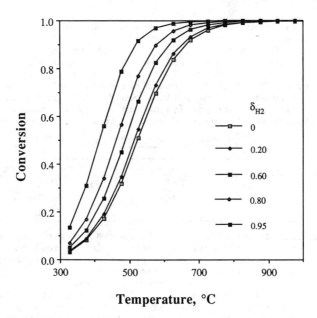

Figure 5. Combined effects of temperature and fraction hydrogen removed on the conversion. Pressure = 1 atm; steam to oil molar ratio = 10.

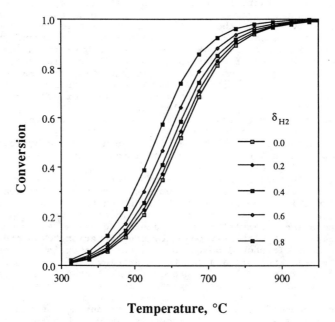

Figure 6. Combined effects of temperature and fraction hydrogen removed on the conversion. Pressure = 1 atm; steam to oil molar ratio = 0.

factor of 3 was used. The LHSV for the reactor is about 0.3 and the residence time is close to 2 seconds.

Reactions 3, 4 and 5 were considered for the model. The rate constants reported by Wenner (25) were used. Sheppard et al. (26) reported that Wenner's kinetic model deviated by 9.6% from manufacturer's data. They proposed other models and more accurate parameters. However, for our work we preferred to begin with Wenner's model because it permits one to employ a simple Arrhenius relationship for the temperature dependence of the rate terms.

Table I Wenner kinetic model

Reaction	A	E_a (Kcal/Kg mole)	Driving Force
3	11.5	24.1	$p_{EB} - (p_{ST}p_{H2}/K_1)$
4	16.6	31.3	$p_{EB}p_{H2}$
5	13	33.2	p_{EB}

Results and Discussion. The plots in Figure 8 represent the conversion obtained in a fixed bed reactor for various catalyst loadings, assuming that no membrane is present. The catalyst density of a catalytic membrane about 5 microns thick would be on the order of 0.011 gr/cc or smaller. The conversion obtained at that catalyst density would not be attractive if a single stage reactor is to be employed. Moreover, if one requires that the catalytic membrane be permselective, the amount of catalyst that would be playing an effective role in the reaction would be smaller than the catalyst contained in the total membrane. The activity of the exposed surface of a permselective membrane is not sufficiently large to permit one to achieve an attractive conversion in a single stage reactor.

The amount of hydrogen that diffuses through the membrane can be adjusted by varying the partial pressure of this component on the sweep side of the reactor. The curves in Figure 9 indicate that the greater the amount of hydrogen removed, the higher is the conversion achieved in the reactor under the same conditions. The fact that reaction 3 is equilibrium limited constrains the performance for a conventional tubular reactor. At 600°C under conditions similar to those used for Figure 9, in a conventional reactor conversion is limited to about 77%. The permselective membrane reactor permits one to by-pass this limitation. With this reactor conversions higher than the thermodynamic "constraint" can be achieved, i.e., for a 80 cm-long reactor for which the partial pressure of the hydrogen on the sweep side is 0.001 atm, the conversion obtained is 0.90 (not shown on the figure).

Figure 10 contrasts the conversion obtained with a conventional reactor to that conversion obtained with a permselective reactor when the partial pressure of the hydrogen on the sweep side is 0.02 atm. It can be seen that the difference in conversion between the two systems increases as the temperature increases. An important operational limitation is imposed by the sintering temperature for the membranes. It should be noted that this factor is not considered since the membranes are assumed to be stable.

The major impact of this process is reflected in the reaction selectivity, which is a very important indicator of reactor performance. The selectivity increases as the recovery of the hydrogen produced increases. This result is due to the fact that reaction 5 is suppressed as hydrogen is removed. Figure 11 contrasts the packed bed and membrane reactors in terms of selectivity at three different temperatures. As the temperature increases the

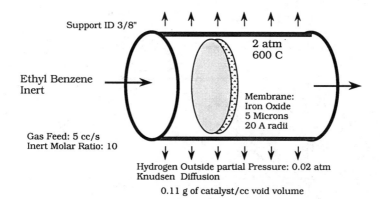

Figure 7. Permselective Membrane Reactor.

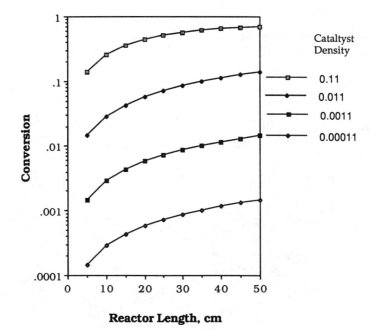

Reactor Length, cm

Figure 8. Dependence of the conversion on the catalyst density (gr-catalyst/cc-void). Base case. No removal of hydrogen occurs by diffusion; Temperature = 600°C.

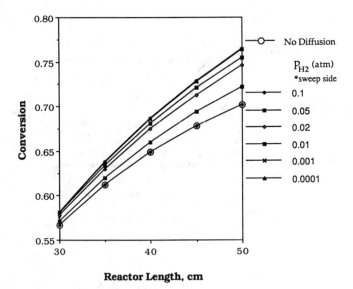

Figure 9. Effect of the partial pressure of the hydrogen on the sweep side (atm) on the conversion.

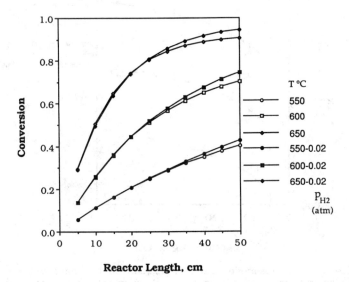

Figure 10. Combined dependence of the conversion on the temperature (°C) and on the partial pressure of the hydrogen on the sweep side (atm).

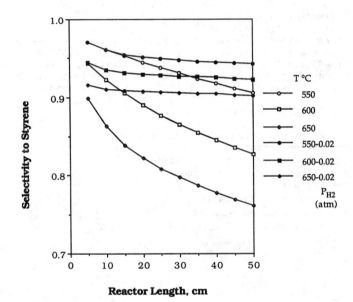

Figure 11.   Combined dependence of the selectivity to styrene on the temperature (°C) and on the partial pressure of the hydrogen on the sweep side (atm).

selectivity decreases. However, the effect of temperature on the selectivity for the permselective membrane reactor is not as drastic as that for the conventional reactor. Combining the results shown on Figures 10 and 11, for operation at 600°C , one predicts that in the best case ca. 15% improvement in the yield can be achieved when a permselective membrane reactor is used.

We are currently in the process of producing a new generation of ceramic membranes which are intended to have smaller pore sizes, higher specific surface areas, and improved permselectivities. These new membranes are being designed to operate at temperatures near 600°C. The use of a permselective membrane for the dehydrogenation of ethylbenzene can have a great impact not only on the conversion achieved but more importantly on the selectivity of the reaction network.

Acknowledgments

The authors gratefully acknowledge financial support received from the United States Department of Energy (DE-AS07-861D12626), (PO#AX079886-1); The National Science Foundation (ECE-8504276); and the U.S. Environmental Protection Agency (R-813457-01-0). One of us (F. T.) also wishes to thank CONACYT-México for financial support.

Legend of Symbols

A	Pre-exponential factor in rate constant
E_a	Activation energy
K_3	Equilibrium constant for reaction 3
p_i	Partial pressure of component i
P_{tot}	Total pressure
r_{H2O}	Steam to oil molar ratio
δ_{H2}	Fraction hydrogen removed
ξ	Extent of reaction 3
Subscripts	
EB	Ethylbenzene
H2	Hydrogen
ST	Styrene

Literature Cited

1. Anderson, M. A; Gieselmann, M. J.; Xu, Q. J. Memb. Sci., 1988, 39, 243.
2. Xu, Q.; Anderson, M. A. Mat. Res. Soc. Symp. Proc., 1989, 132, 41.
3. Scherer, G. W. J. Non-Cryst. Solids, 1988, 100, 77.
4. Pouxviel, J. C.; Boilot, J. P. J. Non-Cryst. Solids, 1987, 94, 374.
5. Kormann, C.; Bahnemann, D. W.; Hoffmann, M. R. J. Phys. Chem., 1988, 92, 5196.
6. Hackley, V. A.; Anderson, M. A. Langmuir, 1989, 5, 191.
7. Ramsay, J.D. F. Mat. Res. Soc. Symp. Proc., 1988, 121, 293.
8. Gieselmann, M. J.; Anderson, M. A. J. Amer. Ceram. Soc., 1989, 72(6), 980.
9. Anderson, M. A.; Gieselmann, M. J.; Villegas, M. A. J. Non-Cryst. Solids, 1989, 110, 17.

10. Leenaars, A. F. M.; Burggraaf, A. J. J. Colloid Interface Sci., 1985, 105(1), 27.
11. Mukherjee, S. P. Ultrastructure Processing of Ceramics, Glasess and Composites, John Wiley and Sons: New York, 1984, 178.
12. Glaves, C. L.; Brinker, C. J.; Smith D. M.; Davis, P. J. Chem Materials, 1989, 1, 34.
13. Uhlheron, R. J. R.; Huis in 't Veld, M. M. B. J.; Keiser, K.; Brurggraaf, A. J. J. Mat. Sci. Lett., 1989, 8(10), 1135.
14. Huis in 't Veld, M. M. B. J.; Uhlheron, R. J. R.; Keiser, K.; Brurggraaf, A. J. Proceedings of the 1st International Conf. on Inorganic Membranes, Montpellier France, 1989.
15. Clough, D. E.; Ramirez, W. F. AIChE Journal, 1976, 22(6), 1097-105.
16. Lee, E. H. Catalysis Reviews, 1973, 8(2), 285-305.
17. Sheel, J. G. P. ; Crowe, C. M. The Canadian Journal of Chemical Engineering, 1969, 47, 183-7.
18. Wang, I.; Wu, J.-C.; Chung, C.-S. Applied Catalysis., 1985, 16, 89-101.
19. Reid, R. C.; Prausnitz, J. M.; Sherwood, T. K. The Properties of Gases and Liquids. McGraw-Hill: New York, 1977.
20. Raymond, M. E. D. Hydrocarbon Process. 1975. 57(4) 139
21. Ito, N.; Shindo, Y.; Hakuta,T.; Yoshitome H. Int. J. Hydrogen Energy. 1984, 9(10), 835-9.
22. Shinji, O.; Misono, M.;Yoneda, Y. Bull. Chem. Soc. Jpn., 1982, 55, 2760-4.
23. Mohan, K. ; Govind, R. Separation Science and Technology, 1988, 23, 1715-33.
24. NAG Library. Numerical Algorithm Group Inc. 1984.
25. Wenner, R. R.; Dybdal, E. C. Chemical Engineering Progress.1948,44 (4), 275-86.
26. Sheppard, C. M.; Maier, E. E. ; Caram, H. S.Ind. Eng. Chem. Process Des. Dev.1986, 25, 207-10.

RECEIVED May 9, 1990

Chapter 20

Novel Oxidative Membrane Reactor for Dehydrogenation Reactions
Experimental Investigation

Renni Zhao, N. Itoh, and Rakesh Govind

Department of Chemical Engineering, University of Cincinnati, Cincinnati, OH 45221

The use of a membrane reactor for shifting equilibrium controlled dehydrogenation reactions results in increased conversion, lower reaction temperatures and fewer by-products. Results will be presented on a palladium membrane reactor system for dehydrogenation of 1-butene to butadiene, with oxidation of permeating hydrogen to water on the permeation side. The heat released by the exothermic oxidation reaction is utilized for the endothermic dehydrogenation reaction.

Application of membranes in reactors to enhance the reaction yield and conversion has been recognized in recent years. Various configurations for membrane reactors have been proposed and studied in the literature(1). One of these configurations involves a permselective membrane which selectively separates the reaction product(s) to enhance the conversion of equilibrium limited reactions, such as hydrogenation or dehydrogenation of hydrocarbons. Other potential advantages include lower reaction temperature thereby reducing effect of by-product reactions resulting in fewer by-products, and lower separation and recycle costs.

In this paper experimental studies have been conducted on an oxidation palladium membrane reactor for dehydrogenation of 1-butene to butadine . Since only hydrogen permeates through a palladium membrane, separation and and hence shifting of the reaction to the products is achieved without any loss of the reactant. Furthermore, palladium is a good conductor of heat, thereby allowing the heat evolved from the exothermic oxidation reaction (permeating hydrogen to water) on the permeation or separation side to flow across the membrane to the reaction side . This heat coupling between an exothermic oxidation reaction and an endothermic dehydrogenation reaction allows the reactor to operate adiabatically.

Background

The use of palladium-based membranes results from the 1866 discovery by Thomas Graham(2) that metallic palladium absorbs an unusually large amount of hydrogen. Hydrogen permeates through Pd-based membranes in the form of highly active atomic hydrogen which can react with other

0097–6156/90/0437–0216$06.00/0

compounds adsorbed on the catalyst surface. The use of palladium membranes gained importance from increased application of membrane based separation in the field of chemical processing, biotechnology, environmental control and natural gas and oil exploration(3). It is reported that palladium-based membrane reactors have been used to some extent in the industrial production of chemicals and pharmaceuticals in the Soviet Union(4).

One of the earliest applications of membrane to shift equilibrium was developed by Wood(5) (1960). He showed that by imposing a non-equilibrium condition on a hydrogen-porous palladium silver alloy membrane, an otherwise stable cyclohexane vapor is rapidly dehydrogenated to cyclohexene.

The idea of conducting hydrogenation and dehydrogenation reactions simultaneous on opposite surfaces of a membrane, which is selectively permeable to hydrogen, was first presented by Grgaznov et a(6)(7).

Itoh et al(8). studied dehydrogenation of cyclohexane in a palladium membrane reactor containing a packed bed of Pt/Al$_2$O$_3$ catalyst. The removal of hydrogen from the reaction mixture using the palladium membrane increased the conversion from the equilibrium value of 18.7% to as high as 99.5%. It was shown that for given rates of permeation and reaction, there is an optimum thickness of membrane, at which maximum conversion is obtained.

In a separate parametric study, Mohan and Govind(1)(9) analyzed the effect of design parameters, operating variables, physical properties and flow patterns on membrane reactor. They showed that for a membrane which is permeable to both products and reactants, the maximum equilibrium shift possible is limited by the loss of reactants from the reaction zone. For the case of dehydrogenation reaction with a membrane that only permeates hydrogen, conversions comparable to those achieved with lesser permselective membranes can be attained at a substantially lower feed temperature.

Ilias and Govind(10) have reviewed the development of high temperature membranes for membrane reactor application. Hsieh(4) has summarized the technology in the area of important inorganic membranes, the thermal and mechanical stabilities of these membranes, selective permeabilities, catalyst impregnation, membrane/reaction considerations, reactor configuration, and reaction coupling.

Recently, Itoh and Govind(11) have reported a theoretical study of coupling an exothermic hydrogen oxidation reaction with dehydrogenation of 1-butene in an isothermal palladium membrane reactor.

Other studies on dehydrogenation reaction, such as decomposition of hydrogen sulfide, have been conducted with porous membranes such as Vycor glass, and alumina(Fukada et al.(12), Kameyama et al.(13)(14)), Raymont(15) has suggested the decomposition of abuntantly available hydrogen sulfide as a possible means for generating hydrogen. It is known that a palladium membrane can enhance the conversion of a thermodynamically limited dehydrogenation reaction.

Model Development

A schematic of a palladium membrane reactor is shown in figure 1. The reversible reaction of 1-butene dehydrogenation occurs on the reaction side of the membrane in which the chrome-alumina catalyst is uniformly packed. The oxidation of hydrogen with oxygen in air occurs in the permeation or separation side on the palladium membrane surface. The palladium membrane acts as a catalyst for the oxidation reaction. The hydrogen produced by the dehydrogenation reaction permeates through the palladium membrane and then reacts with the oxygen on the permeation side. The dimensionless model equations governing the membrane reactor are essentially differential material and energy balances for each side of membrane and are summarized in Table I.

In this model the follow simplifying assumption have been made:

1.   One-dimension plug flow;
2.   Negligible axial diffusion flux of heat and mass;
3.   Negligible radial gradients of temperature and concentration;
4.   Negligible pressure drop on either side of membrane;
5.   The heat and mass transfer resistances, aside from the permeation process itself, are negligible.

Plug flow conditions exist at high Reynolds numbers typically found in flow through packing or small radius tubes. In a permeator, the convection flow dominates over the diffusion flow i.e. a high Peclet number can be expected in a reactor with a membrane. Moreover, the pressure drop in a packed bed is usually a small fraction of the total pressure and can be neglected without significant error.

Experimental Method

In the experimental study, the reactor, schematically shown in Fig.2, consisting of two separate rectangular parts with a 90 mm length, 25 mm width and 25 mm depth groove. The palladium membrane(100 mm × 33 mm and 0.025 mm thickness) held in place by two pieces of gasket which have a 80 mm × 20 mm rectangular hole in the center, is sandwiched between the two reactor parts. Six thermocouples were located along the length of reactor to determine the temperature profile inside the reactor. The reversible reaction of 1-butene dehydrogenation occurred on the reaction side of membrane where chrome-alumina catalyst was uniformly packed. The oxidation of hydrogen with air occurred on the palladium surface on the other side of membrane, referred to as the permeation (or separation) side. The palladium membrane acts as the catalyst during the oxidation. This surface reaction of oxidation decreases the hydrogen concentration on the separation side , thereby increasing the permeation of hydrogen through the membrane. Further, heat liberated by the exothermic oxidation reaction on the separation side flows across the membrane and facilitates the endothermic dehydrogenation reaction, thereby increasing the reaction rates. (See Table II.)

A schematic of the experimental apparatus is showed in figure 3. The feeds of 1-butene, argon and 10% oxygen and nitrogen mixture were supplied from gas cylinders. The flow rates are measured by mass flow meters individually. The down stream flow rate on the permeation side chamber is open to the atmosphere, thereby maintaining the permeation side at atmospheric pressure. The products of dehydrogenation and oxidation were analyzed correspondingly using FID or TCD detector. The detector signal was monitored by an on-line microcomputer.

Results and Discussion

Measurement of reaction rate constant. The disappearance rate of 1-butene can be expressed by the following rate expression([16]):

Table I Basic Equations Developed for the Membrane Reactor System

<u>Mass Balance</u>
  Reaction Side (L>0)

$$\frac{dU_C}{dL} = -Da_r^0 \exp\{\epsilon_r (1 - \frac{1}{\theta_r})\}f_r \tag{1}$$

$$\frac{dU_H}{dL} = Da_r^0 \exp\{\epsilon_r (1 - \frac{1}{\theta_r})\}f_r - T_U^0 \exp\{\epsilon_p(1-\frac{1}{\theta_r})\}(\sqrt{\frac{P_{Hr}}{P_0}} - \sqrt{\frac{P_{Hs}}{P_0}}) \tag{2}$$

$$U_D = 1 - U_C \tag{3}$$

$$U_I = DI_r \quad (constant\ for\ inert\ gas) \tag{4}$$

Separation Side (L>0)

$$\frac{dV_H}{dL} = T_U^0 \exp\{\epsilon_p(1-\frac{1}{\theta_r})\}(\sqrt{\frac{P_{Hr}}{P_0}} - \sqrt{\frac{P_{Hs}}{P_0}}) - 2Da_s^0 \exp\{\epsilon_s(1-\frac{1}{\theta_s})\}f_s \tag{5}$$

$$V_O = V_O^0 + (1 + U_H + V_H - U_C - U_H^0 - V_H^0)/2 \tag{6}$$

$$V_w = 2(V_O^0 - V_O) \tag{7}$$

$$V_I = DI_s \quad (constant\ for\ inert\ gas) \tag{8}$$

<u>Initial Conditions (L=0)</u>
  Reaction Side,

$$U_C = 1, \quad U_I^0 = DI_r = \frac{u_I^0}{u_C^0}, \quad U_D = U_H^0 = 0, \quad \theta_r = 1 \tag{9}$$

Separation Side

$$V_O^0 = \frac{v_O^0}{u_C^0}, \quad V_I^0 = DI_s = \frac{v_I^0}{u_C^0}, \quad V_W = V_H^0 = 0, \quad \theta_s = 1 \tag{10}$$

<u>Rate Equations</u>
  Reaction Side

$$f_r = \frac{P_C^{1/2} - (P_D P_H / K)^{1/2}}{P_r^{1/2}(1 + 1.210\, P_C^{1/2} + 1.263\, P_D^{1/2})^2} \tag{11}$$

Separation Side

$$f_s = \frac{P_H^2 P_O}{P_s^3} \tag{12}$$

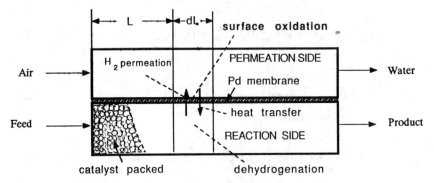

Figure 1   Schematic of Membrane Reactor

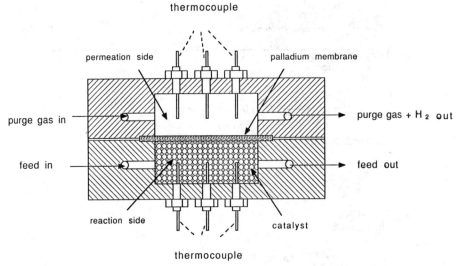

Figure 2   Structure of Membrane Reactor

Table II Experimental System Parameters

Membrane materal	99.9% Palladium Foil
Membrane Dimension	100mm × 33mm × 0.025mm
Type of Catalyst	Chromium (III) Oxide on Alumina
Porosity of Catalyst	0.30 ml/g
Catalyst weight	47.12 g
System Temperature	350 ^{0}C —— 450 ^{0}C
System Pressure	
Reaction side	1 atm
Permeation side	1 atm
Feed Flow Rate	
Reaction side	5 —— 45 ml/min
Permeation side	10 —— 140 ml/min
Feed Composition	
Reaction side	Pure 1-butene or 1-butene and Argon Mixture
permeation side	Argon or 10 % Oxygen Mixture

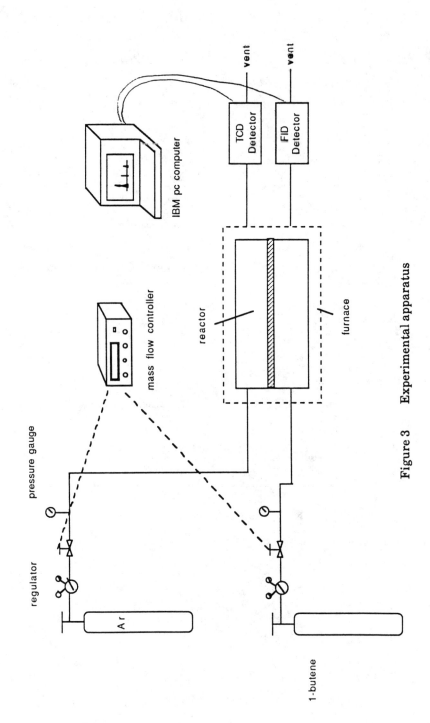

Figure 3   Experimental apparatus

$$r_r = K_f \phi_d [p_C^{1/2} - (p_D p_H / K)^{1/2}]$$

where K_f, ϕ_d and K are respectively the reaction rate constant, adsorption denominator and the reaction equilibrium constant.

$$\phi_d = \frac{1}{(1 + 1.210 p_C^{1/2} + 1.263 p_D^{1/2})^2}$$

and

$$K = \frac{p_D p_H}{(p_{cis-2-butene} + p_{trans-2-butene} + p_{1-butene})}$$

The reaction equilibrium constant is calculated from thermodynamic data([17]).

$$K = 1.966 \times 10^{11} exp[\frac{-30260}{RT}] \qquad Pa$$

The rate constant was independently measured by using a differential reactor packed with chrome-alumina catalyst particles. Before reaction, the catalyst was treated with 10 % oxygen mixture gas for 1 hour and then with argon for about half hour at the reaction temperatures. This pretreatment is to regenerate the catalytic activity of the catalyst which had been deactivated due to coke deposition in the previous run. Either pure 1-butene or mixture of 1-butene and argon was fed into the reactor. The output stream was sampled and analyzed for the composition of the organic portion by a Perkin-Elmer 990 gas chromatograph equipped with a FID detector and a VZ-10 60/80 column. The conversions were calculated based on the mole fraction of butadiene in the exit stream. This data was further analyzed to determine the rate constant. The experiment was carried out at several different input conditions and reaction temperatures. Figure 4 shows the Arrhenius plot of the obtained reaction rate constant for 1-butene dehydrogenation. The rate constant is given by :

$$K_f = 3.199 \times 10^5 exp[\frac{-21600}{RT}]$$

For the case of determining the rate constant of hydrogen with oxygen on the palladium surface, direct reaction of oxygen and hydrogen takes place around 500 °C([18]) and the reaction occurs mostly on the surface of the vessel. Leder and Butt([19]) studied the reaction between hydrogen and oxygen at 100 °C with a dilute platinum catalyst. For investigating the oxidation reaction rate, hydrogen was fed into the reaction side. When the reaction had attained the reaction temperature, the exit valve on the reaction side was shut off, thereby allowing the hydrogen to permeate through the membrane. 10% O_2 mixture gas was introduced into the permeation side. The system pressure and flow rates were controlled and determined by mass flow meters. The products were analyzed by gas chromatograph with molecule series 5A column. Because of the catalytic effect of palladium and high temperature, the oxidation reaction was extremely fast. Hydrogen absorbed on the palladium surface reacted immediately due to the presence of oxygen. The results show that the exit hydrogen flow rate on permeation side is almost zero since the oxidation reaction was completed immediately. Hence, it can be assumed that the concentration of hydrogen on the permeation side is zero ($V_H = 0$) at the reaction operating condition and proper mixture flow rate.

Measurement of palladium membrane permeability. The permeation rate of hydrogen gas through the palladium membrane, Q_H, was assumed to obey the half-power pressure law(20). The permeation flux of hydrogen through the membrane is proportional to the difference between the square roots of the hydrogen partial pressure on the high and low pressure sides of membrane.

$$Q_H = A \, q \, [\, (P_H^h)^{1/2} - (P_H^l)^{1/2} \,] / Z$$

where q is the permeability of hydrogen through the membrane..

The hydrogen flow rate through the palladium membrane at various temperatures and differential pressures is shown in figure 5. The permeation rate increases with increasing pressure differential across the membrane and the temperature. In figure 6, the permeability of hydrogen through the palladium membrane increases with temperature. The permeability of hydrogen through a palladium membrane can be expressed as:

$$q = 2.578 \times 10^{-5} exp [- \frac{2562.7}{T}] \qquad mol/(m \, h \, atm^{-1/2})$$

Comparison of Experimental Data and Simulation Results. For the case of isothermal operating condition without oxidation on separation side, the experimental results and the corresponding calculated curves are shown as plots of conversion versus flow rate of 1-butene, at different flow rate of purge gas, in figures 7-8. The equilibrium shift can be reached at low reactants flow rate. When there is a low reactant flow rate on reaction side, it is possible to double equilibrium conversion. As shown in Fig.9, conversion increases with increasing the purge gas flow rate and decreasing the reactant flow rate. At a fixed feed rate of 1-butene, it can be seen that increasing flow rate of purge gas results in higher conversion. This is due to the increasing permeation rate of hydrogen through the membrane from the reaction side to the separation side. When the purge flow rate is fixed, the conversion increases with decreasing flow rate of 1-butene, due to increasing residence time of the reactant in the reactor.

For the case of isothermal operating condition with oxidation on separation side, computer simulation shows that when the air dilution ratio is less than the theoretical demand (DIs = 4.77), the conversion of reactant is less than 100%. When DIs is greater then 4.77, the conversion attained is 100%. When DIs was increased beyond 10,there was an insignificant increase in the conversion since the oxidation reaction rate became the controlling step. On the other hand, for a given Damkohler number on the separation side, Da_s^0, the conversion increases with increasing air flow rate i.e., increasing value of DIs. In the region of DIs below 4.77, there is a maximum conversion for a given Damkohler number. Beyond this maximum point, there is no effect on the conversion for

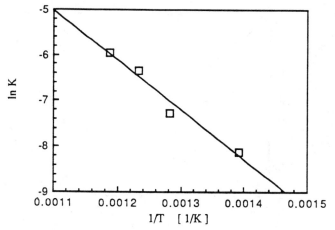

Figure 4   Arrhenius Plot of the Observed Reaction Rate of 1-butene Dehydrogenation

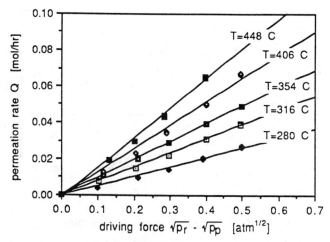

Figure 5   Permeation Rate of Hydrogen through the Palladium Membrane

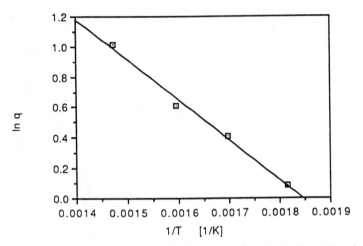

Figure 6   Permeability of Hydrogen through the Palladium Membrane

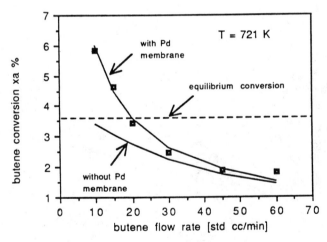

Figure 7   Conversion Shift at T=721 K and 70 std cc/min Purge Gas
Flow Rate

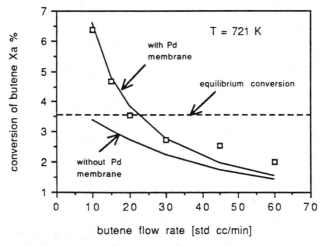

Figure 8   Conversion Shift at T=721 K and 140 std cc/min Purge Gas Flow Rate

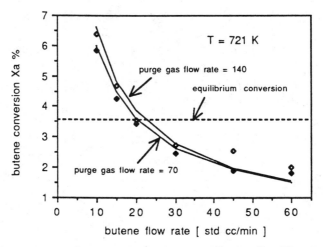

Figure 9   1-butene conversion versus Purge Gas Flow Rate

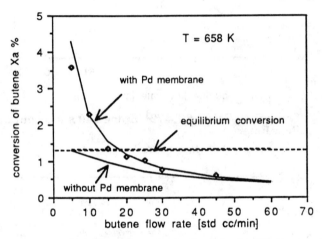

Figure 10   Equilibrium Shift at T=658 K and with Hydrogen Oxidation
             on the separation side

increasing Damkohler number. Figure 10 shows the equilibrium shift at $T = 658K$ with hydrogen oxidation on the permeation side. The solid line is the simulation result assuming that the hydrogen reacted immediately on the permeation side. For low reactant flow rate, equilibrium shift can be achieved. The deviation between experimental data and simulation results at low flow rate is perhaps due to the hydrodynamic resistance due to the catalyst and permeation through the membrane.

Conclusion

The feasibility of the palladium membrane system with an oxidation reaction on the permeation side and 1-butene dehydrogenation reaction on the reaction side in a membrane reactor has been successfully demonstrated. The palladium and its alloy membrane not only can withstand high temperature but also are selectively permeable to hydrogen and has catalytic activity for oxidation and dehydrogenation reactions. The oxidation reaction increases hydrogen flux through the membrane and supplies thermal energy for the endothermic dehydrogenation reaction.

Nomenclature

A	membrane area m^2	
Da_r^0	Damkohler number for reaction side at T_0,	$k_r \cdot G \cdot P_r^{1/2}/u_c^0$
Da_s^0	Damkohler number for separation side at T_0,	$k_s \cdot A \cdot P_s^3/u_c^0$
DI_r, DI_s	dilution ratio, u_1^0/u_c^0, v_1^0/u_c^0	
E	activation energy, J/mol	
G	total weight of catalyst, g	
k_f	rate constant of dehydrogenation, $mol/(g\text{-}cat \cdot s \cdot P_a^{1/2})$	
k_s	rate constant of oxidation, $mol/(m^2 \cdot s \cdot P_a^3)$	
K	equillibrium constant of dehydrogenation of 1-butene, Pa	
L	dimensionless reactor length	
p_i	partial pressure of gas i, Pa	
p_{Hr}	partial pressure of hydrogen on reaction side, Pa	
p_{Hs}	partial pressure of hydrogen on separation side, Pa	
P_r	total pressure on reaction side, Pa	
P_s	total pressure on separation side, Pa	
P_0	reference pressure $= 101325$ Pa	
q	permeability of hydrogen through palladium, $mol/(m \cdot h \cdot atm^{-1/2})$	
Q	permeation rate of hydrogen, mol/h	
R	gas constant, $cal/(mol \cdot K)$	
T	absolute temperature, K	
T_0	absolute temperature at inlet of reactor, K	
Tu^0	dimensionless number relating hydrogen permeation at T_0 and P_0, $D \cdot C_0 \cdot A/(t_m \cdot u_c^0)$	
u_i	flow rate of gas i in reaction side stream, mol/s	
u_i^0	flow rate of gas i in reaction side inlet, mol/s	
U_i	dimensionless flow rate of gas i in reaction side stream, u_i/u_c^0	
U_H	dimensionless flow rate of hydrogen in reaction side stream $u_H/(m \cdot u_c^0)$	
v_j	flow rate of gas j in separation side stream, mol/s	
v_j^0	flow rate of gas j in separation side inlet, mol/s	
V_j	dimensionless flow rate of gas j in separation side stream, v_j/u_c^0	
V_H	dimensionless flow rate of hydrogen in separation side stream, $v_H/(m \cdot u_c^0)$	
z	thickness of membrane, m	

Greek symbols
ε $E/(RT_0)$
θ T/T_0
ϕ_d adsorption denominator

Subscripts
C   reactant of dehydrogenation
D   product of dehydrogenation
H   hydrogen
i   component i in reaction side stream
I   inert
j   component j in separation side stream
O   oxygen
p   permeation
r   reaction side
s   separation side
W   water
0(zero)   indicate $T = T_0$

Superscipts
0(zero)   indicate value at inlet condition
h   high pressure side
l   low pressure side

Literature Cited

1. Mohan, K.; Govind, R. AIChE. J. 1988, 34, 1493.
2. Graham, T. Phil. Trans. Roy. Soc.(London). 1866, 156, 399.
3. Hwang, S. T.; Kammermeyer, K. Membrane in Separations; Wiley-Interscience: New York, 1975
4. Hsieh, H. P.; AIChE. Symposium 1989, 268, 53.
5. Wood, J. B. J. Catalysis 1968, 11, 30.
6. Gryaznov, V. M.; Smirnov, V. S.; Slinko, G. Catalysis; Hightower, J. W., Ed American Elsevier: New York, 1973.
7. Gryaznov, V. M. Platinum Metal Rev. 1986, 30, 68.
8. Itoh, N. AIChE. J. 1987, 33, 1576.
9. Mohan, K.; Govind, R. AIChE. J. 1988, 34, 1493.
10. Llias. S.; Govind, R. AIChE. Symposium Series 1989, 268, 18.
11. Itoh, N.; Govind, R. AIChE. Symposium Series 1989, 268, 10.
12. Fukada, K.; DoKiya, M.; Kameyama, T.; Kotera, Y. Ind. Eng. Chem. 1978, 17, 243.
13. Kameyama, T.; Dokiya, M.;. Fujishige, M; Yokokawa, H.; Fukuda, K. Int. J. Hydrogen Energy 1983, 8, 5.
14. Kameyama, T.; Dokiya, M.;. Fujishige, M; Yokokawa, H. Int. Eng. Chem. Fund. 1981, 20, 97.
15. Raymont, M. E. D. Hydrogencarbon Process, 1975, 54, 139.
16. Happel, H.; Blanck, H.; Hamill, T. D. I&EC Fund. 1960, 5, 289.
17. Aston, J. G.; Szasz, G. J. J. Chem. Phys. 1946, 14, 67.
18. Bohmholdt. G.;Wicke, E.; Z. Physik. Chem. 1967, 56, 133.
19. Jacoson, Encyclopedia of Chemical Reaction; Reinholg Publishing: 1949; Vol. 3, p 625.
20. Leder, F.; Butt, J.B. AIChE. J. 1960, 12, 718.)

RECEIVED May 9, 1990

METAL OXIDE CATALYSTS

Chapter 21

Niobium Oxalate

New Precursor for Preparation of Supported Niobium Oxide Catalysts

Jih-Mirn Jehng and Israel E. Wachs

Zettlemoyer Center for Surface Studies, Department of Chemical Engineering, Lehigh University, Bethlehem, PA 18015

The aqueous preparation of supported niobium oxide catalysts was developed by using niobium oxalate as a precursor. The molecular states of aqueous niobium oxalate solutions were investigated by Raman spectroscopy as a function of pH. The results show that two kinds of niobium ionic species exist in solution and their relative concentrations depend on the solution pH and the oxalic acid concentration. The supported niobium oxide catalysts were prepared by the incipient wetness impregnation technique and characterized by Raman, XRD, XPS, and FTIR as a function of niobium oxide coverage and calcination temperature. The Raman studies reveal that two types of surface niobium oxide species exist on the alumina support and their relative concentrations depend on niobium oxide coverage. Raman, XRD, XPS, and FTIR results indicate that a monolayer of surface niobium oxide corresponds to ~ 19% Nb_2O_5 for an Al_2O_3 support possessing ~ 180 m^2/g. The surface niobium oxide phase is found to be stable to high calcination temperatures.

Supported niobium oxide catalysts have recently been shown to be effective catalysts for many catalytic reactions: pollution abatement, selective oxidation, hydrocarbon conversion, carbon monoxide hydrogenation, etc. [1]. In a previous study [2], it was shown that the presence of the surface niobium oxide phases retards the loss in surface area of the Al_2O_3 and TiO_2 supports and stablizes the V_2O_5/TiO_2 system during high temperature treatments. The surface niobium oxide species on the Al_2O_3 support was also found to possess

0097–6156/90/0437–0232$06.00/0

strong Bronsted acidity [3]. These important properties impart the surface niobium oxide phase on the Al_2O_3 support with a high hydrocarbon cracking activity at elevated temperatures.

Niobium ethoxide [$Nb(OC_2H_5)_5$] has traditionally been used as a precursor for the preparation of supported niobium oxide catalysts. This non-aqueous preparation method requires a controlled enviroment and special procedures to avoid the decomposition of the niobium ethoxide in the presence of moisture. It is well-known that transition metal ions form a stable solution chelate with oxalate groups, and molybdenum oxalate [4] and vanadium oxalate [5] have been widely used for the aqueous preparation of supported molybdenum oxide and supported vanadium oxide catalysts. In the present study, niobium oxalate [$Nb(HC_2O_4)_5$] was investigated as an aqueous precursor for the preparation of supported niobium oxide catalysts.

EXPERIMENTAL METHODS

Materials and Preparation Methods

Niobium oxalate was supplied by Niobium Products Company with the following chemical analysis: 20.5% Nb_2O_5, 790 ppm Fe, 680 ppm Si, and 0.1% insolubles. Niobium oxalate was dissolved into a constant concentration of aqueous oxalic acid solution, and the pH of the solution was varied from 0.50 to 5.00 by adding ammonium hydroxide. The supported niobium oxide on Al_2O_3 catalysts were prepared by the incipient-wetness impregnation method using the niobium oxalate/oxalic acid aqueous solution and Al_2O_3(Harshaw, 180 m2/g). The samples were dried at 110-120°C for 16 hours, and then calcined at 500°C under flowing dry air for 16 hours.

Raman Spectroscopy

Raman spectra were obtained with a Spex Triplemate spectrometer(Model 1877) coupled to an EG&G intensified photodiode array detector cooled thermoelectrically to -40°C, and interfaced with an EG&G OMA III Optical Multichannel Analyzer(Model 1463). The samples were excited with the 514.5nm Ar^+ laser. The beam was focused on the sample illuminator where the sample typically spins at about 2000 rpm to avoid local heating. The Raman scattering was collected by the spectrometer, and analyzed with an OMA III built-in software package. The overall spectral resolution of the spectra is about 2 cm-1.

X-Ray Powder Diffraction (XRD)

The crystalline Nb_2O_5 phase in the supported niobium oxide catalysts was detected by an APD 3600 automated X-ray powdered diffractometer using Cu K_α (45KV, 30MA) radiation. The Nb_2O_5/Al_2O_3 samples were calcined at 700°C to increase the Nb_2O_5 particle size and enhance the XRD signals.

X-Ray Photoelectron Spectroscopy (XPS)

XPS experiments were performed on a Physical Electronic Instruments ESCA/AUGER system. The samples were placed on the sample holder at a 45° angle to the entrance of analyzer and the system was evacuated to 10^{-9} - 10^{-10} Torr. The XPS spectra were calibrated against the Au $4f_{7/2}$ peak using the Mg K_α line as the X-ray exciting radiation.

CO_2 Chemisorption

The CO_2 uptake of supported niobium oxide on Al_2O_3 at different Nb_2O_5 loadings was measured with a Quantasorb BET apparatus using a 1:9 ratio of CO_2/He mixture gases. The samples were degassed at 250°C for 2 hours under flowing He, and the CO_2 chemisorption was performed at room temperature.

RESULTS AND DISCUSSION

Niobium oxide reference compounds

The Raman spectra of several niobium oxide compounds, with their corresponding symmetry and coordination, are shown in Figure 1. The $Nb_6O_{19}^{-8}$ unit is a well-characterized structure which consists of three different types of Nb-O bonds at each niobium center: a short Nb=O terminal double bond, a longer Nb-O-Nb bridging bond, and a very long and weak Nb---O single bond connected to the center of the cage-like octahedral structure [6-8]. From the known structure of $K_8Nb_6O_{19}$ the main frequencies of the $K_8Nb_6O_{19}$ Raman spectrum in Figure 1 can be assigned: Nb=O terminal stretching mode (903, 879, and 831 cm^{-1}), corner or edge-shared octahedral NbO_6 stretching mode (734, 537, and 463 cm^{-1}), Nb=O bending mode (289 cm^{-1}), and Nb-O-Nb bending mode (223 cm^{-1}). The multiple terminal stretching modes present in the high wavenumber region are due to distortions present in the $K_8Nb_6O_{19}$ structure. Niobium pentoxide, Nb_2O_5, possesses a more order octahedral structure with no Nb=O terminal bonds, and a major band appears at 690 cm^{-1} which is characteristic of an octahedral NbO_6 stretching mode as well as Nb-O and Nb-O-Nb bending modes at ~300 cm^{-1} and ~230 cm^{-1}, respectively. For the niobium oxalate precursor a sharp and strong Raman band is present at 958 cm^{-1} due to a Nb=O terminal bond and the associated bending modes appear in the 200-400 cm^{-1} region. The Raman band at

~570 cm^{-1} arises from the bidentate oxalate ligands coordinated to the niobium [9,10]. The Raman frequencies of the reference compounds are tabulated in Table 1.

Niobium Oxalate Aqueous Solutions

Niobium oxalate has a low solubility in aqueous solutions, but its solubility can be dramatically increased by the addition of oxalic acid to the aqueous solutions. At high oxalic acid concentrations, however, the niobium oxalate and oxalic acid precipitate from solution. The solubility curve of niobium oxalate in aqueous solutions is shown in Figure 2 as a function of the oxalic acid concentrations. Figure 3 shows a series of Raman spectra of the niobium oxalate in aqueous oxalic acid solutions with varying pH (0.50 to 5.00). At low pH (<3.00), Two peaks are observed in the 900-1000 cm^{-1} region which are chracteristic of Nb=O terminal stretching modes. The behavior of these two peaks with pH variation suggests that two niobium oxalate species exist in solution. The associated bending modes appear in 200-400 cm^{-1} region. A Nb-O_2-C_2 breathing mode also appears at ~570 cm^{-1}. At high pH (>5.00), two new Raman bands form at ~670 cm^{-1} and ~220 cm^{-1} which indicate the formation of hydrated Nb_2O_5.

It is known that the niobium oxide complexes in oxalic acid aqueous solutions display an equilibria between two ionic species containing 2 or 3 oxalate groups which depend on the solution pH and the oxalic acid concentration [11,12]. Thus, the two Nb=O terminal bonds appearing in the aqueous Raman spectra are assigned to the two different niobium oxalate ionic species present in the solution. The Raman spectra also show that the relative intensity of two Nb=O bands, ~910 cm^{-1} and ~930 cm^{-1}, changes with increasing pH. When ammonium hydroxide is added to the solution, the niobium oxalate species with 3 oxalate groups starts to hydrolyze to 2 oxalate groups as one of the oxalate groups is replaced by OH groups. This results in an increase in intensity of the ~910 cm^{-1} Nb=O band with increasing solution pH. Increasing the pH to about 5.00 by further addition of ammonium hydroxide causes the niobium oxalate species to hydrolyze and coagulate to a hydrated Nb_2O_5 precipitate. The aqueous solution chemistry of niobium oxalate is shown below:

Table 1: Raman frequencies of bulk niobium oxide compounds

Vibrational Modes	Wavenumber (cm^{-1})		
	$K_8Nb_6O_{19}$	Nb_2O_5	$Nb(HC_2O_4)_5$
$\nu(Nb=O)$	903,879,831	-	958
$\nu(NbO_6)$	734 537 463	690	-
$\nu(NbO_2C_2)$	-	-	572
$\delta(Nb-O)$	289	302	284
$\delta(Nb-O-Nb)$	223	238	243

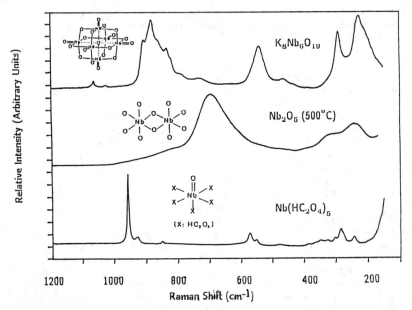

Figure 1: The solubility of nioibum oxalate in solution as a function of oxalic acid added.

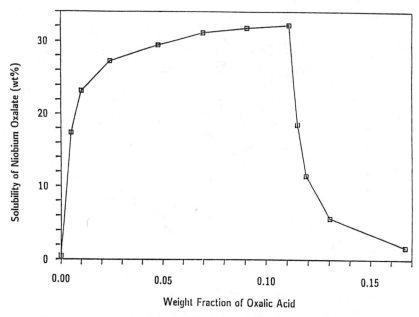

Figure 2: Raman spectra of bulk niobium oxide compounds.

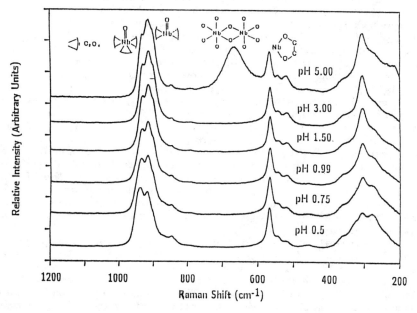

Figure 3: Raman spectra of niobium oxalate in oxalic acid solution as a function of pH from 0.5 to 5.00.

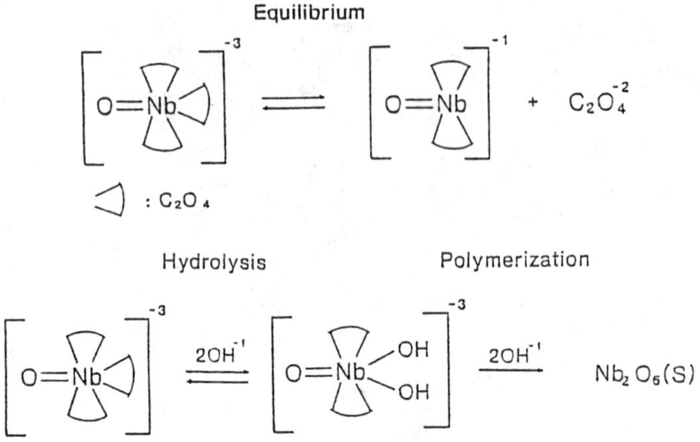

Equilibrium

$$\left[O=Nb \right]^{-3} \rightleftharpoons \left[O=Nb \right]^{-1} + C_2O_4^{-2}$$

◁ : C_2O_4

Hydrolysis   Polymerization

$$\left[O=Nb \right]^{-3} \xrightarrow{2OH^{-1}} \left[O=Nb\begin{smallmatrix}OH\\OH\end{smallmatrix} \right]^{-3} \xrightarrow{2OH^{-1}} Nb_2O_5(S)$$

Supported Niobium Oxide on Alumina

The Raman spectra of supported niobium oxide on alumina are shown in Figure 4 as a function of Nb_2O_5 loading. The nature of the supported niobium oxide phase is determined by comparison of the Raman spectra of the supported niobium oxide samples with those of niobium oxide reference compounds. The Raman features of 1-22% Nb_2O_5/Al_2O_3 samples are different than the bulk niobium oxide compounds due to the formation of a two-dimensional surface niobium oxide overlayer on the alumina support [2]. At low surface coverages (<8% Nb_2O_5/Al_2O_3), the weak and broad Raman band in the 890-910 cm^{-1} region is present due to a distorted octahedral (approaching square-pyramidal) surface niobium oxide species possessing Nb=O bonds, and the mode at ~230 cm^{-1} is characteristic of a Nb-O-Nb linkage. At high surface coverages (>8% Nb_2O_5/Al_2O_3), an additional Raman band at ~630 cm^{-1} is also present due to a slightly distorted octahedral surface niobium oxide species. The Raman studies reveal that two types of surface niobium oxide species exist on the alumina support, and that their relative concentrations depend on the surface niobium oxide coverage.

A series of supported niobium oxide on alumina catalysts, 0-45% Nb_2O_5/Al_2O_3, were further characterized by XRD, XPS, CO_2 chemisorption, as well as Raman spectroscopy in order to determine the monolayer content of the Nb_2O_5/Al_2O_3 system. The transition from a two-dimensional metal oxide overlayer to three-dimensional metal oxide particles can be detected by monitoring the

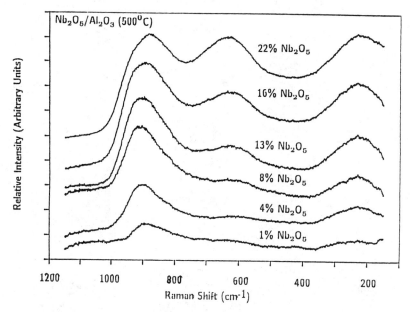

Figure 4: Raman spectra of Nb_2O_5/Al_2O_3 (500°C) as a function of niobium oxide coverage.

$(Nb/Al)_{surface}$ ratios in such systems with XPS because of the vastly different XPS cross-sections of these two phases [13]. The $(Nb/Al)_{surface}$ ratios of the niobium oxide on the alumina support were obtained by integrating the areas of the Nb $3d_{3/2,5/2}$ and the Al 2p photoelectron lines, and the $(Nb/Al)_{surface}$ vs. $(Nb/Al)_{bulk}$ curve is shown in Figure 5. The break in the curve corresponds to ~19% Nb_2O_5/Al_2O_3 and suggests that the transition from a two-dimensional phase to three-dimensional particles, monolayer coverage, occurs at this point. This conclusion is supported by XRD measurements which only detect crystalline Nb_2O_5 particles above 19% Nb_2O_5/Al_2O_3, and CO_2 chemisorption measurements (see Figure 6) which indicate that the basic alumina hydroxyls have been removed by the niobium oxide overlayer [14,15]. The slight increase in the CO_2 chemisorption above 19% Nb_2O_5/Al_2O_3 is due to CO_2 chemisorption on the crystalline Nb_2O_5 particles. The Raman spectra in Figure 7 reveal that the 630 cm-1 band of the surface niobium oxide phase begins to shift towards the 690 cm-1 band of crystalline Nb_2O_5 above 19% Nb_2O_5/Al_2O_3 due to the presence of crystalline Nb_2O_5 particles. Thus, XPS, XRD, CO_2 chemisorption, and Raman all demonstrate that a monolayer of surface niobium oxide on alumina , ~180 m2/g, corresponds to ~19% Nb_2O_5/Al_2O_3.

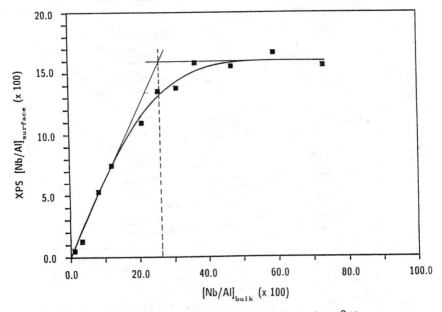

Figure 5: Raman shifts of $Nb_2O_5/Al_2O_3(700^\circ C)$ as a function of niobium oxide coverage.

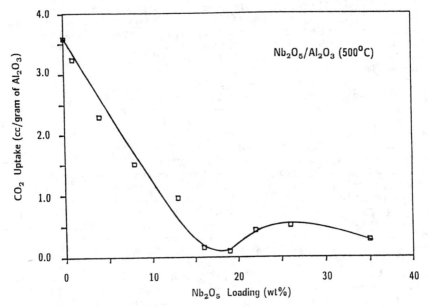

Figure 6: XPS intensity ratios of $(Nb/Al)_{surface}$ as a function of $(Nb/Al)_{bulk}$.

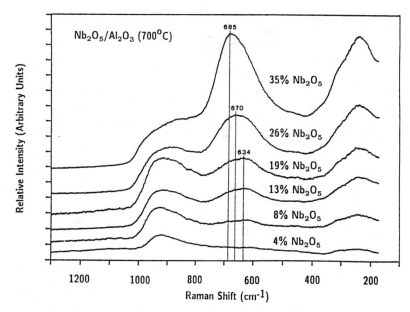

Figure 7: CO_2 uptake of Nb_2O_5/Al_2O_3 (500°C)   as a function of niobium oxide coverage.

ACKNOWLEDGMENTS

The support of Niobium Products Company for this research project is gratefully acknowledged.

REFERENCES

1. Niobium Products Company Inc., <u>Catalytic Applications of Niobium</u>
2. J. M. Jehng, F. D. Hardcastle, and I. E. Wachs, <u>Solid State Ionics,</u> 32/33, 904(1989)
3. L. L. Murrell, D. C. Grenoble, C. J. Kim, and N. C. Dispenziere, Jr., <u>J. Catal.</u> 107, 463, (1987)
4. K. Y. Ng, X. Zhou, and E. Gulari, <u>J. Phys. Chem.</u> 89, 2477, (1985)
5. R. Y. Saleh, I. E. Wachs, S. S. Chan, and C. C. Chersich, <u>J. Catal.</u> 98, 102, (1986)
6. F. J. Farrell, V. A. Maroni, amd T. G. Spiro, <u>Inorganic Chemistry</u> 8(12), 2638, (1969)
7. R.S. Tobias, <u>Can. J. Chem.</u> 43, 1222, (1965)
8. C. Rocchiccioli-Deltcheff, R. Thouvenot, and M. Dabbabi, <u>Spectrochimica Acta</u> 33A, 143, (1977)
9. W. P. Griffith, and T. D. Wickins, <u>J. Chem. Soc.</u> (A), 590, (1967)
10. J. E. Guerchais, and B. Spinner, <u>Bull. Soc. Chim. France</u>, 1122, (1965)

11. E. M. Zhurenkov, and N. Pobezhimovskaya, Radiokhimiya
    12(1), 105, (1970)
12. C. Djordjevic, H. Gorican, and S. L. Tan, J. Less-
    Common Metals II, 342, (1966)
13. Z. X. Liu,Z. D. Lin, H. J. Fan, and F. H. Li, Appl.
    Phys. A 45, 159, (1988)
14. W. S. Milliam, K. I. Segawa, D. Smrz, and W. K. Hall,
    Polyhedron 5, 169, (1986)
15. C. L. O'Young, C. H. Yang, S. J. DeCanio, M. S.
    Patel, and D. A. Storm, J. Catal. 113, 307, (1988)

RECEIVED May 9, 1990

Chapter 22

Redox Cycle During Oxidative Coupling of Methane over PbO–MgO–Al$_2$O$_3$ Catalyst

Alvin H. Weiss[1], John Cook[1], Richard Holmes[1], Natka Davidova[2], Pavlina Kovacheva[2], and Maria Traikova[2]

[1]Department of Chemical Engineering, Worcester Polytechnic Institute, Worcester, MA 01609
[2]Institute of Kinetics and Catalysis, Bulgarian Academy of Sciences, 1040 Sofia, Bulgaria

The partial oxidation of methane to ethane in a great oxygen deficiency was studied at 700, 770, and 820° C at methane to oxygen ratios of 10:1, 20:1, 50:1, 500:1, and infinity (no oxygen). W/F was 7.6, 20.3, and 41.3 ghr/mole in a 10 mm ID quartz tube containing 1 g catalyst mixed with quartz sand. The catalyst was mixed oxide PbO-MgO-Al$_2$O$_3$ precipitated as crystals by a hydrotalcite preparation procedure and then oxidized at 750°C for one hour. BET surface area was 20.8 m^2/g before calcining, 8.7 after. DC Arc Plasma analysis confirmed equiatomic content of the metals. X-ray diffraction analysis showed that the calcined catalyst contained PbO in highly dispersed MgO - Al$_2$O$_3$ matrix. XRD showed that PbO was reduced during reaction to inactive Pb when oxygen was present in the gas feed in quantities insufficient to oxidize hydrogen produced. The oxygen in PbO was used for reaction, thereby deactivating the catalyst. Subsequent oxygen treatment converted the Pb back to PbO and restored catalytic activity, thereby completing the redox cycle. Maximum yield of ethylene plus ethane at 10:1 CH$_4$/O$_2$ was 13.7%. No acetylene was observed, and ethylene/ethane ratios were less than one percent of the value predicted by thermodynamic calculation. This suggests, but does not prove, that ethylene was produced thermally, not catalytically, on this catalyst.

In recent years the shortage of both ethylene and ethylene feedstocks has resulted in a search for alternate sources of these two commodities. Methane is an obvious choice, since it comprises up to 85 mole % of the hydrocarbons in natural gas.

0097–6156/90/0437–0243$06.00/0
© 1990 American Chemical Society

A wide variety of metal oxides were screened by Keller and Bhasin (1) in their pioneering investigation into methane dimerization. Since then, many catalysts have been developed and tested (2-6). However, due to the thermal stability of methane, even in the presence of oxidizers, conversion is extremely low. To date, total yields of C_2 products of 15 to 20% (2, 4) are considered high.

Many different types of catalysts have been employed, including both metal oxides, such as PbO (5-7), which react in a redox cycle to Pb metal, and catalysts containing metals with fixed valence, such as Li promoted MgO (2), which produce active Li^+O^- sites for methane dimerization. Such catalysts were discussed in four papers at the recent Ninth International Congress on Catalysis (8). The most effective catalysts have several common traits. Low surface area has been found to be very important (2, 9) in the conversion of methane. Also of importance is the basicity of the catalyst. (However, not every basic material causes C_2 formation.)

Iwamatsu, et al (9), have shown that, in the absence of PbO, C_2 yield increases from 4.4 to 9.0% over MgO when surface area is calcined from 70 to 17 m^2/g. Similar effects were shown on alkali-doped MgO, and the yields of C_2 over 7 m^2/g 0.2% Na^+-MgO, 2% K^+-MgO, and 2% Cs^+-MgO were 9.8, 13.2, and 8.2%, in that order. (Yield = (2x moles C_2 hydrocarbons produced)/(moles CH_4 in feed)).

Bytyn and Baerns (5) showed that the selectivity obtained using supported PbO in the absence of MgO depended on the acidity of the catalyst support. Yield was best (15%) on SiO_2 - gel with pK_A = +6.8 and minimal (0.5%) on SiO_2 - Al_2O_3 with pK_A = -5.6.

Hinsen, et al (6) showed an optimal effect of PbO on C_2 selectivity, even though there was great loss of surface area during operation at 1013K. Lee and Oyama (10) have published an extensive review on oxidative coupling of methane.

In this present work we show, by starving the reaction for oxygen in the gas phase, that PbO catalyst then provides oxygen. It is reduced to Pb^o; and the catalyst deactivates. The Pb^o is reactivated to catalytic PbO by oxidation, thereby demonstrating the redox cycle.

Experimental

Catalyst. One catalyst, prepared to contain equi-atomic amounts of lead, magnesium, and aluminum, was prepared for this study. Synthesis was accomplished by the hydrotalcite preparation procedure reported by Reichle (12). It is a standard aqueous precipitation and crystallization procedure that avoids filtering and washing problems associated with gel precipitates. A solution of Pb, Mg, and Al nitrates was added under agitation to a solution of NaOH and Na_2CO_3 and maintained at 70°C for thirty hours. This caused precipitation of crystalline material.

Before being used as a catalyst, the material was calcined in air at 750°C for 1.5 hours. After calcination, the color of the catalyst was bright yellow and a 25% weight loss was noted. BET surface areas of uncalcined and calcined materials were 20.7 and 8.7 m^2/g, respectively.

Bulk phase analyses of samples (both calcined and uncalcined) were made using a Spectrametrics, Inc., SMI IV DC Arc Plasma Analyzer and showed nominally 1:1:1 atomic ratios of Pb:Mg:Al (1.00:1.05:0.92 and 1.00:1.15:0.69, respectively). This corresponds to a $PbO:MgOAl_2O_3$ weight ratio of 0.710:0.128:0.162. The calcined catalyst has no hydrotalcite structure, but is really Mg and Al oxides on a support of PbO. This is shown by the x-ray diffraction pattern of Figure 1, obtained using a General Electric XRD-5 X-ray Diffraction Unit. The calcined material showed the existence of two polymorphous modifications of PbO (α-PbO and ß-PbO) and low intensity characteristic signals of α-Al_2O_3, γ-Al_2O_3, and MgO.

Reaction. Purified methane (99.0% min.), and ten percent oxygen in helium were obtained from Matheson Corp.. Both CP helium and air (>99.9%) were supplied by Airco. A Hastings LF-50 mass flow meter was used to monitor the flow of helium into the system. Air flow rate was measured by a Matheson ALL-50 mass flow meter. The methane feed was monitored by a Hastings LF-100 mass flow meter. All gases were passed over Drierite and Ascarite to remove moisture and CO_2.

A ten-inch length of Poly-Flo tubing connected the air-cooled reactor outlet to a water-cooled condenser - receiver vessel. The reactor was a two-foot-long piece of quartz tubing (OD = 12 mm ID = 10 mm). (Stainless steel causes the complete oxidation of methane during the methane activation process (*1*).) A 13.5 inch Lindberg heavy duty tube furnace was mounted vertically around the reactor. Gas flow was downward. Additional insulation was used to protect the surrounding area from the high temperatures generated and to reduce heat loss from the top due to free convection. Experiments flowing helium through the empty reactor showed that a 7 cm isothermal hot zone existed slightly above the center of the reactor. The temperature inside the reactor was within 5°C of that outside. It was in this region that the catalyst bed was placed. No measurements were made during reaction to establish either true bed temperatures or the effects of heat and mass transfer.

For a typical experiment, 1 gram of catalyst sieved to -20 + 40 mesh was diluted in 7 grams of -50 + 70 mesh white quartz sand (Aldrich) to minimize exotherms. The catalyst bed was supported by a small plug of quartz wool. Two inches of quartz wool were also placed above the catalyst bed. An 1/8 inch thermocouple was located at the center of the hot zone but outside of the reactor. The thermocouple was attached to an Omega CN300 temperature controller, which both displayed and controlled the reactor temperature.

The main set of experiments consisted of passing methane and oxygen over the atmospheric pressure catalyst bed at CH_4/O_2 ratios of 10/1, 20/1, 500/1 and infinity. The last value corresponds to no oxygen present in the feed. Three temperatures, 700°C, to 770°C, and 820°C were studied at each ratio. W/F (grams of catalyst/mole of methane fed per hour) was held constant at 7.6 g-hr/mole. (This corresponded to a methane flow rate of 55 cm^3/min). In all tests the total feed rate to the reactor was 104 cm^3/min. This value was maintained through the use of helium as a make-up gas.

A second study examined the effect of varying methane W/F. Again, as the W/F for methane was varied the total flow to the catalyst bed was maintained at 104 cm^3/min. The W/F values examined were 7.6, 20.3 and 41.3 g-hr/mole. Methane to O_2 ratio was held constant at 10/1.

When the reactor was at temperature, methane was diverted to the reactor and the first sample was taken after two minutes. Samples were taken every five minutes on line after that. Once the catalyst had deactivated, methane flow was halted and the system was purged with helium for one hour. The reactor was found to be etched after use.

Air at 26 cm^3/min was fed to the reactor to regenerate the catalyst. Again helium was used to keep the total flow at 104 cm^3/min. Regeneration was allowed to proceed for one hour, after which the reactor was either taken off-line in order to x-ray the catalyst or was again purged with helium to prepare for a second methane reaction. In the latter case, the reactor was purged with helium for one hour to remove oxygen from the lines. After the helium purge was completed, methane was again fed to the reactor.

Blank runs, in which porcelain chips were used to fill the catalyst zone in the quartz reactor, were made at 820°C. Methane at 10/1 CH_4/O_2 ratio (55 cc/min CH_4 + 55 cc/min 10% O_2 in He) was 1.5% converted at 100% selectivity to ethane. No conversion of methane was measured in the absence of O_2. Ethane conversion at 5/1 C_2H_6/O_2 at a total gas rate of 82.5 cc/min over the porcelain was 85.0%; and selectivities were 3.0% CO_2, 86% C_2H_4, and 11% CH_4. In the absence of O_2, C_2H_6 was not greatly different, 78%; and selectivities were 98% to C_2H_4 and 2% to CH_4. This suggests that most of the ethylene in the catalytic methane runs that are the subject of this paper was produced thermally, not catalytically, from ethane formed by the dimerization.

A check was made at 820°C of the behavior of ethane at 5/1 C_2H_6/O_2 over one gram of the catalyst at 82.5 cc/min total. Conversion ranged from 65.5 to 81.8% at practically 100% selectivity to methane. The reason for no observation of C_2H_4 or CO_2 in the product is not known.

Reactor effluent was continuously fed to a 0.25 ml sampling loop located inside a Hewlett-Packard Model 2520 Gas Analyzer. Separation of the products was over two serial columns, 6' x 1/8" Poropak Q 80/100 mesh, followed by 10' x 1/8" molecular sieve 5A 60/80 mesh. Thermal conductivity detection was used with helium carrier gas. Columns were isothermal at 60°C.

Results and Discussions

Figure 2 shows that lower CH_4/O_2 values produce much greater conversions at the three temperatures studied. When the ratio was kept low, 10/1, 20/1, 50/1, there was no significant deactivation in the course of one thousand minutes. However when the ratio was increased to 500/1 and infinity (i.e., a great deficiency of O_2), deactivation of the catalyst was observed in an hour.

X-ray diffraction patterns on Figure 3 show the presence of metallic lead in catalyst deactivated by operation in the absence of

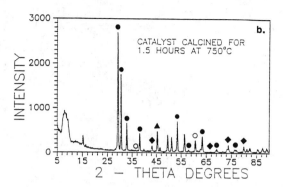

Figure 1.   XRD spectra of the unused catalyst showing PbO and no Pb^O. ◆ = MgO; ○ = α-PbO; ● = β-PbO; ▲ = γ-Al$_2$O$_3$.

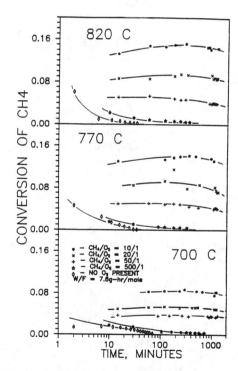

Figure 2.   CH$_4$ conversion vs. time on stream at 820°C, 770°C, and 700°C. Parameters of CH$_4$/O$_2$ ratio at W/F = 7.6 g–hr/mole

oxygen. When catalyst was operated in an excess of oxygen, Pb^O was not visible in the XRD pattern (Figure 4). When deactivated catalyst was reoxidized by an air-helium mixture, activity was restored, albeit at a somewhat lower level than that of the first reaction cycle. Figure 5 shows that after the regeneration of the deactivated catalyst, the metal Pb^O was transformed to ß-PbO. Analogous results were obtained with catalyst operated at 500 CH_4/O_2 (also a deficiency of O_2).

The results are in accord with the observations of Asami, et al (2), which were that the PbO could be cycled in a redox cycle if O_2 and CH_4 were not simultaneously fed to the catalyst.

The yields of products C_2H_4 and C_2H_6 were highest at 820°C. Figure 6 shows the effect of CH_4/O_2 ratio in the feed at W/F = 7.6 g-hr/mole for C_2H_4, C_2H_6, and CO_2. It was rare to observe a trace of CO in the product; and no acetylene was observed in any experiment. Consequently, the reactions that satisfy the stoichiometry for O_2 consumption are:

$$2CH_4 + 1/2 O_2 \longrightarrow C_2H_6 + H_2O$$

$$2CH_4 + O_2 \longrightarrow C_2H_4 + 2H_2O$$

$$CH_4 + 2O_2 \longrightarrow CO_2 + 2H_2O$$

(These reactions should not be regarded as mechanistic, since oxygen comes from PbO produced in the redox cycle).

Table I compares for each temperature and for CH_4/O_2 ratios in the feed of 10/1, 20/1, 50/1, and 500/1, the ratio of oxygen needed to convert the hydrogen produced by reaction to that actually present in the feed. These calculations are the average for the yields shown on Figure 6. When the ratio is equal to unity, exactly a sufficient amount of oxygen is in the feed to maintain the lead redox cycle. When the ratio of needed-to-available oxygen is significantly greater than one, which is the case at 500/1 and no oxygen, then oxygen to combust the H_2 and to produce CO_2 must come from the PbO of the catalyst. Hence the observed deactivation.

Table I. Oxygen Deficiency at High Feed CH_4/O_2 Ratio

CH_4/O_2	O_2 Needed/O_2 Available		
	700°C	770°C	820°C
10/1	1.0	1.0	1.1
20/1	1.0	1.0	1.1
50/1	1.25	1.25	1.3
500/1	2.0	3.0	3.0

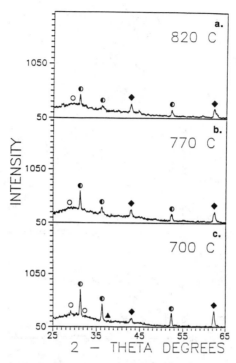

Figure 3. XRD spectra of the deactivated catalyst showing Pb^O after reaction in the absence of O^2 at 820°C, 770°C, and 700°C. Legend the same as Figure 1, plus $\mathbb{O}$ = Pb^O.

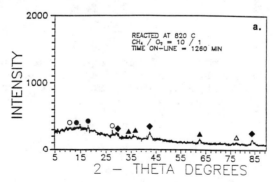

Figure 4. XRD spectra of the non−deactivated catalyst showing PbO, but no Pb^O, after reaction in an excess of oxygen. Legend same as Figure 1.

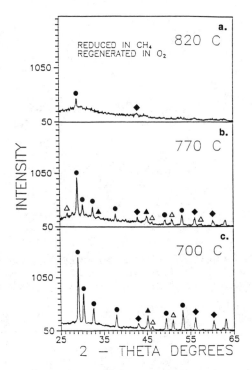

Figure 5.   XRD spectra of the regenerated deactivated
catalyst showing PbO and no Pb^O after reaction in the
absence of O$_2$ at 820°C, 770°C, and 700°C. Legend
same as Figure 1, plus $\triangle = \alpha$-Al$_2$O$_3$.

The maximum yields of C_2 species were obtained at 10/1 methane/oxygen feed ratio and 820°C. Table II lists the effects of varying space time in the range of 7.6 - 41.3 g-hr/mole. The maximum yields of C_2's are listed, along with the corresponding methane conversions and CO_2 yield.

Table II. The Effect of Temperature and Space Time on C_2 Yield

T (°C)	W/F g-hr/mole	Conv (%)	C_2 Maximum Yield(%)	CO_2 Yield(%)
	7.6	15.2	11.7	3.6
820	20.3	16.8	13.3	3.6
	41.3	17.3	13.7	3.6
	7.6	9.9	6.1	3.9
700	20.3	10.8	6.2	4.6
	41.3	11.4	5.5	5.9

C_2H_4/C_2H_6 ratios were calculated at equilibrium for the reaction

$$C_2H_6 \; \xrightleftharpoons{} \; C_2H_4 + H_2$$

both at 1 atm H_2 and at 0.1 atm H_2. These values are plotted in Figure 7 vs. reciprocal temperature. The experimental C_2H_4/C_2H_6 ratios were also calculated from the yield curves of Figure 6 at maximum conversion and plotted. The experimental values on Figure 7 fall into a range that is quite low relative to the equilibrium ratios possible, less than 1%. They show that this particular PbO-MgO-Al_2O_3 catalyst catalyzes neither the dehydrogenation reaction of ethane to ethylene nor the dimerization of methane to ethylene to anywhere near the equilibrium possibility.

If the catalyst does not produce adsorbed ethane, ethane would have to readsorb to react to ethylene. This is not too probable; and it is in accord with our observation of very low ethylene to ethane ratio (at best 1:1) relative to the ratios possible at equilibrium (about 100:1). The mechanism of the reaction over this catalyst appears to be the redox cycle proposed by Keller and Bhasin (1), except that the greater part of the ethane probably desorbs rather than reacts further to ethylene.

Conclusions

When PbO-MgO-Al_2O_3 is degraded by calcination, the result is PbO in a catalytic form. Oxygen from PbO is used to convert the methane produced in the reactions. If this oxygen is not supplied to PbO

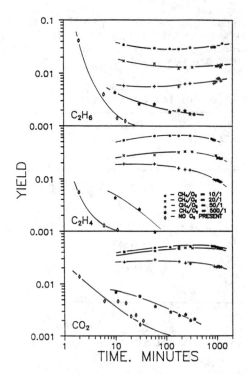

Figure 6.   Yields of C_2H_6, C_2H_4, and CO_2 vs. time on stream at 820°C. Parameters of CH_4/O_2 ratio at W/F = 7.6 g-hr/mole.

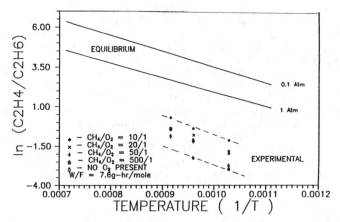

Figure 7.   Calculated equilibrium C_2H_4/C_2H_6 ratios at both 0.1 and 1.0 atm H_2 and experimental C_2H_4/C_2H_6 ratios vs. reciprocal temperature.

from the gas phase during reaction, the PbO is converted to metallic lead. The result is rapid catalyst deactivation. The Pb^O can be readily reoxidized back to the catalytic state. This redox behavior was observed in this study by x-ray diffraction.

The ethane produced in the system is only dehydrogenated to about one percent of its thermodynamic potential, suggesting, even though product ethane to ethylene ratios are about 1:1, that dehydrogenation does not proceed on the catalyst.

Acknowledgments

This work is part of the US-Bulgarian Cooperative Research Program. The authors extend their appreciation both to the United States Army for Lt. Cook's and Capt. Holmes' support and to the National Science Foundation for its grant No. INT-8810539.

Literature Cited

1. Keller, G. E.; Bhasin, M. M. *J. Catal.* 1982, *73(1)*, pp. 9-15.
2. Ito,T.; Wang, J.; Lin, C.; Lunsford, J. *J. Am.Chem. Soc.* 1985, *107*, pp. 5062-5068.
3. Otsuka, K.; *Sekiyu Gakkaishi* 1987, *30(6)*, pp. 385-396.
4. Sofranko, J. A.; Leonard, J. J.; Jones, C. A.; Gaffney, A. M.; Withers, H. P., Preprints, Symposium on Hydrocarbon Oxidation, ACS New Orleans Meeting, **August 30-September 4, 1987**, pp. 763-769.
5. Bytyn, W.; Baerns, M. *J. Appl. Catal.* 1986, *28*, pp. 199-207.
6. Hinsen, W.; Bytyn, W.; Baerns, M., Proceedings of Eighth International Congress on Catalysis, Toyko, Japan, 1984, pp. 581-592.
7. Asami, K.; Shikada, T.; Fujimoto, K.; Tominaga, H. *Ind. Eng. Chem. Res.* 1987, *26*, pp. 2384-2353.
8. Session on Methane Conversion, Proceedings of Ninth International Congress on Catalysis, Calgary, Alberta, 1988, pp. 883-989.
9. Iwamatsu, E.; Moriyama, T.; Nobhiro, T.; Aika, K. *J. Chem. Soc., Chem. Commun.*, 1987, *1*, pp. 19-20.
10. Lee, J. S.; Oyama, S. T. *Catal. Rev. - Sci. Eng.*, 1982, *30*, pp. 249-280.
11. Reichle, W. T. *ChemTech*, 1986, *16*, pp. 58-63.

RECEIVED May 9, 1990

CATALYSTS IN FUEL PRODUCTION

Chapter 23

Carbon Monoxide Hydrogenation over Rh–Molybdena–Alumina Catalysts

Selectivity Control Using Activation Energy Differences

N. A. Bhore, K. B. Bischoff, W. H. Manogue, and G. A. Mills

Center for Catalytic Science and Technology, Department of Chemical Engineering, University of Delaware, Newark, DE 19716

Changes in rates of CO hydrogenation were calculated as a function of temperature using activation energies (Kcal/mole) determined for formation of oxygenates (18.6) and for hydrocarbons (31.2) over novel, high activity Rh-Mo/Al$_2$O$_3$ catalysts. For a 3% Rh-7.5% Mo/Al$_2$O$_3$, selectivity to oxygenates was found to increase from 65% at 250° to 86% at 200°; predicted selectivity is 95% at 160°. Kinetic measurements showed that formation of methanol, higher alcohols and hydrocarbons are inhibited by CO over Rh/Al$_2$O$_3$. But over Rh-Mo/Al$_2$O$_3$, methanol formation is not CO-inhibited. A dual-site mechanism is proposed for Rh-Mo/Al$_2$O$_3$ with high activity/selectivity attributed to hydrogen activation by a MoO$_{3-x}$ site, not inhibited by CO, and CO activation on Rh sites. These data provide guidelines for process optimization and the development of more active and selective catalysts.

Oxygenates are of growing interest for use as automotive fuels because of their clean-burning and high-octane characteristics and because they can be manufactured from non-pertroleum resources (1). Ethyl alcohol and ethers, particularly methyl-*tert* - butyl ether, are being used increasingly in gasoline blends. In addition, methyl alcohol also has great merit as a transportation fuel, particularly if it can be manufactured more economically. A catalyst which could hydrogenate CO to methanol effectively at lower temperature is desirable since this would permit operation at higher conversion per pass (presently thermodynamically limited) and therefore permit less of expensive recycle. Further, a more active catalyst would permit operation at lower pressure and so decrease plant investment.

Multifunctional catalysts offer important opportunities for scientific advances and industrial applications since they are able to activate simultaneously different molecular species such as CO and H$_2$. Of critical interest is the molecular structures of the catalyst responsible for such multiple activations, how the activated species interact, and how the reaction dynamics control activity and selectivity.

Hydrogenation of CO is a widely studied reaction with many practical applications. The catalytic performances of supported Rh catalysts for CO hydrogenation are very dependent on the support and added modifiers (2-6). Of particular interest is the novel Rh-Mo/Al$_2$O$_3$ catalyst system which displays exceptionally high activity for CO hydrogenation and high selectivity for formation of oxygenates (7-12).

Kinetic and characterization tests were carried out using catalysts consisting of Rh on Al$_2$O$_3$ with various amounts of added molybdena. The results are discussed in terms of selectivity enhancement by utilizing differences in activation energies for selective and non-selective reactions. Reaction mechanisms are discussed in terms of a dual-site functionality with implications for design of improved catalysts.

0097–6156/90/0437–0256$06.00/0

Experimental

Catalysts of composition shown in Table I were prepared by impregnation. All contain a nominal 3% Rh deposited from rhodium nitrate solution. The alumina was Catapal. Those containing molybdena were prepared in stages. The support was first impregnated with ammonium molybdate, pH1, followed by drying and air calcination. Then rhodium was deposited. For the 15% Mo catalysts, a dual impregnation was used to overcome solubility limitation. Before testing, catalysts were reduced in flowing H_2 at 500° [All temperatures are °C]. Preformance testing was in a flow reactor system. Data were obtained at 3 MPa, $H_2/CO = 0.5 - 5$, at 3000 to 36,000 GHSV, 200° to 250°. Steady state product analysis was by on-line GC. To make comparisons, space rates were varied at constant temperature to obtain equal conversions generally limited to 6%. This conversion level was selected to provide sufficient product for accurate analysis and yet differential reactor conditions are maintained so that conversions can be used as measures of rates of hydrogenation.

In order to assign and compare catalyst reactivity rates, measured conversions were "normalized" to 3000 GHSV by multiplying observed conversions by the factor : actual GHSV/3000. The normalized conversions were used to specify rates to individual products and rates for overall CO conversion. The reaction has been shown not to be mass or heat transfer limited (12). CO and irreversible H_2 chemisorption were measured at room temperature, the former using a pulse injection system and a thermal conductivity detector, and the latter using a static system. Prior to measurements, catalysts were reduced under the same schedule as for reactor runs.

Results and Discussion

Catalyst Composition - Effect on Performance. Comparison of catalyst selectivities are best made at equal conversions. It was previously shown that selectivity of CO hydrogenation to oxygenates decreases with increasing conversion. For example, selectivity decreased linearly from 35% at 4% conversion to 20% at 18% conversion for Rh/Al_2O_3 catalysts at 250°C.(10)

The rates and selectivities of Rh/Al_2O_3 catalysts are increased greatly by the addition of realtively large amounts of molybdena (7-12). As shown in Table I, CO conversion rate was increased 12-fold by addition of 7.5% Mo. Equal conversion was obtained at 36,000 instead of 3000 GHSV at 250°. At the same time, selectivity to oxygenates was increased to 64.8% from 28.9% at about the same conversion.

Selectivity to oxygenates increased progressivly with addition of Mo to Rh/Al_2O_3. When measured at 250°, % selectivity - % Mo were 29 - 0; 58 - 2.8; 65 - 7.5; 69 - 15, Table I.

Furthermore, molybdena brought a high capability for the shift-conversion reaction. At 6% CO conversion, typically 25% of the converted CO goes to CO_2. The amount of CO_2 observed is consistent with the reaction $CO + H_2O \longrightarrow CO_2 + H_2$ utilizing the amount of water produced from formation of hydrocarbons and higher alcohols.

Temperature — Effect on Catalyst Performance. The rates of formation of some of the products as a function of temperature have been published previously (10) and additional data are shown in Fig. 1. The values of apparent E_{act}, calculated using the Arhennius realtionship, given in Table II for overall CO conversion is the same for Rh/Al_2O_3 and $Rh-Mo/Al_2O_3$. Hence higher reaction rates with the latter catalyst is not due to a lower E_{act}. It should also be noted that for the $Rh-Mo/Al_2O_3$ catalysts E_{act} for oxygenates as a group is 18.6 Kcal/mole, much lower than the 31.2 volume for hydrocarbons. Individual products show smaller but significant differences from these values, for example 32.4 for CH_4 and 27 for C_2H_6. The consequence is that there is a double penalty for operation at higher temperatures. Not only are hydrocarbons increased relative to oxygenates, but also the hydrocarbons consist of larger amounts of less valuable CH_4. It is also significant that $E_{act, C_{2+}oxy.} > E_{act, C_1 oxy.}$. This is believed to result from the requirement for CO dissociation for formation of hydrocarbons and higher alcohols but not C_1 oxygenates.

Table I.   CO Hydrogenation by Catalysts of Various Compositions, $H_2/CO=2$; 3MPa. All contain 3% Rh

Temp.	250	275	225	200	225	235	250	200	225
GHSV	3000	3000	3000	3000	18000	18000	36000	3000	3000
% Conversion of CO, includes CO_2									
Conv.	5.7	12.5	9.0	7.3	4.0	6.0	5.3	6.0	27
to CO_2	1	1	21.5	23.9	23.0	25.6	24.5	25	37
% of CO Converted, excludes CO_2									
CH_4	60	69	34.8	9.4	18.4	23.8	26.8	7.2	10.6
C_2H_6	4	4.1	4.5	3.1	5.8	4.7	5.9	2.4	4.4
C_3H_8	5	5.0	1.5	1.1	2.3	1.9	2.3	0.9	1.6
C_4H_{10}	2	1.4	0.4	0.3	0.6	0.5	0.6	0	0.5
C_5H_{12}		2.0		0	0	0	0		
Total HCs	71	81.5	41.7	13.9	27.1	30.9	35.6	10.5	17.1
MeOH	2	0.8	13.4	37.6	21.7	17.7	15.7	24.0	10.9
MeOMe	1	0.2	15.9	30.0	28.3	26.9	26.5	39.5	56.5
MeCHO	2	2.3	0	0	0	0	0	0	0
EtOH	11	5.2	12.3	5.8	7.4	7.4	6.7	18.0	1?
MeOAc	3	2.6	2.4	1.1	1.1	0.7	0.7	0.2	0.7
HOAc	0	0	0	0	0	0	0	0	0
EtCHO	0.4	0	0	0	0	0	0	0	0
C_3H_7OH	2.7	2.5	2.4	1.9	0.9	2.2	0.7	2.4	2.7
MeOEt	3	21.	12.6	9.9	14.0	14.5	14.5	7.0	11.2
EtOAc	3	3.6	0	0	0	0	0	0	0
C_4H_9OH	0	0.6	0	0	0	0	0	0	0.4
Total Oxy.	28.9	18.9	58.7	86.3	73.3	69.2	64.8	90	83
C_1oxy.	5.1	2.6	34.3	71.4	55.0	49.6	47.3	65.9	71.3
C_2oxy.	20.7	14.2	22.2	13.0	17.4	17.4	16.8	22.8	8.9
C_3oxy.	3.0	2.5	2.4	1.9	0.9	2.2	0.7	2.4	2.6
C_2+oxy. % of oxy.	82.4	86.2	41.6	17.2	21.7	28.1	26.7	27.7	14.0

Table II.   Apparent Activation Energies, CO Hydrogenation

	Product	Kcal/g mol
3% Rh/Al_2O_3	-CO	21.3
3% Rh-7.5% Mo/Al_2O_3	-CO	21.6
	Total Oxy	18.6
	Total HC	31.2
	C_1oxy	17.2
	C_2oxy	24.3
	CH_4	32.3
	C_2H_6	27.5
	C_3H_8	28.6
	C_4H_{10}	26.3

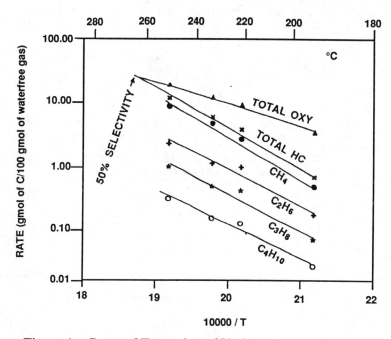

Figure 1. Rates of Formation of Hydrocarbons and Total Oxygenates Over 3%Rh7.5%Mo/Al$_2$O$_3$, 3MPa H$_2$/CO=1. Rates are CO Conversions to Products Shown, Normalized to 3000 GHSV. See Table II for E$_{act}$.

<u>Utilizing Activation Energy Differences for Selectivity Control</u>. The wide differences in E_{act} between formation of oxygenates and hydrocarbons results in a more rapid decrease in the rate of formation of hydrocarbon relative to oxygenates as temperature is decreased. Selectivity is increased. The relative rates for selective, r_o, and non-selective, r_h, reactions are expressed by the relationship

$$\frac{r_o}{r_h} = \frac{\text{selectivity to oxygenates}}{\text{selectivity to hydrocarbons}} = D \cdot e^{\frac{-(E_o - E_h)}{RT}}$$

D is a constant whose value, log D = –5.60, was established from experimental selectivities for Rh-7.5% Mo/Al$_2$O$_3$. The following selectivities to oxygenates represent those calculated and found and those predicted for various temperatures, using E_o=18.6 and E_h=31.2.

Temp.°	273	250	225	200	180	160	140
Calculated %	50	61	75	85	91	95	98
Found %		65	73	86			

One application of this calculation is to provide a prediction of the selectivities which may be obtained with catalysts of sufficient activity to be used at lower temperatures.

The use of lower temperatures to increase selectivity has a penalty — namely loss of conversion rate. The decrease in rate can also be calculated for selective and nonselective reactions as a function of temperatures. This is illustrated by the following:

$$\frac{\text{rate } T_1}{\text{rate} T_2} = e^{\frac{-E(T_1 - T_2)}{RT_1 T_2}}$$

		oxygenates	hydrocarbon
E_{act} cal/mole	18,600	31,200	
			rate loss
50° decrease, 250° – 200°		7 fold	24 fold
90° decrease, 250° – 160°		42 fold	524 fold

The above calculations can provide the initial basis for optimizing process design in which advantages of increased selectivity — improved product value, lower plant and operations costs for separation, and possible longer catalyst life — are calculated and related to disadvantages of lower rates of conversion — larger catalyst inventory and increased reactor investment. Thus an increase of selectivity from 65 to 86% for Rh-7.5%Mo/Al$_2$O$_3$ in going from 250° to 200° may more than compensate for the requirements imposed by a 7-fold increase in catalyst inventory to reach the same conversion level.

<u>Rate Comparisons With Other Catalysts</u>. It is of interest to compare the space-time-yield for Rh-Mo/Al$_2$O$_3$ catalyst and industrial Cu/ZnO/Al$_2$O$_3$ catalysts. The space-time-yield STY, for Rh-7.5%Mo/Al$_2$O$_3$ at 250° at 36,000 GHSV in g/hr/ml catalyst corresponds to 1.0 for all products, 0.76 for oxygenates and hydrocarbons, or 0.4 for oxygenate liquids (0.51 ml/hr/g). Commercial catalysts are said to produce about 0.5 ml methanol/hr/gm cat. Thus

the Rh-Mo/Al2O3 catalyst is more active than a commercial catalyst but not as selective. Rh-Mo/Al2O3 catalysts are by far the most active of supported Rh catalysts identified in a wide survey (6).

Kinetics. The coefficients of the kinetic power-law rate expression for CO hydrogenation

$$Rate_{species} = A \cdot p_{H_2}^x \cdot p_{CO}^y$$

were determined for Rh/Al2O3 and Rh-Mo/Al2O3,Table III(10,12). A negative value for the exponent of pCO for the Rh/Al2O3 catalysts is interpreted to indicate that there is an inhibition of the reaction by preferential adsorption of CO relative to H_2 on the Rh. However, for the Mo-modified catalyst the exponent of pCO is zero for the overall conversion of CO, for MeOH and for CO_2 formation. While CO_2 is mechanistically a secondary product, it follows the power-law because product water is immediately converted to CO_2. Significantly, the exponent of pCO remains negative for the formation of methane and higher alcohols. This is interpreted to mean that formation of CH_4 and higher alcohols is occurring at Rh sites, and that dissociation of CO is involved which is subject to inhibition by CO. Formation of methanol does not involve CO dissociation and is not inhibited by CO.

H2 and CO Chemisorption and Turnover Frequency. The dispersion of Rh in the Rh/Al2O3 catalyst was determined to be 39%, based on H_2 chemisorption and assuming 1H/1Rh. However, for Mo catalysts, H_2 cannot be used for this purpose because of the formation of non-stoichiometric so-called molybdenum bronzes, H_xMoO_y. Therefore, CO was used to measure Rh dispersion for Mo catalysts. H_2 adsorption on Rh/Al2O3 provided an initial calibration point. It was determined that CO does not adsorb appreciably on partially reduced molybdena under the above-mentioned conditions. While CO can adsorb in different forms, as determined by infrared measurements, it is assumed that the stoichiometry of CO chemisorption on Rh does not change with increased Mo and can be used as a measure of Rh dispersion. The amount of CO chemisorbed decreased progressively and substantially with addition of molybdena, Table IV. Also shown is the overall rate of CO conversion, labelled the turn-over-frequency, for each Rh atom in the sample. The TOF shows an increase as increasing amounts of Mo are added. Thus, even though the number of CO adsorption sites decreases, the rate of CO conversion increases. Furthermore, more impressive increases are observed if the comparison is done on the TOF based on each CO adsorption site. Thus at 15% Mo, the overall activity per CO site increased by 150 fold!

Conclusions and Comments

The exponential form of the reaction rate dependence on activation energy and temperature makes rates very sensitive to these variables. As a consequence, differences in activation energies between selective and non-selective reactions can provide for significantly increased selectivities at lower temperatures. Decreasing reaction temperatures from 250° to 200°, for Rh-MoAl2O3 for example, increases selectivity to oxygenates (E_{act} 18.6 Kcal/mole) from 65% to 85% relative to hydrocarbons (E_{act} 31.2 Kcal/mole). Reaction rates are decreased 7-fold. The selectivity is predicted to increase to 95% at 160°. Changes in the distribution of individual hydrocarbons and oxygenates with reaction temperature are also predicted. Such considerations provide a preliminary basis for process optimization through temperature selection.

The greatly enhanced activity and selectivity imparted by Mo addition to Rh/Al2O3 is not explained by activation energy differences alone. Gilhooey, Jackson and Rigby (9) found wide variations in the apparent activation energies and pre-exponential factors for Rh on various supports. They concluded that the compensation effect, which involves the pre-exponential factor, made conclusions on mechanism ambiguous.

Table III. Power Law Coefficients for CO Hydrogenation

$$Rate_{species} = A \cdot p_{H_2}^x \cdot p_{CO}^y$$

Catalyst	Species	X	Y
3%Rh/Al$_2$O$_3$	- CO	0.8	- 0.3
3%RH-15%Mo/Al$_2$O$_3$	- CO	0.72	-0.03±0.09
	+ CH$_4$	1.02	- 0.32±0.09
	+ CH$_3$OH	1.53	- 0.01±0.11
	+ C$_{2+}$OXY.	0.91	- 0.47±0.23
	+ CO$_2$	0.38	- 0.04±0.06

Table IV. CO Chemisorption and Site Reactivity (TOF)
as Function of Mo in 3%Rh,x%Mo/Al$_2$O$_3$

Wt% Mo	CO Chemisorption m moles/g	% Dispersion of Rh	#CO Reacted/site-sec.x 1000**	
			per atom Rh	per site Rh***
0	112	39*	0.4	1
2.8	74	26	4.0	15
7.5	46	16	8.0	50
15.0	28	10	15.0	150

*   Determined by H$_2$ chemisorption.
**   CO Hydrogenation at 225°, H$_2$/CO = 2, 3 mPa.
*** Per CO site.

Examination of the power-law exponents presented here show that the rate of hydrogenation of CO to hydrocarbons and oxygenate is inhibited by CO over Rh/Al_2O_3 but not for methanol formation over $Rh-Mo/Al_2O_3$. Interestingly, the inhibition for CH_4 and higher alcohols formation remains. The implication is that the mechanism of the rate determining step for methanol differs from methane and that the latter is dependent on the Rh.

Based on these results and other characterization tests (10 - 12), it is proposed that $Rh-Mo/Al_2O_3$ catalysts operates by a dual site mechanism in which CO is activated by Rh and hydrogen is activated by $MoO_{(3-x)}$ with migration of activated hydrogen to the activated CO. A major point is that while Rh is capable of activating H_2, its activation is inhibited by CO during CO hydrogenation. In contrast, H_2 activation by MoO_{3-x} is not inhibited by CO. As a consequence of increasing the hydrogenation capability, which is rate-limiting, the overall catalytic activity for CO conversion is greatly accelerated. The increase in oxygenates is also due to increased hydrogenation ability which shows up particularly in methanol formation. The formation of hydrocarbons and higher alcohols involve CO dissociation believed to occur on Rh. As the power-law data show, their formation even over $Rh-Mo/Al_2O_3$ is inhibited by CO which is visualized as strongly occupying the Rh sites.

It should also be noted that the formation of CO_2 (by the shift reaction) is accelerated by $Rh-Mo/Al_2O_3$ but not by Rh/Al_2O_3. The presence of CO_2 could provide an explanation for increased activity of $Rh-Mo/Al_2O_3$ if methanol formation over Rh catalysts is via CO_2 rather than CO. Methanol formation is via CO_2 over $Cu/ZnO/Al_2O_3$ catalysts under usual industrial conditions (14).

Also, the high activity for the shift reaction for molybdena-containing catalysts allows the use of less expensive hydrogen-lean synthesis gas. The enhanced shift reaction also simplifies product separation since water is minimized.

For the practical purpose of achieving higher selectivities at lower temperatures, say 160°, catalysts of increased activity are required. The results discussed here are believed to provide a guide for design of improved dual-site catalysts. The search should be for a structure which provides a H_2 activation site not inhibited by CO. It is speculated that to fulfill this role requires a non-metallic, non-stoichiometric structure such as a partially reduced oxide. For Mo catalysts there is the potential for improvements by use of unusual oxide structures, or of reduction to a MoO_{3-x} of more optimum level, or possibly by use of sulfides instead of oxides. A better knowledge of the interface of Rh and partially reduced molybdena is of great interest as well as the mobility of activated hydrogen to or from the Rh (4,8). The extremely high increase in activity, namely 150-fold, of Rh sites identified by CO chemisorption on the 15% $Mo/Rh/Al_2O_3$ catalyst, illustrates the possibility of catalysts with greatly increased activity. These could be used with great advantage at lower temperatures than present industrial catalyst.

Acknowledgments

Early experimental and conceptual contributions by Dr. C. Sudhakar are recognized. This work was supported by the Department of Energy Grant DE-FG22-84PC70780.

References

1. Ecklund, E.E., Mills, G.A.. CHEMTECH, 19 (9) 549-556, (10) 626-631 (1989).
2. Wilson, T.P., Kasai, P.M., Ellgen, P.C. *J. Catal.* **69**, 193-201 (1981).
3. van den Berg, F.G.A., Glezer, J.H.E., Sachtler, W.M.H. *J. Catal.* **93**, 340-352 (1985).
4. Underwood, R.P., Bell,A.T. *Applied Catal.* **21**, 157 - 168 (1986).
5. Arakawa, H., Hamaoka, T., Takeuchi, K., Matsuzaki, T., Sugi, Y. in *Catalysis: Theory to Practice*, Phillips, M. J., Tiernan, M., Eds. Chem. Inst. Canada, Ottawa, v 2, 602-609 (1988).
6. Wender, I. and Klier, K. Chapter 5 in *Coal Liquefaction — A Research & Development Needs Assessment.* DOE Contract DE-AC01-87ER30110 (1989).

7. Jackson, S.D., Braneth, B. J., Winstanley, D. *Applied Catal.* **27**, 325-333 (1986).
8. Kip, J. B., Hermans, E.G.F., Van Wolput, J.M.H.C., Hermans, N.M.A., van Grondelle, J., Prins, R. *Applied Catal.* **35**, 109-139 (1987).
9. Gilhooey, K., Jackson, S.D., Rigby, S. *Applied Catal.* **21**, 349 - 357 (1986).
10. Sudhakar, C., Bhore, N.A., Bischoff, K.B., Manogue, W.H., Mills, G.A. in *Catalysis 1987*, Ward J.W., Ed., Elsevier Sc. Pub. Amsterdam, 115-124 (1988).
11. Bhore, N.A., Sudhakar, C., Bischoff, K.B., Manogue, W.H., Mills, G.A. in *Catalysis: Theory to Practice*, Phillips, M. J., Tiernan, M., Eds., Chem. Inst. Canada, Ottawa v. 2, 594-601 (1988) v5 (1989).
12. Bhore, N.A. *Modifiers in Rhodium Catalysts for Carbon Monoxide Hydrogenation: Structure – Activity Relationships.* Ph.D. Thesis, University of Delaware, 1989.
13. Hall, W. K.., Proceedings of the Climax Fourth International Conference on the Chemistry and Uses of Molybdenum, Barry and Mitchell, Eds., Climax Molybdenum Co., Ann Arbor, MI, p. 224 - 233 (1982).
14. Chinchin, G. C., Denny, F. J., Parker, D. G., Short, G. D., Spencer, M.S., Waugh, K. and Whan, D. A., Amer. Chem. Soc. Div. Fuel Chem. Preprints, **29**, (5), 178 - 188 (1984).

RECEIVED May 9, 1990

Chapter 24

Photochemically Driven Biomimetic Oxidation of Alkanes and Olefins

J. A. Shelnutt and D. E. Trudell

Fuel Science Division 6211, Sandia National Laboratories, Albuquerque, NM 87185

A photochemical reaction for oxidation of hydrocarbons that uses molecular oxygen as the oxidant is described. A reductive photoredox cycle that uses a tin(IV)- or antimony(V)-porphyrin photosensitizer generates the co-reductant equivalents required to activate oxygen. This "artificial" photosynthesis system drives a second catalytic cycle, mimicking the cytochrome P_{450} reaction, which oxidizes hydrocarbons. An iron(III)- or manganese(III)-porphyrin is used as the hydrocarbon-oxidation catalyst. Methylviologen can be used as a redox relay molecule to provide for electron-transfer from the reduced photosensitizer to the Fe or Mn porphyrin, but appears not to enhance efficiency of the process. The system is long-lived and may be used in time-resolved spectroscopic studies of the photo-initiated reaction to determine reaction rates and intermediates.

Many alkane and olefin oxidation systems that mimic biological oxidation of hydrocarbons by cytochrome P_{450} have been reported. Most use an iron, manganese, or ruthenium porphyrin as the analog of the heme (iron porphyrin) functional group of the enzyme.[1-8] In the great majority of these chemistries a single oxygen atom donor, such as iodosylbenzene or hypochlorite, is used as the oxidant rather than molecular oxygen.[1-4] When molecular oxygen is used as the oxidant, as is the case for cytochrome P_{450}, reducing equivalents must be supplied to reduce the Fe porphyrin causing it to bind and split dioxygen and, subsequently, oxidize the alkane substrate. Several biomimetic systems have been demonstrated using either sodium borohydride, hydrogen/Pt, ascorbate, or zinc metal as the co-reductant.[5-8]

We have been investigating these reactions from the standpoint of stereochemically controlling the reaction at the metal site by designing metalloporphyrins with a shape- and size-selective pocket at the metal center.[9,10] The pockets designed so far are small, and thus

0097–6156/90/0437–0265$06.00/0

require an oxidant, like O_2, that is small enough to enter the cavity. It is also desirable that the system be stable and operate over many hours. We are interested also in the possibility of photo-initiating the reaction so that reaction intermediates can be followed using time-resolved spectroscopic techniques for kinetic studies. This can be accomplished if the reductant is the product of a photoredox cycle.

Here we describe such a photochemically driven system for oxidation of alkanes and olefins. The system is illustrated in Figure 1. The cycle on the left is the photoredox chemistry that produces the reductant, a long-lived tin(IV)-porphyrin radical anion. In the cycle, a tin(IV)- or antimony(V)-porphyrin absorbs a photon of visible light resulting in the formation of the triplet excited state of the porphyrin. The porphyrin photosensitizer in its excited state is reduced by a sacrificial electron donor such as triethanolamine (TEOA).[11-13] The resulting long-lived π anion of the porphyrin has a redox potential low enough to reduce either a Fe(III) or Mn(III) porphyrin, which acts as a catalyst for the biomimetic oxidation of hydrocarbons.[13,14] After reduction of the FeP, the photosensitizer anion (SnP$^-\cdot$) returns to the resting redox state (SnP). Actually, two molecules of the porphyrin anion are required in the biomimetic P_{450} cycle as indicated in Figure 1; the first to reduce Fe(II) to Fe(III), allowing O_2 to bind and a second to split dioxygen to form the reactive O=FeP intermediate. In some cases a molecule such as heptylviologen (HV^{2+}) is used to relay the electron from the SnP anion to the P_{450} cycle. Acetic or benzoic anhydride is sometimes used as an oxygen atom acceptor (replacing H^+) in the splitting of dioxygen in the hydrocarbon (RH) oxidation cycle.

Experimental

A photochemical reaction like that illustrated in Figure 1 was carried out in acetonitrile under an O_2 or air atmosphere. In a typical reaction, 0.24 μmol of Fe(III) tetra(pentafluoro-phenyl)porphyrin (FeTF$_5$PP) chloride, 0.45 μmol of Sn(IV) protoporphyrin IX (SnProtoP) dichloride, 1.1 mmol of TEOA, 1.4 μmol of heptylviologen (N,N'-diheptyl-4,4'-dipyridinium dichloride), and 11 μmol of benzoic (or acetic) anhydride, were added to 1 ml of acetonitrile. Hexane (4.7 mmol) was added as a substrate. The samples, contained in a 1-cm path length cuvette, were irradiated with a tungsten lamp for 1-6 h. Light of wavelengths less than 380 nm was filtered to insure that photosensitization of the reaction was only due to visible light absorption by the porphyrin. 1-, 2-, and 3-Hexanol and 2- and 3-hexanone products were quantified at the end of the run by gas chromatography.

For the dark reactions using Zn amalgam as reductant, the reaction is run for 2 h, but is complete in about 10 min in most cases. Although the yields in some cases represent less than one catalyst turnover, the reaction can be continued by adding more amalgam. In some cases methylviologen is used as a relay molecule, and acetic anhydride is used as an oxygen atom acceptor. The product yield is sensitive to the amount of water in the acetonitrile solvent, and, in addition, acetic acid improved the overall yield and raised the alcohol/ketone product ratio. Presumably, acetic acid aids is the dioxygen lysis step in the reaction.

Results and Discussion

Table 1 gives yields and hexanol-to-hexanone product ratios for typical runs and control experiments. In the presence of the P_{450} catalyst, a generally higher overall yield of products is observed when illumination and other conditions were identical; however, a lower average hexanol to hexanone ratio of 1.3 is observed. However, in the absence of the catalyst FeTF$_5$PP, photosensitized production of hexanols and hexanones is observed in an average ratio of 2.3. In the absence of O_2, light, photosensitizer, or triethanolamine, there is no significant yield of oxidized hexane.

TABLE 1. Oxidation of hexane by air in acetonitrile

Catalyst	Reductant System	-ol/-one	Yield[a] (turnovers/h)
FeTF$_5$PP	SnProtoP, hν, TEOA	1.3[b]	4.3
FeTF$_5$PP	SbProtoP, hν, TEOA	1.0	0.8
FeTF$_5$PP	hν, TEOA	0.8	0.2
MnTPP	SnProtoP, hν, TEOA	0.9	0.2
FeTF$_5$PP	Zn/Hg	0.2-1.1	0.5-1.0[c]
-	SnProtoP, hν, TEOA	2.3[b]	1.7
-	H$_2$ProtoP, hν, TEOA	2.7	0.6
-	SbProtoP, hν, TEOA	2.2	1.4[d]
-	hν, TEOA	-	0.0
-	SnProtoP, hν	-	0.0

a.   Yield is in photosensitizer turnovers (mol product/mol photosensitizer) for selected runs. Catalyst concentrations are about one-half of the photosensitizer concentration.
b.   Hexanol/hexanone value is average for all (~20) runs with turnovers > 1.
c.   Total turnovers under various solvent conditions with O_2 as the oxidant.
d.   Light intensity higher than for SbProtoP/FeTF$_5$PP run, accounting for higher yield in this case.

It is apparent that more than one oxygen activation pathway exists. The excited triplet state of tin porphyrins is known to be quenched in the presence of O_2,[14] suggesting a possible direct mechanism of O_2 activation by the photosensitizer. We have examined reactions of both singlet O_2 and superoxide anion under our reaction conditions. Chemically-produced superoxide (KO$_2$/18-crown-6) is not reactive under our experimental conditions. On the other hand, singlet oxygen, produced by irradiation of free base protoporphyrin (H$_2$ProtoP),[15] is reactive in the presence of tertiary amines and gives about the same hexanol to hexanone ratio (2.7, see Table 1) as is observed in the presence of the SnP photosensitizer.

For the singlet-oxygen photochemical reaction or for the complete reaction mixture when illumination conditions and O_2 pressures cause the singlet-oxygen mediated reaction to dominate, alkane oxidation continues for more than 6 h as shown in Figure 2.

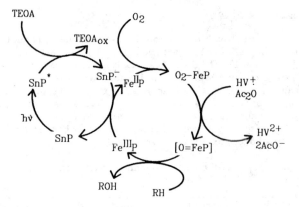

Figure 1. Scheme for photochemical production of co-reductant to drive the oxidation of hydrocarbons by mimicking the cytochrome-P_{450} cycle. The SnP sensitized photoredox cycle is on the left; the P_{450} catalytic cycle is shown on the right.

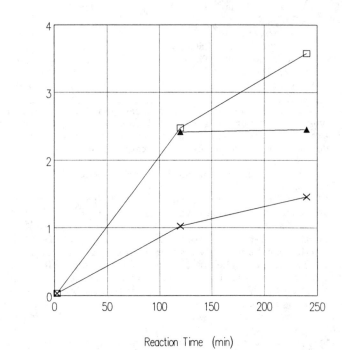

Reaction Time (min)

Figure 2. Oxidation of hexane by singlet oxygen produced photochemically using a SnP sensitizer (PM) in the presence of a tertiary amine. The FeP catalyst and HV are not included in this reaction. Total 1-, 2-, and 3-hexanol yield in photosensitizer turnovers (☐); total 2- and 3-hexanone yield (✗); and hexanol-to-hexanone ratio (▲). PM turnovers = mol product/mol PM.

Sn-, Sb-, and H_2ProtoP all have triplet lifetimes of 10 ms or longer, and form singlet O_2 by intermolecular triplet-triplet annihilation. In fact, the photophysical parameters and singlet oxygen sensitizing properties of SnProtoP[14,16] are similar to metal-free porphyrins.[17] The similarity of photosensitizing characteristics of Sn-, Sb-, and H_2-porphyrins explains the similarity of their properties in the FeP-free reaction (Table 1). However, only the Sn and Sb porphyrins form the stable anions capable of driving the Fe-porphyrin catalyzed reaction.

In the presence of iron or manganese porphyrin, the alcohol/ketone product ratio is modified (-ol/-one ≈ 1) indicating that a competing reaction comes into play. If the FeP-catalyzed reaction is to account for the low product ratio, then this reaction necessarily must give a lower hexanol to hexanone ratio. We can test this hypothesis by determining the product ratio for the P_{450} cycle when driven by addition of a suitable reductant in the absence of light. Table 1 includes the yield and product ratio for the dark reaction of hexane and O_2 catalyzed by FeTF$_5$PP using a Zn/Hg amalgam as the co-reductant.[18] The ranges of yields and product ratios are for a variety of solution conditions. The FeP or MnP catalyst is required for significant yields of oxidized hexane.

Most importantly, when the FeP catalyst is present in the dark reaction the product ratio is one or less. Therefore, the dark reaction appears to compete favorably with the formation of singlet O_2 and the photochemical reaction proceeds as shown in Figure 1. The dark reaction, which also occurs in the the presence of light, results in the observed lowering in the alcohol/ketone ratio and higher yield measured for the light-driven reaction in the presence of the FeP catalyst. Also, viologen appears not to aid the light reaction, since the yield generally remains unchanged or is slightly lowered in its presence (data not shown).

When cyclehexene is the substrate in the dark reaction, the products cyclohexene oxide (20%), 2-cyclohexen-1-ol (44%), and 2-cyclohexen-1-one (36%) are observed in the ratios (relative product yields are given in parenthesis) observed in other dioxygen-based systems that mimic the cytochrome P_{450} reaction.[20,21] For example, Masui *et al.*[20] measured corresponding yields of 27%, 18%, and 54% in their biomimetic system (with styrene present). Single oxygen-donor systems give much higher proportions of cyclohexene oxide and cyclohexen-1-ol than of cyclohexen-1-one (65%, 31%, 4%, respectively).[22] On the other hand, autoxidation by metalloporphyrins generally yields much smaller epoxide yields (3%, 24%, 73%, respectively).[19,20] These results suggest that the biomimetic cytochrome P_{450} cycle contributes significantly; however, alkane oxidation probably does not occur entirely at the metalloporphyrin, but also through a free radical mechanism as in autoxidation.

When Mn tetraphenyl porphyrin (MnTPP) is used as the catalyst, imidazole binding as a fifth ligand acts as a promoter for P_{450} reaction as has been noted in earlier studies.[22,23] Interestingly, the hexanol-to-hexanone ratio increases (0.8 → 1.1) upon replacement of Cl$^-$ with imidazole. Such effects of axial ligand replacement have been noted previously for MnTPP.[22] Both of these results support the contention that the reaction involves the Fe- or

Mn-porphyrin catalyst under these solution conditions, and partially occurs through the biomimetic intermediate.

The photochemically-driven P_{450}-like chemistry produces stable yields of hydrocarbon oxidation products for more than 2 h as shown in Figure 3. In Figure 3, the yield of products is given in terms of catalyst turnovers (mol product/mol FeP catalyst) as a function of reaction time. Also plotted in Figure 3 is the average yield per hour, which degrades over the 4 h reaction time. The alcohol/ketone ratio is not a function of the reaction time; the ratio is slightly above unity, indicating that the FeP-catalyzed reaction pathway dominates rather than the singlet oxygen reaction. The decrease in alkane oxidation rate could be explained by degradation of the FeP-catalyzed cycle.

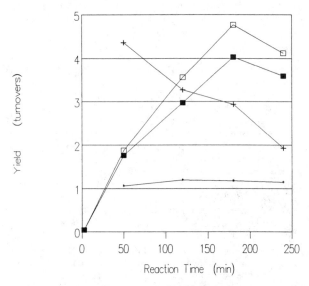

Figure 3. Hexanol and hexanone yields and product ratio as a funtion of reaction time for the reaction illustrated in Figure 1. Total 1-, 2-, and 3-hexanol yield in catalyst turnovers (▢); total 2- and 3-hexanone yield (■); hexanol/hexanone ratio (*); catalyst turnovers/h (+).

Conclusions

A photochemically driven reaction that mimics biological photosynthesis, electron-transfer, and hydrocarbon-oxidation reactions is described. The reaction occurs at room temperature and uses O_2 as the ultimate oxidant. Most importantly, the reaction can be run for hours without significant degradation. This means that the oxidation of low molecular weight alkanes by O_2, which proceeds at a lower rate than for hexane, can be investigated. Further studies are underway to determine the detailed reaction mechanisms involved in the photochemical reaction and the relative contributions of various oxidative pathways. Transient absorption and Raman spectrocopic techniques will also be applied to determine reaction rates.

Acknowledgments

This work performed at Sandia National Laboratories and supported by the United States Department of Energy Contract DE-AC04-76DP00789.

Literature Cited

1. For recent reviews see: (a) Ortiz de Montellano, P. R., Ed. "Cytochrome P-450, Structure, Mechanism, and Biochemistry" (Plenum: New York) 1986. (b) Guengerich, F. P.; Macdonald, T. L., Acc. Chem. Res. 1984, 17, 9.
2. Groves, J. T.; Nemo, T. E.; Myers, R. S., J. Am. Chem. Soc. 1979, 101, 1032.
3. Groves, J. T.; Quinn, R., J. Am. Chem. Soc. 1985, 107, 4343.
4. Colman, J. P.; Brauman, J. I.; Meunier, T.; Hayashi, T.; Kodadek, T.; Raybuck, S. A., J. Am. Chem. Soc. 1985, 107, 2000.
5. Tabushi, I.; Koga, N., J. Am. Chem. Soc. 1979, 101, 6456.
6. Perree-Fauvet, M.; Gaudemer, A., J. Chem. Soc. Chem. Commun. 1981, 874.
7. Tabushi, I.; Yazaki, A., J. Am. Chem. Soc. 1981, 103, 7371.
8. Fontecave, M.; Mansuy, D., Tetrahedron 1984, 40, 2847.
9. Shelnutt, J. A., Stohl, F. V.; Granoff, B., Preprints, Fuel Chem. Div., ACS, Los Angeles, CA, September 25-30, 1988.
10. Quintana, C. A.; Assink, R. A.; Shelnutt, J. A., Inorg. Chem. 1989, 28, 3421.
11. Shelnutt, J. A., J. Am. Chem. Soc. 1983, 105, 7179.
12. Shelnutt, J. A., J. Phys. Chem. 1984, 88, 6121.
13. Shelnutt, J. A., U. S. Patent 4,568,435, 1986.
14. Kalyanasundaram, K.; Shelnutt, J. A.; Grätzel, M., Inorg. Chem. 1988, 27, 2820.
15. (a) Cannistraro, S.; Jori, G.; Van de Vorst, A., Photobiochem. Photobiophys. 1982, 3, 353. (b) Cox, G. S.; Whitten, D. G., Adv. Exp. Med. Biol. 1983, 160, 279.
16. Land, E. J.; McDonagh, A. F.; McGarvey, D. J.; Truscott, T. G., Proc. Natl. Acad. Sci. USA 1988, 85, 5249.
17. Gouterman, M., in "The Porphyrins" Vol. 3, Dolphin, D., ed. (Academic: New York) Chpt. 1, 1978.
18. Karasevich, E. I.; Khenkin, A. M.; Shilov, E., J. Chem. Soc., Chem. Commun. 1987, 731.
19. Paulson, D. R.; Ullman, R.; Soane, R. B.; Closs, G. L., J. Chem. Soc., Chem. Commun. 1974, 186.
20. Masui, M.; Tsuchida, K.; Kimata, Y.; Ozaki, S., Chem. Pharm. Bull. 1987, 35, 3078.
21. Leduc, P.; Battioni, P.; Bartoli, J. F.; Mansuy, D., Tetrahedron Lett. 1988, 29, 205.
22. Nappa, M. J.; McKinney, R. J., Inorg. Chem. 1988, 27, 3740.
23. Battioni, P.; Renaud, J-P.; Bartoli, J. F.; Mansuy, D., J. Chem. Soc., Chem. Commun. 1986, 341.

RECEIVED May 9, 1990

Chapter 25

Oligomerization of Isobutene with an Improved Catalyst

S. Börje Gevert, Peter Abrahamsson, and Sven G. Järås

Department of Engineering Chemistry I, Chalmers University of Technology, Göteborg, S–412 96, Sweden

Oligomerization of iso-butene was used as a model reaction for testing of improved catalysts based on phosphoric acid on silica. The catalysts were produced by impregnation of silica spheres (2-3 mm diameter) with ortho-phosphoric acid, drying and calcining under various conditions.
The silica spheres used had a high pore volume (1-1.3 cm^3/g) and large pore diameter (300-600 Å). One catalyst was made by impregnation, calcining at 350°C, re-impregnation and drying at 170°C. This catalyst showed a conversion of 98% by weight of iso-butene in a continuous tube-reactor at 160°C, atmospheric pressure and a weight hourly velocity of 0.543 gram/cm^3. The degree of conversion is effected by the calcining temperature after the first impregnation wich changes the water content and the bindings between the phosphoric acid and the silica.

In conversion processes in oil refining, LPG (C_3 and C_4) are formed as byproducts. When catalytic cracking is used the LPG will have a high content of unsaturated components. The low price of the LPG, compared to gasoline, makes oligomerization of the molecules in LPG worthwhile for the industry. There are mainly two processes used in the refineries for this purpose. In alkylation, liquid sulphuric or hydrofluoric acids are employed and in the other process phosphoric acid on silica or diatomeric earth. In the refining industry the latter process is normally called polymerization, although dimerization is the dominating and most wanted reaction. In the alkylation process iso-butene is reacted with either propane or butane. If using the less strong phosphoric acid, two alkenes must be used as reactants and at a higher temperature. The mounted phosphoric acid catalyst leaks less acid to down-stream processing and therefore gives less severe corrosion problems than with alkylation (1). The chemistry of the oligomerization can be shown as:

$$C = C - C \ + \ H^+ \longrightarrow C - C^+ - C$$

$$C - C^+ - C \ + \ C = C - C \longrightarrow \ \substack{C \\ C} C - C \substack{C \\ C}$$

$$\substack{C \\ C} C - C \substack{C \\ C} \ \xrightarrow{-H^+} \longrightarrow C_6H_{12} - \text{isomers}$$

0097–6156/90/0437–0272$06.00/0

The formed carbonium ions and hexenes can react further to trimers and higher.

The phosphoric acid on silica catalyst is also an important catalyst for the production of petrochemical intermediates like nonene and alkylated aromates.

Light alkenes can be produced by dehydrogenation of corresponding alkanes to increase the basis for gasoline production in a refinary (2). To produce an environmental better excepted gasoline, the high alkene content in the oligomerization product can be reduced by hydrogenation. This will however slightly reduce the octane number of the product.

The traditional method of producing the mounted phosphoric acid on silica-catalyst is to start from kiesel guhr and phosphoric acid. The water content of the phosphoric acid should be low so that the final catalyst contains 65-70% by weight of H_3PO_4. The mixture is extruded or bricketted, dried, calcined and finally hydrated to a preferred level, usually with steam. There are several proposals in the patent literature to improve the mechanical strength of the final product. One method is to steam treat the mixture after calcning (3). It is also possible to add bentonite, montmorillonite, halloysite, or other compounds to the mixture before extrusion. The acidity of the catalyst can also be reached by using an aluminum silicate or zeolite instead of the above mentioned acids (4,5). The kiesel guhr carrier can be replaced by a silica hydrogel (6-12). Recently Bernard et al. (13) proposed to extrudate the silica carrier before impregnating with the phosphoric acid. Little interest has been shown in the development of new or improved catalysts for oligomerization of propene and butene in the recent literature. It is therefore of interest to use modern analytical instruments to explain and improve the catalyst. In this paper, an improved catalyst based on mounted phosphoric acid on silica will be presented.

EXPERIMENTAL

The used silica carriers were spheres with a diameter of 2.2 mm, recieved from Shell in London. In Table I, more details of pore volumes and pore sizes are presented. The phosphoric acid used was of pro analysi quality from Kebo Lab (Sweden) and the iso-butene from AGA Gas Inc. (Sweden).

Table I. Silica spheres. Pore volume and average pore size according to manufactures, surface area BET, and bulk density

ID	Pore volume cm^3/g	Average pore size Å	Surface area m^2/g	Bulk density g/cm^3
I	1.0	300	79	0.43
II	1.0	600	62	0.40
III	1.3	600	60	0.36

The catalysts were made by combining the following steps in different sequences:

* WI - Wet impregnation with ortho-phosphoric acid of 42% by weight.
* D - Drying at 170°C for 5 hours in a muffle furnace.
* C - Calcining in a muffle furnace at various temperatures.
* DI - Dry impregnation with phosphoric acid of 42% by weight.
* S - Steaming in 100% steam at 100°C for 5 hours.

Testing of the activity of the catalysts was done in a 23 cm^3 stainless steel reactor (15 mm inner diameter) with iso-butene in vapour phase as feed. In Figure 1, a simple flow diagram of the reactor system is presented. Temperature, pressure and gas flow were kept constant at 160°C, 1 atm and 12.5 gram/h of iso-butene respectively. The yields were determined by measuring the weight of liquid products and the amount of iso-butene feed. The liquid products were analyzed on GC with FID detector (Hewlet Packard 5880A) and the simulated distillation was made according to ASTM D2887-3. BET surface area was measured with Micromeritics Digisorb 2600 and pore distribution was measured with mercury intrusion. Crushing strength was measured on the spheres by crushing 15 samples. The short and total acid were determined by grinding, leaching with water, and titrating with caustic at room temperature and at the boiling point, respectively. Brom cresol green (first true end point) was used as indicator.

RESULTS AND DISCUSSION

The pore volume of the silica sphere III is higher than pore volume of the other spheres (Table I). This is also reflected in the larger average pore size distribution (Fig. 2).

A catalyst produced from silica spheres III with the sequence: WI, D, C, (350°C) DI, D reduced the surface area from 60 m^2/g to 5 m^2/g. The phosphoric acid had filled most part of the pores. Crushing strenght was maintained at 1.6 kg before and after preparation and after testing.

In Figure 3, the conversion as a function of time is presented for three different catalysts produced from the three different silica spheres by the same procedure. Calcination was done at 425°C. The catalyst produced from silica spheres III, with both larger pores and pore volume compared with silica spheres II and I, gives the highest activity. The conversion was 78% after 50 h of experiment. The catalyst produced from silica sphere II, gives a better conversion than the catalyst produced from silica sphere I. Thus the silica carrier should have a large average pore size and a large pore volume. It is important to use a high concentration of the acid in order to reach a H$_3$PO$_4$ concentration of 65-70% in the final catalyst. On the other hand, too high viscosity, which is found in concentrated phosphoric acid, reduces the impregnation speed. In Figure 4 the effect on activity of calcining at 350°C and 425°C is presented. The higher temperature gives a lower activity and calcining at 350°C gives a catalyst with a conversion of 97.6%. This catalyst also maintained the conversion level in an experiment of over 70 h. Calcining after impregnation effects the bindings between the phosphoric acid and the silanol groups and also the dehydration of phosphoric acid (Table II). In a separate experiment silica spheres were calcined at 550°C, before

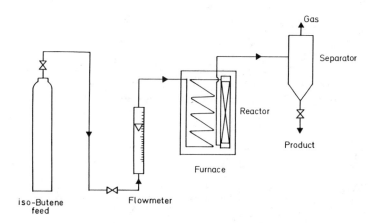

Figure 1. Simple flow diagram of reactor system.

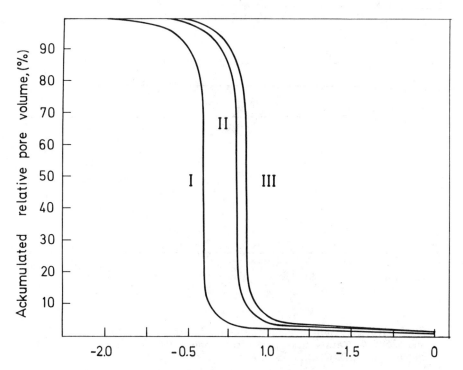

Figure 2. Distribution of pore volume for silica spheres I , II , and III by mercury intrusion.

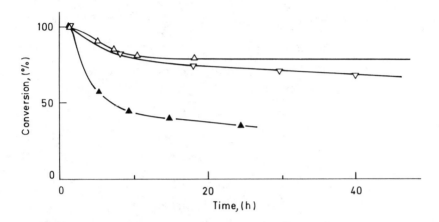

Figure 3. Conversion in weight percent of catalysts prepared with silica spheres I▲, II▽, and III△ by the following sequence: WI, D, C, DI, and D. Calcination was done at 425°C.

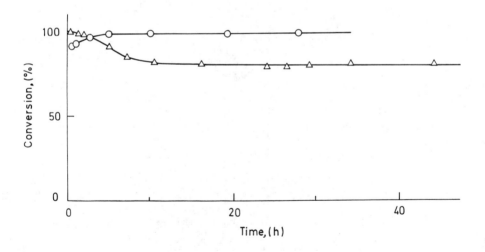

Figure 4. Conversion in weight percent of catalysts calcined at 350°C ○ and 425°C △. Catalysts prepared from sphere III by the following sequence: WI, D, C, DI, and D.

impregnation without affecting the conversion of the final product. As no effect of calcining at 550°C before impregnation was noted, the silanol groups of the silica remained intact.

If only one wet impregnation followed by drying is carried out, the activity of the final catalyst drops quickly after a few hours of the experiment (Table II). When water is added to the feed, or the catalyst is wetted with water, the activity rises again for a short period. This indicates that the phosphoric acid dehydrates rather fast. The high sensitivity of water content in the feed makes the catalyst, which has been impregnated once only, impractical for commercial use.

Table II. Effects of Preparation Procedure on Conversion

Sphere	Preparation	Conversion (% w/w)
III	WI, C, DI, D	80
III	WI, C, DI, D, C, S	78
II	WI, C, DI, D	75
II	WI, C, DI, D, C	16
III	WI, D	10
III	WI, D + waterinject.	100

In Figure 5 the activity of commercial catalysts is plotted as a function of time. The ground catalyst shows a much higher conversion rate (66%) after 20 h of experiment than the unground catalyst (18%). In the small reactor used, the unground catalyst will give wall effects and channelling due to the small ratio of reactor diameter to unground catalyst. The ground catalyst has a lower activity compared to the best of the catalysts produced in this study. Futhermore, agglomeration was observed of the commersial catalyst but not of the catalysts produced from the spheres. In the production of gasoline a dimer or trimer is wanted, while the formation of tetramers is not wanted, since they normally fall outside the gasoline boiling range. The selectivity of some catalysts is presented in Table III.

Table III. Selectivities

Catalyst	Time on stream h	% w/w in liquid di-	tri-	tetra-	Conversion % w/w
Spheres III, calc 425	3	74.9	22.1	3.0	88.6
Spheres III, calc 425	23	84.0	14.9	1.1	79.6
Spheres III, calc 425	47	84.4	14.6	1.0	79.6
Spheres III, calc 350	27	73.3	22.7	4.0	97.6
Commercial unground	49	68.5	27.2	4.3	16.9
Commercial ground	49	76.5	20.7	2.8	65.4

For catalysts with high conversions, the formation of tetramers is also high. The catalyst calcined at 350°C shows an acceptable selectivity (dimer + trimer= 96%) and at the same time high conversion 98%. Short and total acid

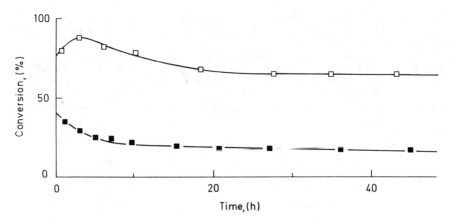

Figure 5. Conversion of commercial catalyst in weight percent.
Unground ■ and ground □.

was determined to 15 and 67 weight percent respectivily of the catalyst. The
unground commercial catalyst shows 4.3% tetramers in the liquid, while the
ground catalyst only shows 2.8%. The higher production of tetramers is
probably an effect of long pore diffusion times of the primary dimer in the
unground catalyst. The yield of trimeres follows the same pattern as the yield
of tetrameres.

CONCLUSIONS

A high activity oligomerization catalyst based on phosphoric acid on silica has
been produced by impregnation of silica spheres. A calcining temperature of
350°C is better than calcining at 425°C or only drying at 170°C. The catalyst
has a low sensitivity to water concentration in the feedstock provided it has
been impregnated twice with calcination between. The activity of the catalyst
produced by impregnation is higher than that of the commercial catalysts used
in this study.

LITERATURE CITED

1.   Eglott, G.; Welnert, P.C., Proc. 3rd World Petroleum Congress, The
       Hague, 1951, Section IV, 201-14.
2.   Guaaon, S.; Spence, D.C.; White E.A. Oil and Gas Journal, 1980, Dec. 8.
3.   Engel, W.F. Dutch Pat. 87999, 1959.
4.   O'Connor, C.T.; Fasol, R.E.; Fouldes, G.A. Fuel Process. Technol, 1986,
       13 (1), 41-51.
5.   Minachev, Kh.M.; Bondarenko, T.N.; Kondratév, D.A.; lzv. Akad. Nank,
       SSSR, Ser. Khim. 1987, 6, 1225-30, Chem Abstr. 108 (15): 130956 a.
6.   Morell, J.C. U.S. Patent 3 044 964, 1962.
7.   Morell, J.C. U.S. Patent 3 050 472, 1962.
8.   Morell, J.C. U.S. Patent 3 050 473, 1962.
9.   Morell, J.C. U.S. Patent 3 170 885, 1965.
10.   Morell, J.C. U.S. Patent 3 213 036, 1965.
11.   Wilshier, B.G.; Smart, P., Western, R., Mole, T., Behrsing, T. Applied
       Catalysis, 1987, 31, 339-459.
12.   Corner, E.S.; Lynch, C.S. US Patent 2 824 149, 1958.
13.   Bernard, J.; Le Page, F. Ger. offen 1 954 326, 1970.

RECEIVED May 9, 1990

Chapter 26

Hydrous Titanium Oxide-Supported Catalysts

Activation for Hydrogenation, Hydrodesulfurization, and Hydrodeoxygenation Reactions

Robert G. Dosch[1], Frances V. Stohl[1], and James T. Richardson[2]

[1]Sandia National Laboratories, P.O. Box 5800, Albuquerque, NM 87185
[2]Department of Chemical Engineering, University of Houston, Houston, TX 77204-4792

Catalysts were prepared on hydrous titanium oxide (HTO) supports by ion exchange of an active metal for Na^+ ions incorporated in the HTO support during preparation by reaction with the parent Ti alkoxide. Strong active metal-HTO interactions as a result of the ion exchange reaction can require significantly different conditions for activation as compared to catalysts prepared by more widely used incipient wetness methods. The latter catalysts typically involve conversion or reduction of an active metal compound dispersed on the support while the HTO catalysts require the alteration of electrostatic bonds between the metal and support with subsequent alteration of the support itself. In this paper, we discuss the activation, via sulfidation or reduction, of catalysts consisting of Co, Mo, or Ni-Mo dispersed on HTO supports by ion exchange. Correlations between the activation process and the hydrogenation, hydrodeoxygenation, and hydrodesulfurization activities of the catalysts are presented.

The objective of this work is to develop new catalysts for use in direct coal liquefaction processing. We are currently developing catalysts that are supported on hydrous titanium oxide (HTO) ion exchange materials (1-3). The ultimate goal of the program is an improved coal liquefaction catalyst produced in quantities that allow testing in a large-scale process development unit such as the Advanced Coal Liquefaction R & D Facility (4) at Wilsonville, Alabama. This paper gives an overview of Sandia's catalyst development program and presents results of studies of HTO supported catalyst preparation and pretreatment procedures.

Catalysts currently employed in process development units for coal liquefaction are hydroprocessing catalysts developed for petroleum refining (5,6). They are composed of combinations of Mo or W with Co, Ni or other promoters dispersed on alumina or silica-alumina supports. When used in liquefaction, these catalysts deactivate rapidly (6-9) causing decreases in product yield and quality and problems with process operability. Thus the

0097–6156/90/0437–0279$06.00/0

existing generation of supported catalysts does not adequately meet the demanding requirements for use in coal liquefaction processes.

Past efforts in developing coal liquefaction catalysts have focused on alumina-supported systems and, except for exploratory studies, little attention has been given to systematic development of novel formulations. A particularly promising approach to the development of new catalysts specifically designed for coal liquefaction processes lies in the formulation of multicomponent systems that, in comparison to work on single or bimetallic systems, are essentially unexplored. Use of multimetallic systems offers the possibility of multifunctional catalysts that are needed to perform the many different reactions encountered in coal processing. Because of its versatility for the preparation of multimetallic catalysts, the HTO system is an excellent candidate for further development.

Hydrous titanium oxide ion exchange compounds exhibit a number of properties ($\underline{2}$) that make them desirable as substrates for active metals: 1) ions of any active metal or mixture of metals can be atomically dispersed over a wide range of concentrations by an easily controlled process; 2) the ion exchange capacity of the materials is large, permitting high loadings of active metals; 3) solution chemistry can be used to provide control of the oxidation state of the active metal; 4) catalyst acidity can be modified by ion exchange; 5) the materials have high surface areas; 6) the ion exchanger substrates are stable in oxidizing and reducing atmospheres, and over a wide pH range in aqueous solution; and 7) the ion exchangers can be prepared as thin films on a wide variety of supports. The latter property offers the potential for tailor-made catalysts with chosen chemical, physical, and mechanical characteristics. The capability of making thin film catalysts could yield a wide range of support acidities and physical properties that might result in less catalyst deactivation. It is the versatility of the HTO system that prompted our investigations of the applicability of HTO catalysts for coal liquefaction. The ion exchange properties combined with high surface areas produce high initial catalyst activities.

HTO catalysts have been evaluated in several fossil fuel applications including direct coal liquefaction ($\underline{3},\underline{10},\underline{11}$), hydropyrolysis ($\underline{12}$), and coprocessing ($\underline{13}$). Initial batch microreactor tests ($\underline{10}$) using equal weights of Shell 324M (a NiMo/ Al_2O_3 catalyst that is commonly used in direct coal liquefaction), Ni HTO, Mo HTO, and Pd HTO catalysts with Illinois #6 coal and SRC-II heavy distillate showed that the HTO catalysts, even at low active metals loadings of 1 wt%, are equally effective for conversion of coal to low molecular weight products as Shell 324M, which contains 15 wt% active metals. In addition, for the same oil yield, the HTO catalysts used less hydrogen than the commercial catalyst. HTO catalysts have also been evaluated at the Pittsburgh Energy Technology Center in bench-scale tests using a feed consisting of a 1:1 mixture of distillate solvent and deashed residuum. The two feed components were obtained from the Wilsonville Advanced Coal Liquefaction R & D Facility. Results ($\underline{11}$) showed that a CoNiMo HTO catalyst gave conversions to cyclohexane solubles, H/ C product ratios, and hydrodesulfurization activities that were similar to those obtained with Shell 324M. The CoNiMo HTO catalyst had not been optimized for hydrotreating coal-derived liquids. Studies ($\underline{12}$) of hydropyrolysis using a coal coated with a Pd HTO catalyst (0.7 wt% Pd on daf coal basis) showed a 50% increase in tar yield compared to a reaction performed with the addition of ground $NiMo/Al_2O_3$ catalyst (3.6 wt% active metals on daf coal basis). HTO catalysts have also been evaluated for use in coprocessing ($\underline{13}$). Results showed that the performances of some Mo, Pd, Ni, Co, and Co-Mo catalysts compared favorably with a commercial Co-Mo catalyst in terms of oil yield, conversion of tetrahydrofuran insolubles, and gas formation. These results all indicate that HTO catalysts have potential in coal liquefaction processes.

Experimental

HTO Catalyst Preparation. Hydrous titanium oxide ion exchangers are amorphous inorganic compounds synthesized in the form of salts of weak acids represented by the empirical formula $C(Ti_xO_yH_z)_n$, where C is an exchangeable cation. HTO catalysts can be prepared by a technique that consists of synthesis of sodium hydrous titanate ion exchange material followed by exchanging active metal ions for the sodium. The synthesis involves three steps:

(1) Reaction of tetraisopropyl titanate with an alkali or alkaline earth metal hydroxide in alcohol solution to form a soluble intermediate:

$$Ti(OC_3H_7)_4 + NaOH \xrightarrow{\quad CH_3OH \quad} \text{Soluble Intermediate}$$

(2) Hydrolysis of the soluble intermediate in acetone/water mixtures to form the HTO exchange material:

$$\text{Soluble Intermediate} \xrightarrow{\quad Acetone, H_2O \quad} NaTi_2O_5H$$

(3) Ion exchange of the alkali or alkaline earth metal by active metal ions in aqueous solution to form the catalyst:

$$2NaTi_2O_5H + Ni^{+2}(aq) \longrightarrow 2Na^+ + Ni(Ti_2O_5H)_2$$

Steps 2 and 3 contain empirical representations of the x-ray amorphous HTO material. Co, Mo, Pd, and NiMo HTO catalysts are prepared by similar procedures. Co, Ni, and Pd were exchanged using metal nitrates; Mo was added by ammonium paramolybdate. HTO-Si support materials were made by adding tetraethyl orthosilicate to the tetraisopropyl titanate prior to addition of NaOH.

Activity Testing. All as-prepared HTO catalysts, which were subsequently tested for hydrogenation (HYD) and hydrodeoxygenation (HDO) activities, were pressed at 6800 psi into disks measuring 1.125 in diameter and about 0.125 in thickness using a Carver hydraulic laboratory press. The disks were then ground to -30 + 40 mesh size material to simplify subsequent catalyst calcination and sulfidation. Most catalysts were calcined in air by heating at 5°C/min to 450°C followed by a 1°C/min increase to 500°C with 1 h at this temperature. Prior to activity testing, all catalysts were sulfided in an atmospheric pressure flow reactor at 425°C for 2 h using a 10 mol% H_2S/H_2 mixture. Catalysts tested for hydrodesulfurization (HDS) activities were calcined using several different temperatures that ranged from 350 to 600°C.

HYD activities were determined by measuring the rate of hydrogenation of pyrene to dihydropyrene (14). Experiments were performed at 300°C in 26 cm^3 batch microreactors that were loaded with 100 mg pyrene, 1 g of hexadecane as a solvent, and 500 psig H_2 cold charge pressure. Reaction times were 20 min and catalyst loadings, which were varied depending on the activity of the catalyst, ranged from 10 to 25 mg. The concentrations of pyrene and dihydropyrene in the products were determined using gas chromatography (GC). All HYD activity testing was performed with -100 mesh catalysts.

HDO activities were evaluated using the rate of disappearance of dibenzofuran (DBF). Experiments were performed at 350°C in 26 cm^3 batch microreactors that were loaded with 100 mg DBF, 25 mg catalyst, 1 g hexadecane, and 1200 psig H$_2$ cold charge pressure. Reaction times were 15 min and products were analyzed by GC. HDO testing was performed with -200 mesh catalysts.

HDS activities were measured in a flow reactor system using thiophene as the model compound. Reaction rates for the HDS of thiophene to butene were determined at 325°C and atmospheric pressure. HDS testing used -60 mesh catalyst.

Results of these activity tests are reported on a weight of active metals basis. The HYD and HDO results are compared to results obtained with Shell 324M (12.4 wt% Mo, 2.8 wt% Ni on an alumina support).

Results and Discussion

Because the HTOs are new catalytic materials with properties that are significantly different from well-known catalysts such as those supported on alumina, a thorough and systematic approach to their development must be taken. The elements included in this approach are 1) process definition; 2) determination of catalyst requirements; 3) optimization of catalyst preparation and pretreatment procedures; 4) characterization of catalysts; 5) catalyst testing; 6) scaleup of catalyst preparation procedures; and 7) process demonstration unit evaluation. The relationship among the program elements is shown in Figure 1. The aspects of this program that are discussed in this paper involve studies of the catalyst preparation and pretreatment procedures that are aimed at enhancing catalyst activity for direct coal liquefaction.

Effects of Catalyst Composition on Activity. Seven NiMo HTO catalysts were prepared at acidification pHs of 4. This series of catalysts (Table I) was

Table I. Composition of NiMo catalysts

Catalyst	Ti:Si[a]	Pretreatment	Wt % Active Metals Mo	Ni	Na (Wt %)
1	No Si	Sulfide	12.31	3.23	0.66
2	No Si	Calcine+sulfide	12.31	3.23	0.66
3	3	Sulfide	9.69	3.03	0.97
4	3	Calcine+sulfide	9.69	3.03	0.97
5	5	Calcine+sulfide	9.62	2.67	1.08
6	5	Calcine+sulfide	9.77	4.08	0.10
7	5	Calcine+sulfide	4.74	0.81	0.19
Shell 324M		Sulfide	12.40	2.80	

[a] Mole ratio

analyzed to determine the effects on catalyst activity of calcination, Si addition to the HTO support, decreased Na content in the catalyst, and lower active metals concentration. The HYD activity results for these catalysts are compared to Shell 324M in Figure 2. Analysis of the HYD activities of the first two pairs of HTO catalysts (catalysts 1-4) shows that the calcined catalyst in each pair has significantly greater activity than the noncalcined catalyst indicating that calcining these materials is important in catalyst activation. Of the calcined and sulfided HTO catalysts, catalyst 2 without Si had the lowest HYD activity. The results of this study show that Si addition causes increased activity, but that too much Si (Ti:Si= 3) causes a decrease in activity. The increased activity observed with Si addition is in agreement with previous (unpublished) results that indicated Si could be added to HTOs via its alkoxide, and the presence of Si tended to reduce the degree of sintering thus yielding higher surface areas (see below). The first five catalysts in Table I contained small amounts of Na (0.6 to 1.1 wt %), whereas the catalyst used in run 6 had < 0.1 wt%. Comparison of activity results from runs 5 and 6 suggests that decreasing the Na content increases the catalyst HYD activity; catalyst 6 has 90% of the HYD activity of Shell 324M. These results are in agreement with studies that showed Na poisoned catalyst sites on Mo HTO catalysts. Additional studies will be performed to determine more definitively if Na is detrimental to the activity of NiMo HTOs. The catalyst used in run 7 had a HYD activity that was 73% greater than that of Shell 324M. The only difference between this catalyst and that used in run 6 was the lower active metals content. We don't know why a decrease in active metals would yield an enhancement in catalyst HYD activity; additional studies are being performed to determine the cause of this effect.

The results of the HDO activity tests for catalysts 1-6 are plotted against the HYD activities (from Figure 2) in Figure 3. There is a direct correlation (correlation coefficient = 0.98) between HYD and HDO activities for these HTO catalysts and Shell 324M. In contrast, HTO catalyst 7 has an HDO activity of 0.116 g^{-1} (metal) sec^{-1}, which is the same as Shell 324M, although its HYD activity is 1.77 g^{-1} (metal) sec^{-1}, 73% higher than Shell 324M. It is not known why the relationship between the HYD and HDO rate constants is different for the catalyst with the low active metals content as compared to the higher active metals containing catalysts. Experiments will be conducted to determine the cause of this difference.

Impacts of Catalyst Preparation and Activation on Surface Area. Studies have been performed to determine the effects of catalyst preparation and activation procedures and Si addition to the support on catalyst surface area. This work was performed with Pd HTO catalysts, which also show an increase in activity with Si addition to the support. Figure 4a shows the effects of the Ti:Si ratio in the support and the acidification pH during catalyst preparation on the surface areas of Pd HTO catalysts that have been calcined at 300°C in air. The presence of a 6:1 Ti:Si mole ratio results in a 2- to 3-fold increase in surface area with respect to Pd HTO without Si addition. When the Si content is increased to give a Ti:Si ratio of 2, any discernible enhancement in surface area compared to catalyst without Si is small and is limited to materials formed at the lower pHs. The effect of pH on the surface areas of catalysts prepared on the same supports is more difficult to discern. There appears to be a general trend whereby the surface areas of samples acidified at pH 4 and 5 are slightly enhanced with respect to materials that either were not acidified (pH>6) or were prepared at pH 2.

Increasing the calcination temperature to 500°C (Figure 4b) shows similar trends to those observed at 300°C. As would be expected, the higher temperature resulted in decreases in catalyst surface areas. The results of this

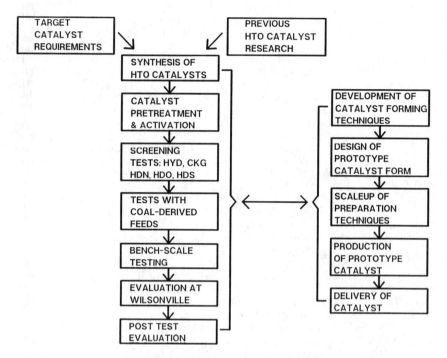

Figure 1. Overview of Sandia's catalyst development program.

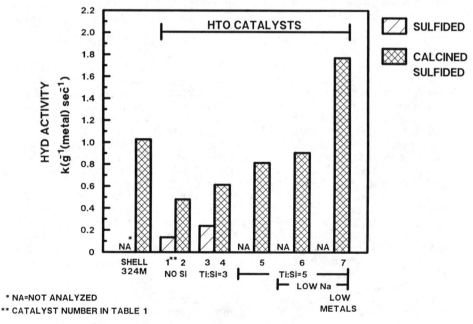

* NA=NOT ANALYZED

** CATALYST NUMBER IN TABLE 1

Figure 2. HYD activity of NiMo catalysts reported on a weight of active metals basis.

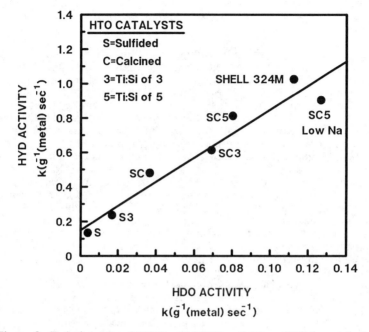

Figure 3. Relationship of HDO and HYD activities of NiMo catalysts.

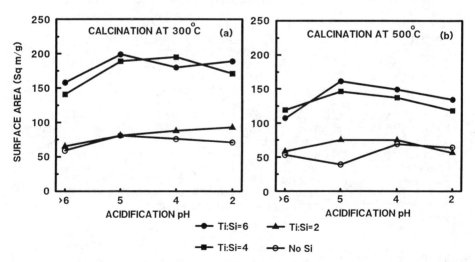

Figure 4. Effects of support composition, calcination temperature, and catalyst preparation conditions on Pd HTO surface areas.

study on Pd HTO catalysts suggest that it may be possible to increase the activity of the NiMo HTO catalysts by decreasing the calcination temperature, optimizing the Ti:Si ratio, and using a somewhat higher pH during acidification.

Evaluation of Precalcination Technique. Optimization of the calcination process for these materials is aimed at maintaining the dispersion of the active metals in the as-prepared material while giving a catalyst that is stable under reaction conditions. The usual procedure for preparing alumina supported catalysts involves calcining after the metals have been impregnated onto the support. This procedure has been used in all previous studies of HTO catalysts. However, a different method, calcination of the support prior to ion exchange, may be possible with these materials. The potential benefit of this procedure, which will subsequently be referred to as precalcining, is that the metal is not present during calcination so that it will not sinter, while the possible disadvantage is that the ion exchange capacity of the calcined material may be too low to enable preparation of a good catalyst. The impacts of precalcining temperature on the ion exchange capacities of HTOs are shown in Figure 5. The results obtained on two sets of catalysts, one exchanged with Co and the other with Mo, show that the exchange capacity decreases with increasing precalcination temperature. For Co, about 2/3 of the HTO's noncalcined exchange capacity is maintained between 400°C and 700°C. In contrast, Mo exchange shows a sharp decrease due to precalcining at 200°C and a more gradual decrease with increasing precalcination temperature. After a 500°C precalcination, less than 1/4 of the initial Mo exchange occurs. These results suggest that formation of a CoMo HTO catalyst using the precalcining technique may require two separate exchange steps: the Mo could be exchanged at room temperature followed by calcination and subsequent exchange with Co. Studies are currently underway to compare activities of catalysts prepared using precalcination and calcination after ion exchange.

Impact of Calcination Temperature on HDS Activity. Studies of HTO calcination after ion exchange are also being performed. Recent work (Figure 6) has shown the effect of calcination temperature on the thiophene HDS activities of two NiMo HTO catalysts. With calcination above 400°C, both catalysts show a significant increase in HDS activity. Differential thermal analyses of these catalysts showed that a thermal event occurred between 400 and 500°C, which may be related to this activity enhancement. The NiMo HTO catalyst with the lower total metals loading maintains this higher activity up to a calcination temperature of at least 600°C. The NiMo HTO with the higher metals loading shows additional activity enhancement up to a calcination temperature of 500°C and then an activity decrease. The reasons for these phenomena are not yet known. Differential scanning calorimetry and thermal gravimetric analysis techniques are currently being performed in conjunction with activity testing to learn more about calcination and activation procedures for these materials.

Conclusions

Hydrous titanium metal oxide catalysts are extremely versatile materials that have promise as direct coal liquefaction catalysts. Previous studies have shown that they perform well in both batch and bench-scale coal liquefaction tests.

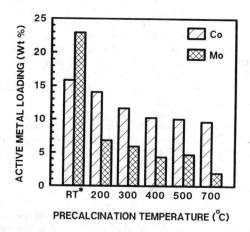

Figure 5. Relationship between HTO precalcination temperature and Co and Mo ion exchange capacities.

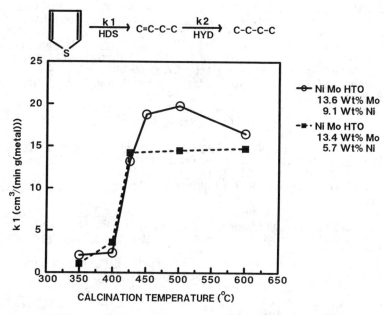

Figure 6. Effects of calcination temperature on thiophene HDS activity for two NiMo HTO catalysts without Si addition.

Studies of preparation and activation procedures for these catalysts have identified several ways for improving catalyst activity: optimizing the calcination temperature, adding Si to the support, decreasing the Na that remains in the catalyst after ion exchange, and using low active metal concentrations. It may also be possible to maintain the high atomic dispersion of the metals at reaction conditions by calcining the titanate prior to ion exchange of the metals onto the support. Current research is addressing these areas. In addition, flow reactor studies are being performed to evaluate how these catalysts perform under different reaction conditions with both model compounds and coal-derived feeds.

Future research will emphasize additional areas that may yield improvements in catalyst performance. These will include varying the catalyst surface area and pore volume distribution, adding additional active metals and promoters, varying the catalyst acidity, and coating the HTO on existing supports.

Acknowledgment

This work was supported by the U. S. Department of Energy at Sandia National Laboratories under Contract DE-AC04-76DP00789 and at the University of Houston by subcontract to Sandia.

Literature Cited

1. Dosch, R. G.; Stephens, H. P.; Stohl, F. V. U.S. Patent 4 511 455, 1985.
2. Dosch, R. G.; Stephens, H. P.; Stohl, F. V.; Bunker, B. C.; Peden, C. H. F. Hydrous Metal Oxide-Supported Catalysts: Part I. A Review of Preparation Chemistry and Physical and Chemical Properties; Sandia National Laboratories Report SAND89-2399, February 1990.
3. Dosch, R. G.; Stephens, H. P.; Stohl, F. V. Hydrous Metal Oxide-Supported Catalysts: Part II. A Review of Catalytic Properties and Applications; Sandia National Laboratories Report SAND89-2400, February 1990.
4. Nalitham, R. V.; Lee, J. M.; Davies, O. L.; Pinkston, T. E.; Jeffers, M. L.; Prasad, A. Fuel Proc. Tech. 1988, 18, 161-83.
5. Stiegel, G. J.; Shah, Y. T.; Krishnamurthy, S.; Panvelker, S. V. In Reaction Engineering in Direct Coal Liquefaction; Shah, Y. T., Ed.; Addison-Wesley: Reading, MA, 1981; Chapter 6.
6. Derbyshire, F. J. Catalysis in Direct Coal Liquefaction: New Directions for Research; IEA Coal Research Report IEACR/08: London, England, 1988.
7. Stohl, F. V.; Stephens, H. P. I&EC Research 1987, 26, 2466-73.
8. Stiegel, G. J.; Tischer, R. E.; Cillo, D. L.; Narain, N. K. Ind. Eng. Chem. Prod. Res. Dev. 1985, 24, 206-13.
9. Moniz, M. J.; Nalitham, R. V. Proc. of the Tenth Annual EPRI Contractors' Conf. on Clean Liquid and Solid Fuels, 1985, p. 19-1.
10. Stephens, H. P.; Dosch, R. G.; Stohl, F. V. Ind. Eng. Chem. Prod. Res. Dev. 1985, 24, 15-9.
11. Stohl, F. V.; Dosch, R. G.; Stephens, H. P. Proc. Direct Liquefaction Contractors' Review Mtg.: U. S. Dept. of Energy, Pittsburgh Energy Technology Center, 1988, p. 490-503.
12. Snape, C. E.; Bolton, C.; Dosch, R. G.; Stephens, H. P. Preprints Fuel Div. Amer. Chem. Soc. 1988, 33(3), 351-6.
13. Monnier, J.; Denes, G.; Potter, J.; Kriz, J. F. Energy & Fuels 1987, 1, 332-8.
14. Stephens, H. P.; Kottenstette, R. J. Preprints Fuel Div. Amer. Chem. Soc. 1985, 30(2), 345-53.

RECEIVED May 21, 1990

Chapter 27

Dispersed-Phase Catalysis in Coal Liquefaction

Bruce R. Utz, Anthony V. Cugini, and Elizabeth A. Frommell

Pittsburgh Energy Technology Center, Department of Energy, P.O. Box 10940, Pittsburgh, PA 15236

The specific reaction ("activation") conditions for the conversion of catalyst precursors to unsupported catalysts have a direct effect on the catalytic activity and dispersion. The importance of reaction intermediates in decomposition of ammonium heptamolybdate and ammonium tetrathiomolybdate, and the sensitivity of these intermediates to reaction conditions, were studied in coal liquefaction systems. Recent results indicate that optimization of activation conditions facilitates the formation of a highly dispersed and active form of molybdenum disulfide for coal liquefaction. The use of the catalyst precursors ammonium heptamolybdate, ammonium tetrathiomolybdate, and molybdenum trisulfide for the conversion of coal to soluble products will be discussed.

The use of an unsupported dispersed-phase catalyst for direct coal liquefaction is not a novel concept and has been employed in many studies with varying success. Dispersed-phase catalysts have been introduced via impregnation techniques (1-4), as water-soluble (5-8) and oil-soluble (9-12) salts, and as finely divided powders (1,2). While some methods of catalyst introduction give higher dispersion of the catalyst and greater activity for the liquefaction of coal, all of the techniques allow the formation of a finely dispersed inorganic phase. The use of dispersed-phase catalysts in direct coal liquefaction offers several advantages. Since they could be considered "once-through" catalysts, deactivation problems are reduced in comparison to supported catalysts, and catalytic activity remains high. Diffusion limitations are minimized owing to the high surface area of small catalyst particles. Maximum interaction of coal, vehicle, and gaseous hydrogen can occur on the catalyst surface with a highly dispersed catalyst.

One of the more popular techniques for producing dispersed-phase catalysts involves the use of water- or oil-soluble catalyst precursors. Small amounts of the water-soluble catalyst precursor are added to the coal-vehicle feed and are subsequently converted to a highly dispersed, insoluble catalytic phase. The reaction conditions that convert the soluble catalyst precursor to a highly active and dispersed-phase catalyst are critical. The objective of this paper is to identify techniques that will convert a molybdenum catalyst precursor to a highly dispersed MoS_2 phase that has higher activity for coal liquefaction than most previously identified dispersed-phase MoS_2 catalysts. The catalyst precursors studies include ammonium heptamolybdate, ammonium tetrathiomolybdate, and molybdenum trisulfide. The bulk of the work was done with the sulfided precursors.

EXPERIMENTAL.

Catalyst screening tests were conducted in 40-mL microautoclave reactors. The liquefaction conditions were the following: temperature, 425°C; reaction pressure, 1000 psig H_2 (pressure at room temperature); residence time, 1 hour; and solvent/coal, 2/1. The coal used was Illinois No. 6 hvBb from the Burning Star Mine, properties of the coal used are summarized in Table I.

Table I.   Analysis of Illinois No. 6 (Burning Star) coal

Proximate Analysis, wt% (As Received)	
Moisture	4.2
Volatile Matter	36.9
Fixed Carbon	48.2
Ash	10.7
Ultimate Analysis, wt% (Moisture Free)	
Carbon	70.2
Hydrogen	4.8
Nitrogen	0.9
Sulfur	3.1
Oxygen (Difference)	9.9
Ash	11.1
Sulfur Forms, wt%	
Sulfatic	0.03
Pyritic	1.2
Organic	1.9

The vehicle used was tetrahydronaphthalene (tetralin). Rapid heat-up rates were obtained by immersing the microautoclave in a preheated fluidized sand bath at 425°C. The microautoclave reached reaction temperature in 1-2 minutes. Slow heat-up rates were obtained by immersing the microautolcave in the fluidized sand bath at room temperature and gradually heating the sand bath to reaction temperature in 3/4 hour to 1 hour. Catalyst metal loadings were 1000 ppm, based on the weight of coal. Coal conversion was measured by the solubility of coal-derived products in methylene chloride and heptane using a pressure filtration technique(13).

Continuous catalytic coprocessing and liquefaction experiments were conducted in a computer-controlled 1-liter bench-scale continuous unit of the type diagrammed in Figure I. The feedstocks included Illinois No. 6 coal and Maya atmospheric tower bottoms (ATB) in coprocessing or with a heavy hydrotreated 650°F-975°F recycle distillate (V-1074) from Run 257 at the Wilsonville Advanced Coal Liquefaction Facility in liquefaction experiments. Properties of the two vehicle oils are summarized in Table II.

Table II. Properties of Vehicle Oils

Ultimate Analysis, wt%	Maya ATB (650°F+)	Wilsonville V-1074
Carbon	84.5	89.1
Hydrogen	10.6	9.8
Oxygen (Direct)	0.3	0.9
Nitrogen	0.5	0.4
Sulfur	4.0	0.04
Ash	0.1	<0.1
oAPI	8.8	-
975°F (Vol%)	30	65
Heptane Insols, wt%	20	0.2
M_n(daltons)	720a	360b
H^*_{ar}	0.07	0.15
f_a	0.33	0.43

aVPO, pyridine, 80°C. bVPO, THF, 40°C.

The continuous unit includes provision for injection of an aqueous catalyst stream into the feed slurry, temperature staging of the preheater and reactor, and use of a mixed H_2/H_2S reducing gas if required. Rapid catalyst heat-up rates were obtained by injecting the total feeds slurry (including catalyst) with the feed gas directly into the reactor. For this study the unit was operated as a single stage reactor system. Catalyst metal loadings were varied from 0-1000 ppm based on coal to study the effect of catalyst concentration on conversions. Coal conversion was measured by the solubility of coal-derived products in methylene chloride and heptane using a pressure filtration technique (13). A modified D-1160 procedure was used to measure distillable product yields.

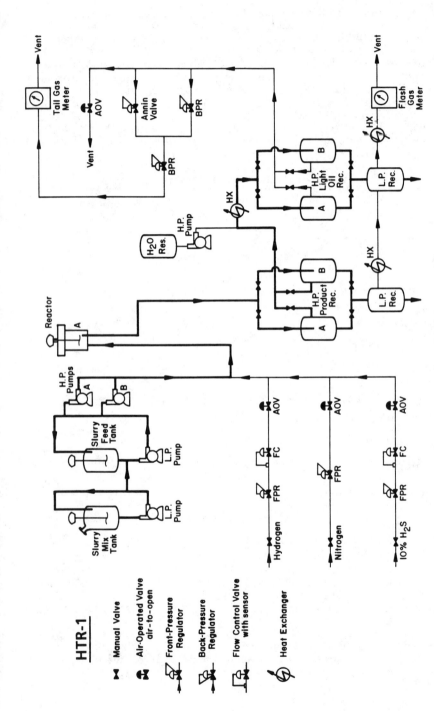

Figure I.   Schematic diagram of the bench-scale continuous-flow units.

Molybdenum trisulfide (MoS_3) was prepared by acidifying a solution of ammonium tetrathiomolybdate with 24 wt% formic acid ([14]). All other reagents were ACS grade.

X-ray diffraction studies were conducted with a Rigaku computer-controlled diffractometer equipped with a long fine-focus Cu X-ray tube, a receiving graphite monochromator to provide monochromatic Cu-K_a radiation, and a scintillation detector.

RESULTS AND DISCUSSION.

Conversion of an aqueous solution of ammonium heptamolybdate [$(NH_4)_6Mo_7O_{24}\cdot4H_2O$] to an active and high-surface-area catalyst is dependent on a number of factors. Gaseous hydrogen sulfide (H_2S) is required to convert ammonium heptamolybdate (AHM), which is essentially an oxide salt, to a series of oxysulfide salts ([15]) and ultimately to molybdenum disulfide (MoS_2), as shown in Figure II. The ratio of ammonium ion to molybdenum may also be important, since studies have shown that increased NH_3/Mo ratios result in higher hydrogen consumption for the conversion of petroleum to upgraded products when AHM is used as a dispersed-phase catalyst ([7,15]). Petroleum upgrading studies demonstrated that the heat-up rate for the conversion of AHM to MoS_2 is extremely important ([15]). Slower heat-up rates resulted in a gradual transition of AHM to MoS_2 and significantly higher conversions of petroleum to distillate products.

Information on the dispersion of MoS_2 was obtained from x-ray diffraction measurements, which are sensitive to the degree of stacking and dispersion of the MoS_2 layers (Figure III). The diffraction pattern of single layers of MoS_2 shows the (100) and (110) bands, but no (002) band, as in the middle curve of Figure III. These crystallites are considered two-dimensional, since there is no growth in the third dimension. When only a very small number of MoS_2 layers are in multilayer stacks, a weak (002) band is present, as in the top curve with the catalyst precursor, MoS_3. When many, multilayer stacks of MoS_2 are present, a strong (002) band can be seen, as in the lower curve. The pattern is that obtained for three-dimensional crystallites. The width-at-half-maximum can be used to estimate the size of the MoS_2 crystallites. The MoS_2 crystallites formed during a gradual heat-up of AHM, in the absence of coal, are three-dimensional, as shown in the bottom curve of Figure III. Coal added to AHM under the same reaction conditions prevents the MoS_2 layers from growing in the third dimension. Since the crystallite size in the plane of the MoS_2 layers is about the same in both cases (150-200 angstroms), the addition of the coal produces a more highly dispersed, minimally stacked MoS_2. All of these results suggest that many factors can affect the extent of MoS_2 dispersion. The factors that have been identified include heat-up rate of the catalyst precursor during the conversion to MoS_2; the NH_3/Mo ratio; the H_2S pressure; and other reagents, such as coal, that might affect the transition of AHM to MoS_2.

In the activation of AHM to MoS_2, one of the intermediates that is observed is ammonium tetrathiomolybdate (ATTM). Previous studies by Lopez et al. ([16]) show that small amounts of ammonium tetrathiomolybdate [$(NH_4)_2MoS_4$] are produced during the decomposition of AHM, which represents an intermediate of a minor decomposition pathway. One advantage of using ATTM as a dispersed-phase catalyst in coal liquefaction is that an external source of H_2S is not required because the catalyst precursor already exists as a water-soluble sulfide salt. It is known that ATTM thermally decomposes to MoS_3 and subsequently to MoS_2 (17). Eggertsen et al. (14) examined the reaction conditions for the thermal decomposition of MoS_3 to MoS_2 and determined that rapid heat-up (direct introduction of MoS_3 into a stream of hydrogen at 450°C) gave MoS_2 having a surface area of 85-158 m²/gm, while gradual heat-up (25 min to 450°C) resulted in MoS_2 surface areas of less than 5m²/gm. Naumann and coworkers (17) applied the results of Eggertsen to the catalyst precursor ATTM, since MoS_3 was considered to be an intermediate in the decomposition of ATTM (Figure IV). The results show that high-surface area MoS_2 is formed (88 m²/gm) if the thermal transition of ATTM to MoS_2 is rapid, and the surface area is low if the thermal transition is gradual.

Studies at the Pittsburgh Energy Technology Center examined the conversion of coal to methylene chloride- and heptane-soluble products using ATTM as a catalyst precursor. Experiments were performed using both rapid and gradual heat-up rates for activation of the dispersed-phase catalyst and were compared with results using AHM as a catalyst (Figure V). The results demonstrate that rapid heat-up of ATTM resulted in coal conversions to methylene chloride- and heptane-soluble material that were higher than conversions from experiments done at a slower heat-up, consistent with expectations for surface area studies of pure compounds. Experiments were also performed with MoS_3, wherein MoS_3 was suspended in the same amount of water used for the water-soluble precursor, ATTM. Results show that conversion of coal to methylene chloride- and heptane-soluble products is greater for rapid heat-up experiments and is comparable to those using AHM (Figure V). Experiments that verify the microautoclave studies were also performed in a 1-liter stirred autoclave, using both batch and continuous modes of operation.

Experiments were also performed with moisture-free solid MoS_3. Elimination of water has several advantages. The addition of water causes a decrease in hydrogen partial pressure within the reactor and is more energy intensive because water is being heated. Experimental results presented in Figure V demonstrate that the addition of the dry catalyst precursor, MoS_3, produced coal conversions comparable to those obtained with the water-soluble catalyst precursors. Therefore, MoS_3 may represent an ideal choice in the preparation of dispersed-phase catalysts. The x-ray diffraction pattern of MoS_2, from MoS_3, showed a very small (002) band, indicating minimal stacking. This implies the presence of a well-dispersed, high-surface-area material that provides comparable conversions to the water-soluble catalyst precursors. Scanning electron microscopy (SEM) was not able to detect the presence of MoS_2, suggesting that the particle size was less than 1000 angstroms. These results support the existence of highly dispersed MoS_2, which has resulted in the high conversion to solvent-soluble products.

Based on the success of the batch autoclave experiments, continuous liquefaction and coprocessing tests using [$(NH_4)_2MoS_4$] were made. High coal conversions to solvent-soluble and distillate products were observed at 450°C. Typical continuous unit results in both liquefaction and coprocessing modes at 450°C are presented in Table III.

$$(NH_4)_6Mo_7O_{24} \cdot 4H_2O + H_2S \rightarrow (NH_4)_xMoO_yS_z$$

$$(NH_4)_xMoO_yS_z + H_2S \rightarrow MoO_xS_y + NH_3$$

$$MoO_xS_y + H_2 + H_2S \rightarrow \text{"MoS}_2\text{"} + H_2O$$

Figure II. Conversion of ammonium heptamolybdate to molybdenum.

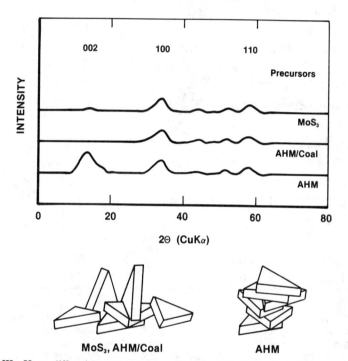

Figure III. X-ray diffraction patterns for molybdenum disulfide (h, k, l) generated from Mo(VI) precursors reacted at coal liquefaction conditions.

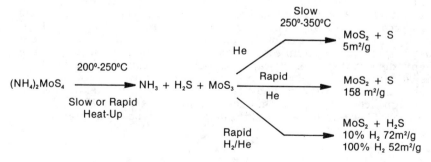

Figure IV. Decomposition of ammonium tetrathiomolybdate.

Table III. Effect of Solvent Type on Product Yields at 450°C Using
Ammonium Tetrathiomolybdate

Coal	Illinois No. 6	Illinois No. 6
Solvent	Maya ATB	V-1074
Temperature(°C)	450	450
Pressure (psig H2)	2500	2500
Residence Time (h)	1.5	1.5
Catalyst Conc. (ppm Mo)	1000	1000
Yields (wt% of Feed)		
Non-Hydrocarbon Gases	5.1	3.7
C1-C4	8.2	5.1
C5-850oF	76.8	76.4
850oF+	13.9	18.5
(C7 Insols)	(8.4)	(4.8)
(CH2Cl2 Insols)	(1.4)	(2.4)
Hydrogen	-4.0	-3.7
maf Coal Conversion to Heptane Solubles (%)	83	93

At 450°C in the presence of the dispersed catalyst, results are remarkably similar with respect to distillate yields, coal conversions, and net hydrogen consumption. Use of a superior hydrogen donor solvent in the liquefaction experiments does lead to a consistently lower gas make and slightly higher coal conversion to heptane solubles than in the coprocessing runs under comparable processing conditions.

A series of continuous unit coprocessing runs were performed with tetrathiomolybdate to determine the change in product yield structure with catalyst concentration. The results obtained in coprocessing Illinois No. 6 coal with Maya ATB are given in Figure VI. High yields of heptane- and methylene chloride- solubles, and C_5-850°F distillates were sustained at molybdenum concentrations of 200 ppm, indicative of the highly dispersed state of the catalyst. Below 200 ppm catalyst distillate yields declined and hydrocarbon gas yields rose toward those obtained under thermal processing conditions. Also noteworthy was the lack of any significant coking in this coprocessing system at 450°C in the presence of low concentrations of the dispersed molybdenum catalyst.

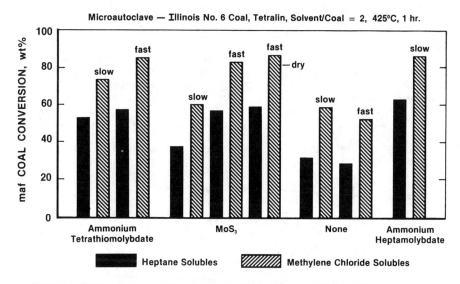

Figure V. Effect of heat-up rate on coal conversion using presulfided catalyst precursors.

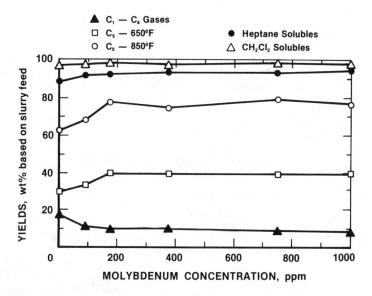

Figure VI. Product yield versus catalyst concentration. (Continuous unit runs; 450°C, 1.5 hours, 2500 psig H_2, Maya ATB and Illinois No. 6, Mo added as ammonium tetrathiomolybdate.)

CONCLUSIONS.

Two catalyst precursors have been identified that result in high conversions of coal to solvent-soluble products when heated rapidly to reaction temperature. The use of MoS_3 and ATTM as catalyst precursors, rather than AHM as a catalyst precursor, offers a number of advantages. Both catalyst precursors are in a sulfided form, and therefore additional H_2S is not required, while AHM requires the addition of H_2S in order to form the oxysulfide intermediates and the final product, MoS_2. Both of the sulfided precursors are highly active and highly dispersed when heated rapidly to reaction temperature, while AHM requires a gradual heat-up, and therefore activation of AHM is much more energy intensive. Results from experiments made in the continuous unit indicate that conversions and yields are maintained down to a catalyst concentration of 200 ppm using ATTM as the catalyst. The ultimate goal is to identify a dry, highly dispersed, catalyst precursor or catalyst that can be added to a coal-vehicle feed without the addition of water, yet results in yields of coal-derived products comparable to those produced using water- or oil-soluble catalyst precursors. Possibly MoS_3 is the catalyst precursor that satisfies those requirements.

ACKNOWLEDGMENTS.

The authors thank Raymond Bernarding for conducting the liquefaction experiments and Fred Vinton for preparing the catalyst precursors. The authors would also like to thank Sidney S. Pollack for his assistance in the x-ray diffraction study.

DISCLAIMER.

Reference in this report to any specific commercial product, process, or service is to facilitate understanding and does not necessarily imply its endorsement or favoring by the United States Department of Energy.

LITERATURE CITED.

1. Schlesinger, M.D.; Frank, L.V.; Hiteshue, R.W. Bureau of Mines Report of Investigations No. 6021 1961.
2. Weller, S.W.; Pelipetz, M.G. Ind. Eng. Chem. 1951, 43(5), 1243.
3. Clark, E.L.; Hiteshue, R.W.; Kandiner, H.J. Chem. Eng. Prog. 1952, 48, 14.
4. Derbyshire, F.J.; Davis A.; Lin, R.; Stansberry, P.G.; Terrer, M.T. Fuel Proc. Tech. 1986, 12, 127.
5. Moll, N.G.; Quarderer, G.J. Chem. Eng. Prog. 1979, 75(10), 46.
6. Moll, N.G.; Quarderer, G.J. U.S. Patent 4,090,943, 1978.
7. Lopez, J.; McKinney, J.D.; Pasek, E.A. U.S. Patent 4,557,821, 1985.
8. Cugini, A.V.; Ruether, J.A.; Cillo, D.L.; Krastman, D.; Smith, D.N.; Balsone, V. Preprints, Div. Fuel Chem., Am. Chem. Soc. 1988, 33(1), 6.
9. Hawk, C.O.; Hiteshue, R.W. Bureau of Mines Bulletin No. 622 1965, 42.

10. Gatsis, J.G.; Gleim, W.K.T. U.S. Patent 3,252,894, 1966.
11. Anderson, R.R.; Bockrath, B.C. Fuel 1984, 64, 329.
12. Aldridge, C.L.; Bearden, R., Jr. U.S. Patent 4,298,454, 1978.
13. Utz, B.R.; Narain, N.K.; Appell, H.R.; Blaustein, B.D. "Coal and Coal Products: Analytical Characterization Techniques" (Ed. E.L. Fuller, Jr.), Am. Chem. Soc. Symp. Ser. 205, 1982, 225.
14. Eggertsen, F.T.; Roberts, R.M. J. Phys. Chem. 1959, 63, 1981.
15. Lopez, J.; Pasek, E.A.; Cugini, A.V. U.S. Patent 4,762,812, 1988.
16. Lopez, J.; Pasek, E.A.; Cugini, A.V. Unpublished results.
17. Naumann, A.W.; Behan, A.S.; Thorsteinson, E.M. Proc. Fourth Int. Conf. Chemistry and Uses of Molybdenum (H.E. Barry and C.N. Mitchell Eds.)1982, 313 .

RECEIVED May 9, 1990

NEW   TECHNIQUES

Chapter 28

Measuring Surface Tension of Thin Foils by Laser Interferometer

First Step in Generating Surface Phase Diagrams

G. A. Jablonski, A. Sacco, Jr., and R. A. Gately

Department of Chemical Engineering, Worcester Polytechnic Institute, Worcester, MA 01609–2280

Surface segregation is a phenomenon which catalytic scientists have struggled to understand and control for many years. Bulk non-equilibrium ternary phase diagrams have been used to control the surface composition of Fe, Ni, and Co foils at elevated temperatures (>900 K) under reaction conditions. To extend this approach to low temperature reaction systems, surface phase diagrams must be used. The thermodynamic development of surface phase diagrams requires surface free energy data for the solid components for the system of interest. A laser technique has been developed to determine the surface tension of thin foils under their own vapor pressure at temperatures between their melting point and the Tammann temperature (0.5 T_m).This surface tension data is then used to calculate the surface free energy. The laser technique uses the zero creep method with a modified Michelson interferometer. This approach has been successfully tested by measuring the surface tension of Al foils at 71 % (1196 d/cm) and 82 % (1080 d/cm) of the melting point under vacuum. The future application of this technique to the study of Fe, Ni, and Co foils will be discussed.

Carbon deposition on metal surfaces is a common problem in the chemical process industry. Catalyst fouling and deactivation as well as corrosion of process equipment (e.g., metal dusting) can occur as a result of this type of deposition. A particularly interesting form of carbon is filamentous carbon, which in addition to its undesirable formation under some conditions, has found use in high strength-light weight composite materials(1). This "spaghetti-like" form of carbon has a metallic crystallite at its head. The metallic crystallite, which originates from the metal surface, is believed to serve as the growth center for the filament(2). An

0097–6156/90/0437–0302$06.00/0

understanding of the mechanisms of filamentous carbon formation and growth would greatly aid in controlling its growth.

Bulk C-H-O phase diagrams have been used to study filamentous carbon initiation and growth mechanisms over Fe, Ni, and Co foils at elevated temperatures from gas mixtures of CO, CO_2, CH_4, H_2O, and H_2(3-6). Through the use of the phase diagram, the solid surface phase can be controlled by controlling the gas phase composition. By controlling the solid surface phase composition, the catalytic properties of a specific phase or phases can be investigated. At elevated temperatures, bulk phase diagrams can be used to control the surface phases because transport limitations between the solid surface and the bulk are minimal. This is not the case at lower reaction temperatures, where transport between the solid surface and the bulk is much slower. To account for this variation of the surface composition from that of the bulk, the surface free energy of the possible metallic species must be incorporated in the phase diagram development. The surface free energy is calculated by the product of the surface tension, and an estimate of the surface area per mole of pure species *(7).

Techniques to measure the surface tension of solids are notoriously difficult and known for their inaccuracies. Reliable surface tension data requires not only a reliable measurement technique but careful control over parameters such as sample purity and the gaseous atmosphere in which the experiments are conducted. The zero creep technique is considered one of the most accurate and reliable of these techniques since it requires only a simple length measurement(8). Samples can be either wires or thin foils. Hondros(9) has postulated that the use of thin foils increases the sensitivity of the technique and thus allows more accurate measurements. The thinner the foil, the more it approximates a surface. Wire gauges are limited due to the loads required to strain the sample. Table I lists some of the results obtained using the zero creep foil technique. It should be pointed out that the terms surface tension and surface free energy are often used interchangeably, though they are not equivalent(9,10).

Creep only occurs appreciably at temperatures greater than the Tammann temperature(19) [$\approx 0.5\ T_m$] of the metal. Most of the investigations shown in Table I are done at 0.85 T_m or greater. The reason for this is that the level of strains which occurs as the temperature approaches the Tammann temperature is not easily measured. In addition, the method used to measure the sample strain (length change) on foils [X-ray shadowgraphs] has large errors associated with it [$\pm$ 10 - 20 μm]. The objective of this work is to develop a technique to determine the surface free energy of thin foils under their own vapor pressure and under reaction conditions at temperatures approaching the Tammann temperature. These values will be used in the generation of surface phase diagrams. The approach will be to utilize the zero creep technique and laser interferometry to allow more accurate sample strain measurements at lower temperatures. Using a modified Michelson interferometer, strain measurement approaching $\lambda/2$ will be possible (λ is the wavelength of the energy source used). The energy source being used is a red HeNe laser [λ= 0.6328 μm].

---

* The surface tension(i.e., surface free energy) decreases substantially with physisorption and chemisorption(2-3 orders of magnitude). Thus, even under reaction conditions, the pure metal surface tension dominates.

Methodology

The technique used to measure the surface tension of foils in this work couples two well known technologies: the zero creep technique for foils, and the technique of laser interferometry. The theory behind each of these techniques will be discussed briefly in order to develop the expressions necessary to generate the desired surface tension data which will be used to calculate the surface free energy.

The Zero Creep Technique. The zero creep technique was developed by Udin, Shaler, and Wulff(8) to measure the surface energy of Cu wires. The technique was later extended for use with thin foils by Hondros(16). Very thin foils, approximating a surface, are readily available. When shaped into a cylinder, the sample will tolerate large loads without necking. Since necking does not occur, the stress can be considered constant throughout the experiment. Figure 1 shows a schematic of a foil and the associated stress under an applied load

Table I. Summary of investigators using the zero creep technique including their method of length measurement

Metal	Temp ($\% T_m$)	γ (dynes/cm)	Length Measurement Method	Ref
Ni	72	2280	micrometer stage	11
Au	90-95	1490	micrometer slide	12
Fe,Ni, Co	95	2525,2595, 2490	micrometer microscope	13
Ni	86 - 100	1850	travelling microscope	14
Ag	91, 97	1140	travelling microscope	15
Cu	88 - 97	1420	travelling microscope	8
Fe	92	2320	X-ray shadowgraphs	16
Au	85,97	1480,1300	X-ray shadowgraphs	17
Fe,Fe-P	88,94	2100	X-ray shadowgraphs	18

(after Hondros,18). For a foil of width, w, length, L, average grain size, a, thickness, θ, with an applied load, F, the principal stresses can be written as:

$$\sigma_x = F/w\theta - 2\gamma/\theta - \gamma_b/a \tag{1}$$
$$\sigma_y = -2\gamma/\theta - \gamma_b/a \tag{2}$$
$$\sigma_z = -2\gamma_b/a - 2\gamma/w \tag{3}$$

where: γ = bulk surface tension
γ_b = grain boundary surface tension

The last term in the expression for the stress in the z-direction is very small in

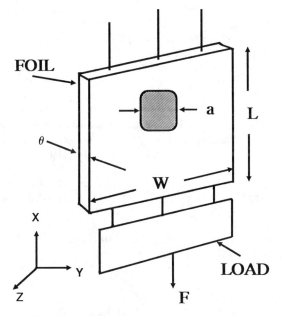

Figure 1. Schematic of a zero creep foil.

comparison to the first term because the width of the sample is 3-4 orders of magnitude greater than the average grain size. For this reason, it can be neglected. The sample strain rate in the x-direction is much greater than in either the y or z-directions since this is the direction in which the force is being applied. Consideration of the strain rate in the x-direction [$\dot{\epsilon}_x$] at zero strain rate coupled with the principal stresses leads to a relationship between the force on the sample, and the bulk and grain boundary surface tension terms(18):

$$F_0 = \gamma w - \gamma_b \theta w/2a \tag{4}$$

This condition is the zero creep point, when the surface forces represented by the terms on the right hand side of equation (4) are exactly balanced by the force due to the load imparted to the sample (F_0). Physically, there would be neither contraction nor elongation when this situation occurs. For this reason, the force at or approaching the zero creep point is difficult to measure directly. Therefore, the force at the zero creep point is determined by measuring the strains of several loaded samples and plotting these strains against the stress which caused them. The force at the zero creep point is determined by interpolation.

The grain boundary surface tension term (γ_b) is typically determined by measurement of the dihedral angle between two grains. This dihedral angle occurs at the intersection of two grains (of different orientation) due to slightly different surface tension values for the grains involved. The angle is measured through the use of interference microscopy. The magnitude of the grain boundary surface tension term can be approximated to be 0.33γ , which has been substantiated by several authors(18,22). However, the relative effect of the grain size also must be considered when approximating the value of γ_b. The total expression involving γ_b ($\gamma_b \theta w/2a$) increases substantially with decreasing grain size. Between 25 - 35 μm, the approximate error due to neglecting γ_b is 10 - 15 %. The error decreases substantially with increasing grain size (i.e., < 7% for grain sizes > 50 μm). Therefore, the sample grain size may be one major source of difference between reported surface tension(surface free energy) data by various investigators and must be kept in mind when comparing results.

Laser Interferometry. A Michelson interferometer consists of two optical path "legs" which are the result of splitting the incident beam using a beamsplitter(20): a sample leg whose optical path changes as the sample length changes, and a reference leg whose optical path length is fixed. The electric field vector, E_i, for each leg of the interferometer can be written as:

$$E_i = A_i \exp i[w_i t + \phi_i] \tag{5}$$

where: w_i = angular frequency
              ϕ_i = phase
              A_i = constant
              t = time

When the two beams recombine at the beamsplitter from which they originated, they interfere. For two coherent beams which interfere, the electric field vector, E_t, can be written as:

$$E_t = E_s + E_r \qquad (6)$$

where: E_s, E_r = electric field vector for the sample and reference beams, respectively

The interference which occurs can be either constructive, manifested by bright fringes, or destructive, manifested by dark fringes. The intensity, I_i, of the interferogram can be written as:

$$I_i = E_t E_t^* \qquad (7)$$

where: E_t^* = complex conjugate of E_t. Therefore,

$$I_i = A_s^2 + A_r^2 + A_s A_r [expi(\phi_s - \phi_r)] + A_s A_r [exp(-i)(\phi_s - \phi_r)] \qquad (8)$$

where: A_s, A_r = intensity of the sample and reference beams
ϕ_s, ϕ_r = phase of the sample and reference beams

Equation (8) can be simplified to:

$$I_i = A_s^2 + A_r^2 + 2A_s A_r \cos(\Phi_t) \qquad (9)$$

where: Φ_t = interference phase

The interference phase, Φ_t, can be represented by:

$$\Phi_t = \phi_s - \phi_r = 2\pi \Delta P/\lambda \qquad (10)$$

where: ΔP = optical path difference

The length change of the sample is directly proportional to the change in the interference phase. As the sample optical path length changes due to changes in the sample length (strains) caused by the load imparted to the sample, the interference fringes shift relative to the speed of the optical path length change (and thus the sample length change). By measuring the intensity of the beam and plotting intensity versus time, the sample length change with time can be recorded. One complete phase shift corresponds to a sample length change of $\lambda/2$. In this way, the sample length change is continuously measured and sample strains can be easily calculated based on the overall length of the sample.

Experimental

The overall schematic of the zero creep/laser interferometer system is shown in Figure 2. The system consists of the reactor in which the sample and reference mirrors are located, the interferometer, and the data collection system. The

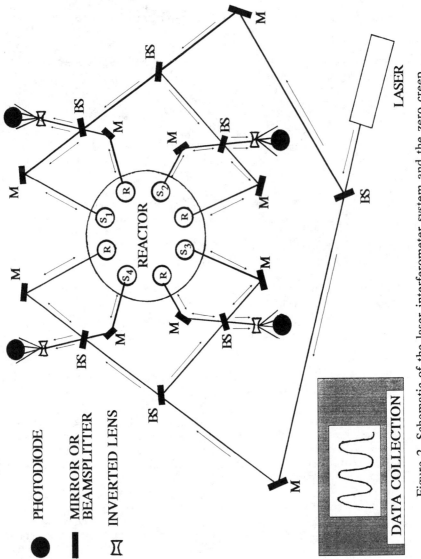

Figure 2. Schematic of the laser interferometer system and the zero creep reactor.

system components are mounted on a vibration isolation table [NRC RS-46-18], effectively isolating them from outside vibrations. The isolation table is completely encased in a plexiglass enclosure to control convection currents and to prevent dust and particulates in the air from soiling the optical components.

Within the enclosure, the 316 stainless steel reactor is sealed in a vacuum tight glass vessel at the bottom of which is a 5.08 cm window through which the incident and resultant laser beams pass. Up to 4 samples can be run at any time. The temperature in the reactor is controlled to $\pm 0.3^\circ$K at 800 K using a current proportional temperature controller [Honeywell UDC 5000]. The glass vessel containing the reactor is evacuated by a roughing pump initially, followed by a diffusion pump.

The foil sample is held in a cylindrical configuration connected to a 1.27 cm diameter mirror through the use of a set of locking collars. The locking collars effectively pinch the sample between them, preventing any slippage and holding the 1.27 cm diameter object mirror in place. The reference mirror sits on a ledge machined out of the stainless steel reactor. The relative positions of the sample mirror and the reference mirror are approximately the same. This was done to minimize any temperature variations between the sample and the reference optical paths. The sample and reference mirrors [Precision Optics Corp.] are high purity quartz substrates with a reflective dielectric coating. The mirrors are able to withstand temperatures up to 1200 K. The temperature of the sample is measured by a thermocouple which is inserted into the center of the sample via a hole in the sample holder. The thermocouple is placed at the interface between the sample holder and the sample itself.

Each sample has its own independent interferometer associated with it. A 5 mW red HeNe laser [Spectra Physics] is used as the energy source for the four interferometers that are in the system. The incident beam is split twice in order to provide incident beams to each of the independent interferometers. Each interferometer consists of a 50/50 beamsplitter, four mirrors [including two mirrors at 45° under the reactor to direct the beam into the reactor], and a beam expander in addition to the sample and reference mirrors located in the reactor. All of the optics have a flatness specification of $\lambda/10$. The mirrors which make up the interferometer components outside of the reactor are enhanced aluminum. The 45° mirrors have adjustment screws so that the sample and reference beams can be aligned from outside the system once a run is started.

The data collection system consists of a photodiode for each interferometer, an amplifier, A -> D board [Metrabyte], and a personal computer for collecting the data in real time. With the exception of the sample and reference mirrors located in the reactor, all of the other components for measuring the sample strains are external to the reactor. Thus, it is not necessary to disturb the sample in order to make these *in-situ* sample strain measurements.

The zero creep/laser interferometer system was evaluated using Al foils [0.002 cm thick, 1.905 cm long]. No thickness tolerance was furnished by the manufacturer, however, no discernible variations were observed using a micrometer accurate to ± 0.1 μm. The foils were washed in acetone and isopropyl alcohol before being mounted in the reactor. The reactor was evacuated before the sample was heated to temperature. The operating vacuum was $7 * 10^{-6}$ atm at temperature. Typical runs were from 3 - 4 days in length, and data was collected every 5 seconds.

Results and Discussion

A typical intensity versus time plot for the Al system at 743 K is shown in Figure 3. The breadth of the plot is an indication of the random error due to background vibration and drift from thermal currents. The absolute stability of the system was determined to be $\lambda/2$ over a 24 hr. period at 775 K. This was done by fixing the sample and reference optical path lengths. Figure 3 represents a recording of the total length change of the sample. One complete phase change corresponds to a sample length change of $\lambda/2$. From this plot, the sample strains can be easily calculated if the initial overall sample length is known. Figure 4 illustrates a typical strain - time plot. The plot shows a very sharp rise in sample strain followed by a period of constant strain. Initially, sample strain is very high due to the large deformations that are occurring in the sample due to the load on the sample at elevated temperature. These deformations can be grain growth or diffusion processes. Eventually, the strain rate starts to decrease until its value is constant. This is typical of a creep curve(21). The constant strain rate represents secondary creep. The constant strain rate is then plotted against the stress which caused it. The stress - strain rate plots for 71 % T_m and 82 % T_m are shown in Figure 5a and 5b, respectively. Samples were loaded such that positive and negative sample strains were observed and recorded. The zero creep point, where the surface forces are exactly balanced by the force due to the load imparted to the sample, is determined by interpolation. Ignoring the grain boundary surface tension, the bulk surface tension at 71 % T_m was determined to be 1034 ± 80 dynes/cm. The bulk surface tension at 82 % T_m was determined to be 934 ± 70 dynes/cm. The average grain size of the Al after reaction was found to be 25 μm by consideration of scanning electron micrographs of the sample . The grain boundary surface tension has been found to be 0.33 γ (18,22). Therefore, the corrected surface tension for Al is 1196 80 dynes/cm at 743 K and 1080 ± 70 dynes/cm at 813 K. The area of the Al was estimated to be 7.8 Å^2/molecule by consideration of the molar area for individual planes(i.e., (100), (110), etc.) and weighting their contributions based on the relative intensity of their XRD peaks. We recognize that XRD is a bulk technique, however, it allows us to approximate the relative contributions of the crystallographic planes. The molar areas of the planes were calculated using a bond breaking model(23). An overall molar area was calculated to be $4.7 * 10^8$ cm^2/mole. This compares well with a value of $4.24 * 10^8$ cm^2/mole calculated by Overbury et al.(24). Using this data, the surface free energy of Al was calculated to be $5.6 * 10^{11}$ ergs/mole(13.4 kcal/mole) and $5.1 * 10^{11}$(12.2 kcal/mole) at 743 K and 813 K, respectively. The surface tension values for Al were reported by Westmacott et al.(25) [1140 ± 200 dynes/cm] determined at 23 - 32 % T_m using a void annealing technique. If we linearly extrapolate the surface tension values obtained using our technique to the temperatures investigated by Westmacott et al., we can compare our values to those reported by Westmacott. Linear extrapolation was used based on the results of Hondros(18) and Udin et al.(8) who showed an approximate linear relationship between the surface tension and temperature. A surface tension of 1643 dynes/cm was estimated versus 1140 ± 200 as reported by Westmacott. The value reported by Westmacott is lower than would be expected, thus it is possible that their sample may have oxidized. Oxidation would tend to lower the

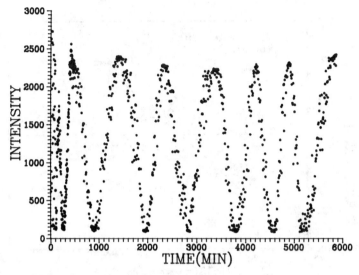

Figure 3. Typical intensity vs. time plot for the determination of the surface tension of Al at 743 K.

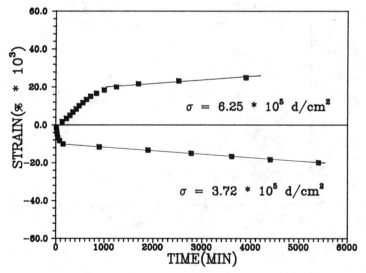

Figure 4. Typical strain (%) vs. time plots for Al in a vacuum at 743 K.

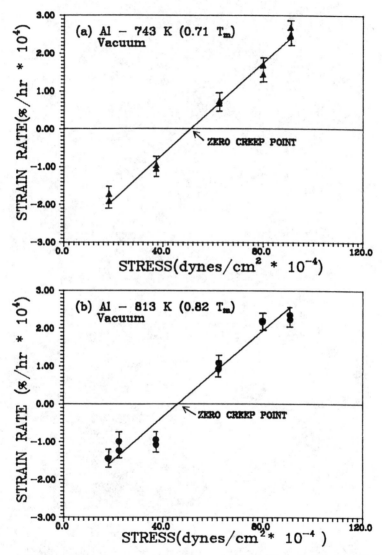

Figure 5. Strain rate vs. stress for Al at (a) 743 K and (b) 813 K in a vacuum.

measured surface tension. In reporting this data, Westmacott et al. suggest that because of the tenacity of the Al surface oxide, it is not possible to use the zero creep technique to measure the surface tension of pure Al. The authors did not disclose the atmosphere that was used to heat their samples in order to create the voids nor did they analyze the sample after any of the heat treatments to determine if a surface or a bulk oxide was present. It is likely that a surface oxide was present on their samples. The partial pressure of O_2 necessary to oxidize Al at 775 K is 10^{-57} atm. Therefore, it is likely that a surface oxide is present on any Al foil. In addition, the presence of any surface oxide will retard the further growth of the oxide since the oxygen must diffuse through the surface oxide layer in order to oxidize the Al metal. This is a very slow process that is not likely to occur in the time frame of the experiment. X-ray diffraction of the Al samples after reaction showed only the presence of Al. No oxide was detectable. This is not surprising since the Al will form a protective surface oxide layer instantaneously at room temperature. The fact that no bulk oxide is present does not preclude the possibility that the surface tension being measured is being influenced by the probable presence of a surface oxide.

The surface tension of thin Sn foils was also determined in air. No attempt was made to prevent the samples from oxidizing. The surface tension of the Sn samples was determined to be 452 dynes/cm at 88 % T_m. This value is lower than the values reported in the literature of 500 dynes/cm(26,27). This lower value is consistent with a sample that was partialy oxidized as evidenced by XRD. No attempt was made to calculate the surface free energy of the Sn due to the presence of the bulk oxide.

Conclusions

The surface free energy of thin foils can be determined at temperatures between the melting point and approaching the Tammann temperature of the metal using the zero creep/laser interferometer technique. The use of the laser interferometer allows smaller sample strains to be measured with a higher level of confidence.

The surface tension of Al under a vacuum was determined to be 1196 dynes/cm at 743 K and 1080 dynes/cm at 813 K. The surface tension of Sn was determined to be 452 dynes/cm at 478 K in air.

Future Developments

In order to apply the phase diagram approach to the study of filamentous carbon initiation and growth mechanisms over Fe, Ni, and Co foils at low reaction temperatures, surface phase diagrams for these materials must be generated. The zero creep/laser interferometer technique will be used to determine the surface tension and thus the surface free energy of these metals. Using these data, the surface phase diagrams can be generated. The effect of the addition of the surface free energy will be to shift the position of the solid phase boundaries. The amount and direction of the shift will be determined by the relative magnitude of the surface free energy terms. Figure 6 shows a C-H-O phase digaram for the Co system at 800 K and 1 atm where the boundaries for: the graphite-gas, the Co_b-Co_3C_b-gas, the Co_s-Co_3C_s-gas, and the Co-CoO-gas equilibria have been

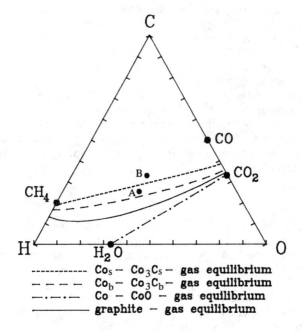

Figure 6. Co-C-H-O phase diagram, 800 K and 1 atm.

plotted. The subscripts b and s signify bulk and surface species, respectively. The Co_s-Co_3C_s-gas phase boundary is hypothetical, while the other phase boundaries shown are based on real thermodynamic data.The relative positions of the two metal carbide curves illustrate the effect that surface free energy can have on the solid phases that are stable. Consider a point A on the phase diagram. If the Co_s-Co_3C_s-gas phase boundary is ignored, the phase diagram is the bulk phase diagram for the system. If a foil were reacted in a gas composition represented by point A, bulk Co_3C would be expected to form with the potential for the formation of β-graphite if the Co_3C is a catalyst for its formation. However, if the surface free energy of the metallic components are included in the thermodynamic development, the result is a shift in position of the metal-metal carbide equilibrium boundary. The curve has shifted upward because of the effect of the surface free energy on the thermodynamics of the system. In reality, there has been a segregation of the metal to the surface. The equilibrium no longer favors the formation of the carbidic phase under the gas composition represented by point A. The addition of the surface free energy terms results in a surface phase diagram. In order to form a carbide phase, a gas composition above the new Co_s-Co_3C_s-gas line would have to be used [i.e., point B] with the new phase diagram. In this case, the carbide would be a surface carbide and it would also co-exist with β-graphite.

The use of surface phase diagrams to control the solid surface phase[s] represents a powerful *in-situ* technique for use in heterogeneous catalytic studies. Knowledge of the surface phase composition allows the catalytic properties of specific metallic phases to be evaluated. This type of *a-priori* control is not possible with other *in-situ* analytical techniques

Acknowledgments

The authors acknowledge the machining talents of Mr. Giacomo P. Ferraro, Jr. and Mr. Severin J. Ritchie, II. In addition, helpful discussions with Dr. Ryszard J. Pryputniewicz are gratefully acknowledged. Also, the fabrication of the high temperature mirrors by Precision Optics Corporation is greatly appreciated. This work was supported by NSF grant # CBT - 8722636.

Literature Cited

1. Tibbetts, G.G.; Endo, M.; Beetz, C.P. SAMPE Journal 1986, 22, 30.

2. Ruston, W.R.; Warzee, M.; Hennaut, J.; Waty, J. Carbon 1969 7, 47.

3. Sacco, Jr., A.; Reid, R.C. A.I.C.H.E. Journal 1979, 25, 839.

4. Sacco, A., Jr.; Thacker, P.; Chang, T.N.; Chiang, A.T.S. J. Catal. 1984, 85, 24.

5. Manning, M.P.; Garmirian, J.E.; Reid, R.C. Ind. Eng. Chem. Process Des. and Dev. 1982, 21, 404.

6. Sacco, A., Jr.; Geurts, F.W.A.H.; Jablonski, G.A.; Lee, S.; Gately, R.A. J. Catal 1989, 119, 322.

7. M. Modell and R.C. Reid Thermodynamics and its Applications, 2nd ed., Prentice-Hall, Inc., Englewood Cliffs, N.J., 1983, p.407.

8. Udin, H.; Shaler, A.J.; Wolf, J.; Trans. A.I.M.E 1949, 185, 186.

9. Hondros, E., Techniques of Metals Research, ed. R.A. Rapp, Interscience Publishers, New York, 1970, p.293.

10. Vermaak, J.S.; Mays, C.W.; Kuhlmann-Wilsdorf, D. Surface Science 1968, 12, 128.

11. L.E. Murr; O.T. Inal; G.I. Wong Proceedings of the Fifth International Materials Symposium, eds. G. Thomas; R.M. Fulrath; R.M. Fisher, Univ. of California Press, Berkely, CA, 1972; p 417.

12. C.E. Bauer; R. Speiser; J.P. Hirth Met. Trans. A 1976, 7A, 75.

13. T.A. Roth Mat. Sci. and Eng. 1975, 18, 183.

14. E.R. Hayward; A.P. Greenough, J. of Inst. of Metals 1960, 88, 217.

15. F.H. Buttner; E.R. Funk; H. Udin J. Phys. Chem. 1952, 56, 657.

16. Hondros, E.D. Acta. Metal. 1968, 16, 1377.

17. Hondros, E.D.; Gladman, D. Surf. Sci. 1968, 9, 471.

18. Hondros, E.D. Proc. Royal Soc. 1965, A286, 479.

19. Tammann, G. Z. Angew. Chemie. 1920, 39, 869.

20. Michelson, A.A. Phil. Mag. 1882, 13, 236.

21. Reed-Hill, R.E. Physical Metallurgy Principles, 2nd ed., Van Nostrand Co., New York, N.Y., 1973; p 873.

22. McLean, M. J. Mat. Sci. 1973, 8, 571.

23. Lee, S., M.S. Thesis, W.P.I., 1989.

24. Overbury, S.H.; Bertrand, P.A.; Somorjai, G.A. Chem. Rev. 1975, 75, 547.

25. Westmacott, K.H.; Smallman, R.E.; Dobson, P.S. Met. Sci. J. 1968, 2, 177.

26. Greenhill, E.B.; McDonald, S.R. Nature 1953, 171, 37.

27. Rice, C.M.; Eppelsheimer, D.S.; McNeil, M.C. J. Appl. Phys. 1966, 37, 4766.

RECEIVED May 9, 1990

Chapter 29

Raman Spectroscopy of Vanadium Oxide Supported on Alumina

G. Deo[1], F. D. Hardcastle[1], M. Richards[1], A. M. Hirt[2], and Israel E. Wachs[1]

[1]Departments of Chemistry and Chemical Engineering, Zettlemoyer Center for Surface Studies, Lehigh University, Bethlehem, PA 18015
[2]Materials Research Laboratories, Inc., 720 King Georges Post Road, Fords, NJ 08863

The molecular state of vanadium oxide supported on different alumina phases (γ, δ-θ, and α) was investigated with Raman spectroscopy. The supported vanadium oxide was found to form a molecularly dispersed overlayer on the different alumina phases. The molecular state of the surface vanadium oxide phase, however, was dependent on the nature of the alumina support. This variation was primarily due to the presence of surface impurities, in particular sodium oxide. The surface sodium oxide content was found to increase with the calcination temperature required to form the different transitional alumina phases (α, δ-θ, γ). The surface vanadium oxide phase consists of polymeric tetrahedra and distorted octahedra on γ-Al_2O_3, monomeric tetrahedra and distorted octahedra on δ,θ-Al_2O_3, and monomeric tetrahedra on α-Al_2O_3.

Recent studies of supported vanadium oxide catalysts have revealed that the vanadium oxide component is present as a two-dimensional metal oxide overlayer on oxide supports (1). These surface vanadium oxide species are more selective than bulk, crystalline V_2O_5 for the partial oxidation of hydrocarbons (2). The molecular structures of the surface vanadium oxide species, however, have not been resolved (1,3,4). A characterization technique that has provided important information and insight into the molecular structures of surface metal oxide species is Raman spectroscopy (2,5). The molecular structures of metal oxides can be determined from Raman spectroscopy through the use of group theory, polarization data, and comparison of the

0097–6156/90/0437–0317$06.00/0

Raman spectra with spectra of known molecular structures (6).

In the present investigation, the interaction of vanadium oxide with different alumina phases (γ, δ-θ, and α) is examined with Raman spectroscopy. Comparison of the Raman spectra of the supported vanadium oxide catalysts with those obtained from vanadium oxide reference compounds allows for the structural assignment of these supported species. The present Raman data demonstrate that the molecular structures of the surface vanadium oxide phases are significantly influenced by the presence of surface impurities on the alumina supports and this overshadows the influence, if any, of the alumina substrate phase.

EXPERIMENTAL

The γ phase (Harshaw, $180m^2/gm$), δ,θ phase (Harshaw, $120m^2/gm$), and α phase (ALCOA, $9.5m^2/gm$) of Al_2O_3 were used in this study. δ,θ-Al_2O_3 was prepared by heating the starting γ-Al_2O_3 at 950°C in O_2. X-ray diffraction was used to confirm the presence of the respective alumina phases. The vanadium oxide catalysts were prepared by incipient-wetness impregnation of vanadium tri-isopropoxide (Alfa) using methanol as the solvent. Due to the air and moisture sensitive nature of the alkoxide precursor, the catalysts were prepared and subsequently heated in N_2 at 350°C. The catalysts were finally calcined in O_2 at 500°C for 16 hrs. All samples are noted as weight percent V_2O_5/Al_2O_3. Surface areas were measured with a Quantachrome BET apparatus using single point nitrogen adsorption. The arrangement of the laser Raman spectrometer has been described elsewhere (5,7). Surface impurities on different alumina phases were measured on a Model DS800 XPS surface analysis system manufactured by KRATOS Analytical Plc, Manchester UK. Specimens were prepared by pressing the different alumina phases between a stainless steel holder and a polished single crystal silicon wafer. Measurements were done at 5.10^{-9} Torr with an hemisherical electron energy analyzer used for electron detection. Mg-Kα x-rays at a power of 360 W were employed in this study and the data were collected in 0.75 eV segments for a total of 1 hour. A pass energy of 80 eV was used for each of the specimens. The electron spectrometer was operated in the fixed analyzer transmission (FAT) mode. Elemental identification from each spectrum were done by comparing the measured peak energy to tabulated values and concentration estimates were made using typical normalization procedures (8).

RESULTS AND DISCUSSION

The major vibrational region of interest in the Raman spectra of vanadium(V) oxide structures lies in the

1200-100 cm^{-1} range. For vanadium oxide systems this region can be primarily divided into three parts. The V-O terminal stretching which occurs at 770-1050 cm^{-1}, the V-O-V stretching region at 500-800 cm^{-1}, and the bending mode at 150-400 cm^{-1}. Lattice vibrations of crystalline compounds may be also present below 150 cm^{-1}.

Vanadium(V) Oxide Reference Compounds

It is known that under ambient conditions supported vanadium oxide exists as a +5 cation (9). For this reason the Raman spectra of some pentavalent vanadium oxide reference compounds were studied. The vanadium(V) oxide reference compounds can be primarily categorized as tetrahedral or octahedral compounds. The structures of these reference compounds have been previously determined and only a brief discussion will be given here. Raman spectra of the vanadium(V) oxide reference compounds are shown in Figures 1-2.

   Spectra of tetrahedral vanadium(V) oxide compounds are shown in Figure 1 with different degrees of polymerization of the monomeric VO_4 unit. The Raman band associated with the terminal V-O bond increases with increasing extent of polymerization : ~ 830 cm^{-1} for monomeric VO_4^{3-}, ~ 880 cm^{-1} for dimeric $V_2O_7^{4-}$, and ~ 940 cm^{-1} for a chain composed of VO_4 units. Due to the presence of V-O-V linkages in the dimeric and polymeric species the bond length of the bridging bonds increase which gives rise to a higher order of the terminal V-O bonds and consequently a higher frequency. The monomeric unit in $Pb_5(VO_4)_3Cl$ gives rise to Raman bands at ~ 830 cm^{-1} (symmetric stretch), ~ 790 cm^{-1} (antisymmetric stretch), and bending modes at 320-350 cm^{-1}. Distortions imposed on the VO_4^{3-} unit will also increase the bond order and shift the Raman band to higher frequencies. For example, $AlVO_4$ possesses three different, highly distorted monomeric VO_4^{3-} units, and the corresponding Raman spectrum exhibits a triplet in the 980-1020 cm^{-1} region. The presence of V-O-V linkages is readily identified with Raman spectroscopy since they give rise to new modes at 200-300 cm^{-1} (bending), and 400-500 cm^{-1} (symmetric stretch), and 600-800 cm^{-1} (antisymmetric stretch). The Raman spectra of tetrahedral vanadium oxide species in aqueous solution also have similar features (6).

   Spectra of octahedral vanadium(V) oxide compounds are presented in Figure 2. Undistorted vanadium(V) oxide compounds do not exist, and all vanadia octahedra are highly distorted, which give rise to short V-O bonds. This is reflected in the Raman band position for the distorted octahedral vanadium oxide compounds which always occur in the 900-1000 cm^{-1} region. The vanadia structure in V_2O_5 approaches a square-pyramidal coordination and the individual vanadia are linked

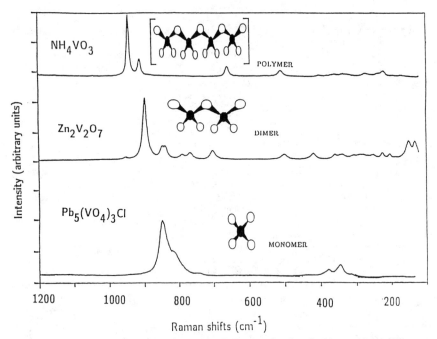

FIGURE 1. Raman Spectra of Tetrahedral Vanadium(V) Oxide Reference Compounds.

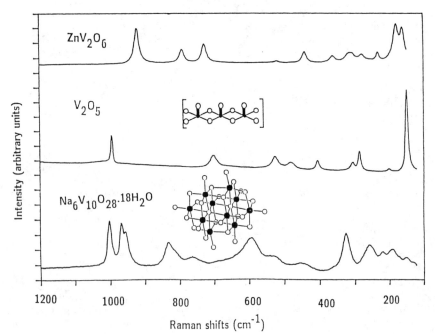

FIGURE 2. Raman Spectra of Octahedral Vanadium(V) Oxide Reference Compounds.

together to form infinite sheets (10). The short V-O bond in this structure is responsible for the band at 997 cm^{-1} (10). Many bands in the 200-800 region are due to the V-O-V linkages, and the strong band at ~ 144 cm^{-1} arises from the lattice vibrations (11). The decavanadate ion in $Na_6V_{10}O_{28}.18H_2O$ is made up of three distinctly different and distorted vanadia octahedra (12). Two of the vanadia octahedra approach a square-pyramidal coordination and the third has two short bonds cis to each other. The presence of three types of terminal V-O bonds in the decavanadate ion can be seen from the Raman spectra which give rise to strong bands at ~ 1000, 966 and 954 cm^{-1}. Due to the presence of numerous V-O-V linkages in the decavanadate structure, a number of strong Raman bands in the V-O-V bending (150-300 cm^{-1}) and V-O-V stretching (500-800 cm^{-1}) regions are present. The vanadia octahedral ion present in ZnV_2O_6 is not as distorted as those found in V_2O_5 and $Na_6V_{10}O_{28}.18H_2O$ and consequently exhibits a strong stretching band at 910-920 cm^{-1} (13). Numerous V-O-V associated bands are also present in the 150-800 cm^{-1} region.

The Alumina Supports

The Raman spectra of γ, δ,θ, and $\alpha-Al_2O_3$ are shown in Figure 3. For $\gamma-Al_2O_3$ there are no Raman bands in the 150-1200 cm^{-1} region. For $\delta,\theta-Al_2O_3$ several Raman bands are observed: two bands of medium intensity at ~ 837 and ~ 753 cm^{-1}, and a strong band at ~ 251 cm^{-1}. The α phase of Al_2O_3 possesses Raman bands at 742, 631, 577, 416, and 378 cm^{-1}. The surface compositions of the alumina supports were determined by X-ray Photoelectron Spectroscopy and are shown in Table I. In addition to the impurities present on $\gamma-Al_2O_3$, δ,θ and $\alpha-Al_2O_3$ show the presence of sodium and flourine ions. These impurities may result from the manufacturing process (e.g. Bayer process) or incomplete purification of the ore (14).

V_2O_5/ Al_2O_3

The supported vanadium oxide on $\gamma-Al_2O_3$ is present as a well dispersed phase for 3-20% V_2O_5/Al_2O_3 (Figure 4). The Raman spectra of $V_2O_5/\gamma-Al_2O_3$ have been discussed before (11), and the vanadium oxide structure has been classified into three main regions. Above 20% V_2O_5/Al_2O_3, the Raman spectra show the formation of V_2O_5 crystallites. At 5% V_2O_5/Al_2O_3 and below, the Raman band is present at ~ 940 cm^{-1} which is typical of the terminal stretching mode found in alkali metavanadates which possess polymeric chains of VO_4 units (Figure 1). The corresponding V-O-V vibrations of metavanadates are also present in these samples. Above 5% V_2O_5/Al_2O_3, new bands

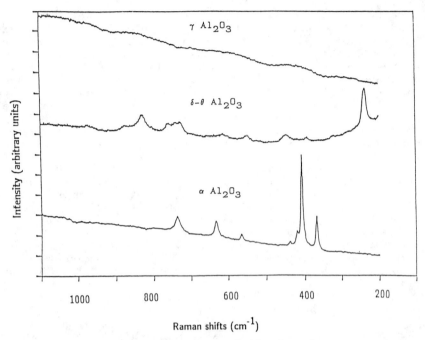

FIGURE 3. Raman Spectra of Alumina Supports.

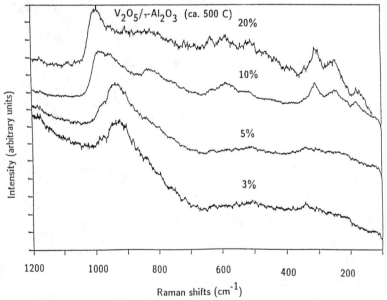

FIGURE 4. Raman Spectra of Vanadium Oxide Supported on
γ-Al$_2$O$_3$.

Table I. Surface concentrations on different alumina
supports (in atomic %)

Surface Atom	Alumina Support Phase		
	α	δ, θ	γ
C	5.5	4.7	7.5
O	57.1	59.6	58.2
F	0.27	0.28	----
Na	0.98	0.18	----
Al	35.5	34.9	33.5
Cl	0.6	0.4	0.86

appear near the 1000 cm^{-1} region which are due to a vanadium oxygen double bond and are associated with the highly distorted octahedral enviroment of the vanadium oxide species (Figure 2). Solid state ^{51}V NMR studies of these samples confirm that tetrahedral surface vanadia species are exclusively formed in the 0-5% $V_2O_5/\gamma-Al_2O_3$ range and that distorted octahedral surface species are predominately present above 5% $V_2O_5/\gamma-Al_2O_3$ (4). The 10% $V_2O_5/\gamma-Al_2O_3$ also has broad bands at 950-1000, at 810-830, 550-580, ~ 500, 290-300, ~ 250, and ~ 180 cm^{-1}. There is a striking similarity between the position of these Raman bands and the Raman bands found in $Na_6V_{10}O_{28} \cdot 18H_2O$ (Figure 2). Similar Raman bands are also present for 20% $V_2O_5/\gamma-Al_2O_3$. Thus, the supported vanadium oxide phase seems to be present with units similar to the decavanadate ion in 10-20% $V_2O_5/\gamma-Al_2O_3$.

The Raman spectra for 1-10% $V_2O_5/\delta,\theta-Al_2O_3$ are shown in Figure 5. Crystalline bands appear at 13% $V_2O_5/\delta,\theta-Al_2O_3$ indicating that monolayer coverage of surface vanadium oxide has been exceeded. 1-5% $V_2O_5/\delta,\theta-Al_2O_3$ samples possess Raman bands of the surface vanadium oxide overlayer as well as weak bands of the $\delta,\theta-Al_2O_3$ support. The Raman bands of the support, however, quickly diminish as the vanadium oxide content is increased. This is due to the absorption of the laser light by the yellow-colored vanadium oxide overlayer. The Raman bands in the 990-1000 cm^{-1} region are characteristic of the distorted octahedra (decavanadate unit). The Raman bands in the 770-790, 530-540, and 150-300 cm^{-1} region are, however, much stronger than the corresponding Raman bands associated with the decavanadate ion and suggest the presence of a second surface vanadium oxide species on the $\delta,\theta\text{-}Al_2O_3$ support. A similar set of Raman bands have recently been observed in a bismuth vanadate (Bi:V=25:1) reference compound which exhibits Raman bands at ~ 790, 530, 350-250, and 150 cm^{-1}. Solid state ^{51}V NMR experiments have shown this structure to contain tetrahedral vanadium oxide species (15). The decrease in the Raman stretching frequency, from ~ 830 cm^{-1} for monomeric VO_4^{3-} to ~ 790 cm^{-1} for the bismuth vanadate (Bi:V=25:1), at the first approximation suggests, that the terminal V-O bonds have been slightly lengthened in this bismuth vanadate structure. Similar to the case of bismuth vanadium oxide compounds, the sodium vanadium oxide compounds (NaVO$_3$→Na:V=1:1; Na$_3$VO$_4$→Na:V=3:1) also show a decrease in Raman terminal stretching frequencies from ~920 to ~820 cm^{-1} (16). Due to the presence of sodium ions on the surface, see Table I, it can be concluded that part of the surface vanadium oxide species are coordinated to surface sodium ions to form monomeric tetrahedral vanadia species. It is also possible to conclude that more than three sodium ions are coordinated per monomeric tetrahedral vanadium oxide species as the Raman band occurs below 820 cm^{-1}. Solid state ^{51}V NMR studies also confirm that two distinctly different surface vanadia species are present on the $\delta,\theta\text{-}Al_2O_3$ support: a perfect tetrahedral structure and a distorted octahedral structure (17). Thus, the supported vanadium oxide phase on $\delta,\theta\text{-}Al_2O_3$ consists of distorted octahedra (decavanadate-like) and monomeric tetrahedra.

The Raman spectra of supported vanadium oxide on $\alpha\text{-}Al_2O_3$ exhibit different structural features than on the previously described alumina phases, as can be seen by comparing Figure 6 (0.7-1.7% V$_2$O$_5$/Al$_2$O$_3$) with Figures 4 and 5. Broad Raman features are present at ~770, ~690, ~550, ~470 and a strong Raman band at ~290 cm^{-1}. As described previously, the ~770 cm^{-1} band position is specific to the monomeric tetrahedral species with lengthened V-O bonds. It should be noted that the sodium ion concentration on the surface of $\alpha\text{-}Al_2O_3$ is much higher than on other alumina supports (see Table I) and, hence, is most probably responsible for the formation of the monomeric tetrahedral species. Thus, the surface vanadium oxide phase on $\alpha\text{-}Al_2O_3$ consists primarily of monomeric tetrahedral vanadia species. Similar spectra have been observed in our laboratory using different types of $\alpha\text{-}Al_2O_3$ supports.

As mentioned earlier, sodium is usually present as an impurity in alumina. However, the absence of sodium on the surface of $\gamma\text{-}Al_2O_3$ implies its presence inside the bulk. To produce α and $\delta,\theta\text{-}Al_2O_3$ it is required to heat $\gamma\text{-}Al_2O_3$ to higher temperatures. As a result of heating γ-Al$_2$O$_3$, sodium migrates to the surface. This can be observed in Table I where the sodium concentration increases from γ to $\alpha\text{-}Al_2O_3$. The migration of sodium changes the acid/base charateristics of the surface. An

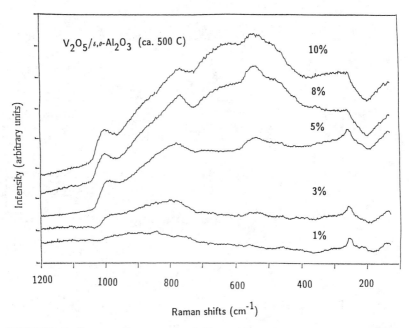

FIGURE 5. Raman Spectra of Vanadium Oxide Supported on $\delta, \theta - Al_2O_3$.

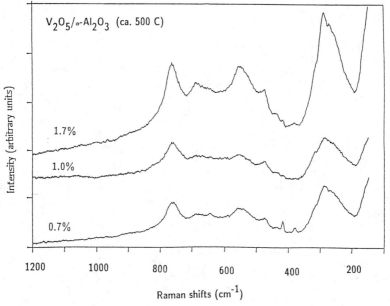

FIGURE 6. Raman Spectra of Vanadium Oxide Supported on $\alpha - Al_2O_3$.

increase in the sodium ion concentration produces a more basic surface. Thus, the basic strength of the surface should be expected to vary as follows:

$$\alpha\text{-Al}_2\text{O}_3 > \delta,\theta\text{-Al}_2\text{O}_3 > \gamma\text{-Al}_2\text{O}_3$$

← more basic

Comparison of this information with the surface vanadium oxide species present on the different alumina phases:

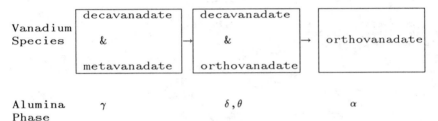

| Vanadium Species | decavanadate & metavanadate | → | decavanadate & orthovanadate | → | orthovanadate |

| Alumina Phase | γ | δ,θ | α |

implies that acidic surfaces favor the decavanadate species, whereas, basic surfaces favor the orthovanadate species. Under ambient conditions, the conditions under which the V_2O_5/Al_2O_3 spectra were taken, the surface of the support is hydrated and there is a similarity in the behavior of the vanadium oxide surface species on Al_2O_3 and the pH dependence of aqueous vanadium oxide structural chemistry (10).

The effect of the impurities besides sodium do not seem to play an important role in determining the structure of the surface vanadium oxide species. Flourine is the only other impurity whose concentration varies appreciably in the different alumina phases. Flourine cannot be responsible for influencing the vanadium oxide structures since flourine is present in similar concentrations in α and δ,θ-Al_2O_3 while the vanadium oxide structures are changing. It may therefore be concluded that surface impurities other than sodium oxide do not contribute significantly to the change in the structure of the surface vanadium oxide species on the different alumina phases.

It should be noted that all of the above studies were performed under ambient conditions where the surface vanadium oxide species are known to be hydrated due to adsorbed moisture. It has been shown that dehydration can alter the structures of the surface vanadium oxide species. For example, dehydration of surface vanadium oxide species on γ-Al_2O_3, which initially possesses vanadia octahedra and tetrahedra, tranforms most of the surface vanadia species to a tetrahedral coordination (4). Dehydration studies on the $V_2O_5/\delta,\theta$-Al_2O_3 and V_2O_5/α-Al_2O_3 samples are currently underway.

CONCLUSIONS

The Raman spectra of model vanadium(V) oxide compounds are very sensitive to vanadium oxygen coordination. A distinct trend in the Raman spectra of the reference tetrahedral and octahedral vanadium(V) oxide compounds is observed. Vanadium oxide is found to interact differently with the different alumina supports. On the $\gamma-Al_2O_3$ support, the surface vanadium oxide overlayer consists of polymeric tetrahedra and distorted octahedra. On the $\delta,\theta-Al_2O_3$ support, the surface vanadium oxide overlayer is composed of monomeric tetrahedra and distorted octahedra. On the $\alpha-Al_2O_3$ support, the surface vanadium oxide overlayer contains monomeric tetrahedra surface species. The interactions, however, appear to be primarily due to the level of sodium impurity present on the surface and not the alumina phase in particular. It is therefore imperative to take into consideration the surface impurities present when describing the molecular structures of the supported vanadium oxide phases.

ACKOWLEDGMENTS

We would like to thank Jih-Mirn Jehng for obtaining the Raman spectra of some of the samples. This study has been supported by the National Science Foundation grant # CBT-8810714.

REFERENCES

1. (a) F. Roozeboom, T. Fransen, P. Mars, and P. J. Gellings, Z.anorg. allg. Chem. 449, 25-40 (1979). (b)F. Roozeboom, M. C. Mittelmeijer-Hazeleger, J. A. Moulijn, J. Medema, V. H. J. de Beer, and P. J. Gellings, J. Phys. Chem. 84, 2783 (1980). (c) L. R. Le Coustumer, B. Taouk, M. Le Meur, E. Payen, M. Guelton, and J. Grimblot, J. Phys. Chem. 92, 1230 (1988).
2. (a) R. Y. Saleh, I. E. Wachs, S. S. Chan, and C. C. Chersich, J. Catal. 98, 102 (1986).(b) I. E. Wachs, R. Y. Saleh, S. S. Chan, and C. C. Chersich, Appl. Catal. 15, 339 (1985).
3. (a) J. Haber, A. Kozlowska, and R. Kozlowski, J. Catal. 102,52 (1986).(b) H. Miyata, K. Fujii, T. Ono, and Y. Kubokawa, J. Chem. Soc., Faraday Trans. 1, 83, 675-685 (1987).(c) G. Bergeret, P. Gallezot, K. V. R. Chary, B. Rama Rao, and V. S. Subrahmanyam, Appl. Catal., 40, 191 (1988).(d) J. Haber, A. Kozlowska, and R. Kozlowski, Proc. 9th Intl. Congr. Catal., 1481-1488 (1988).
4. (a) H. Eckert, and I. E. Wachs, Mat. Res. Soc. Symp. Proc. Vol. 111, 459 (1988).(b) H. Eckert, and I. E. Wachs, J. Phy. Chem., 93, 6796 (1989)

5. I. E. Wachs, F. D. Hardcastle, S. S. Chan,
    Spectrosc., 1(8), 30 (1986).
6. W. P. Griffith, and P. J. B. Lesniak, J. Chem. Soc.
    A, 1066 (1969).
7. I. E. Wachs, and F. D. Hardcastle, Mat. Res. Soc.
    Symp. Proc., Vol. 111, 353 (1988).
8. H. Eckert, G. Deo, I. E. Wachs, and A. M. Hirt, J. of
    Colloids and Surfaces (in press).
9. I. E. Wachs, R. Y. Saleh, S. S. Chan, and C. C.
    Chersich, CHEMTECH, Dec., 756 (1985).
10. R. J. H. Clark, The Chemistry of Titanium and
    Vanadium, Elsevier Publishing Co., 1968.
11. J. M. Jehng, F. D. Hardcastle, and I. E. Wachs, Solid
    State Ionics, 32/33, 904 (1989).
12. H. T. Evans, Inorg. Chem., 5, 967-977 (1966).
13. H. N. Ng, and C. Calvo, Canad. J. Chem., 50, 3619-
    3624 (1972).
14. C. Misra, Industrial Alumina Chemicals, ACS monograph
    184, 1986.
15. F. D. Hardcastle, I. E. Wachs, H. Eckert, and D.
    Jefferson, to be published.
16. F. D. Hardcastle and I. E. Wachs, to be published.
17. H. Eckert, G. Deo, and I. E. Wachs, to be published.

RECEIVED May 9, 1990

Chapter 30

Scanning Electron and Field Emission Microscopy of Supported Metal Clusters

T. Castro[1], Y. Z. Li[1], R. Reifenberger[1], E. Choi[2], S. B. Park[2], and R. P. Andres[2,3]

[1]Department of Physics and [2]School of Chemical Engineering, Purdue University, West Lafayette, IN 47907

STM measurements of the shape of gold clusters supported on flat gold substrates and FEM measurements of the melting temperature of gold clusters supported on tungsten tips are presented. The size dependence of these cluster properties is obtained for cluster diameters ranging from 12 nm to 2 nm.

Understanding the unique properties of supported metal clusters, often consisting of only a few metallic atoms, is the key to understanding and improving a wide class of catalytic processes. Over the past decade, researchers have developed myriad techniques for synthesizing small atomic clusters (1). There also exist a variety of experimental techniques such as scanning electron microscopy (SEM and STEM), field emission and field ion microscopy (FEM and FIM), scanning tunneling microscopy (STM) and atomic force microscopy (AFM), that permit one to study individual supported clusters. In this paper we discuss preliminary studies aimed at combining these research advances. The ultimate goal of this effort is to probe the properties of supported metal clusters of catalytic importance and to measure the dependence of these properties on both the support material and adsorbed molecules.

[3]Author to whom all correspondence should be addressed.

0097–6156/90/0437–0329$06.00/0

Production of Nanometer-sized Clusters

A synthesis approach, that has proved fruitful in molecular-beam studies of clusters, is to vaporize the material of interest into a stream of cold inert gas which acts as a heat bath and promotes aggregation of the condensable species to form clusters. This gas aggregation technique has the desirable characteristics of not being material specific and of being easily adapted for production of large quantities of clusters. It has the undesirable characteristics of typically producing a broad distribution of cluster sizes and compositions and of yielding an aerosol in which the clusters are mixed with uncondensed atoms and inert gas. We have developed an adaptation of the gas aggregation method at Purdue, which is capable of producing nanometer-sized metal clusters having a controlled mean size and a relatively narrow size distribution (2-5). We have also developed a reliable technique for stripping uncondensed atoms from streams of uncharged clusters (3-5).

The multiple expansion cluster source (MECS) used to prepare cluster samples for the present study is described in detail in the thesis of S.B. Park (5). Metal is evaporated from a crucible placed inside a resistively heated graphite oven tube (UT6ST, Ultra Carbon). The evaporation rate is controlled by inert gas pressure in the oven, oven temperature, location of crucible in oven and hole size in the crucible. The metal vapor, primarily composed of individual atoms and diatoms, is entrained in hot inert gas (He or Ar) flowing in the oven tube and expands through a sonic orifice in the oven wall. This subsaturated gas mixture is immediately mixed with room temperature inert gas in a quench region. The supersaturated stream then flows into a tubular, fast flow, condensation reactor. To provide a well defined residence time in the condensation reactor, the flow in this region is laminar and annular guard flows are introduced around the free jet coming from the quench region.

The clusters grow by condensation of metal atoms onto the diatoms originally present in the quench region. Formation of additional diatoms, i.e. nucleation of new clusters, is negligible at the low pressures maintained in the condensation region. The pressure in the condensation reactor is kept at less than ~50 Torr by means of a mechanical pump. When there is sufficient cooling and dilution in the quench step, cluster/cluster agglomeration in the condensation reactor can be largely eliminated and cluster growth is a pure birth kinetic process.   In these situations the cluster size distribution, plotted as a function of cluster diameter, is normal and exhibits a full width at half maximum of approximately 0.5. nm and a mean size that can be controlled by controlling the residence time, the initial monomer concentration and the initial monomer/dimer ratio in the condensation reactor (5). If cluster/cluster agglomeration is prohibited, the clusters grow, with increasing residence time, to a maximum size given by the monomer/dimer ratio in the quench region.   For Au this ranges from 0.5 nm to 3.0 nm in the present MECS.   Clusters larger than this can be grown by allowing these small clusters to agglomerate.   In this manner, multiply twinned, essentially spherical, Au clusters as large as 20 nm in diameter have been grown.

In order to produce supported samples for STM or FEM study, clusters formed on the centerline of the condensation reactor are extracted through a 1 mm diameter capillary into a vacuum chamber typically kept at 10^{-5} Torr.   The resulting supersonic free jet flow is collimated to form a molecular beam of metal clusters, uncondensed metal atoms and inert gas atoms.

Production of Supported Cluster Samples

Free jet expansion from the capillary produces a molecular beam in which the clusters are all traveling approximately at the sonic velocity of the gas in the capillary.   Thus, cluster velocity ranges from 3×10^4 cm/s if Ar is used in the MECS to 1×10^5 cm/s if He is used.

A gas cell is used to decelerate the clusters before they are deposited on a substrate and to strip uncondensed metal atoms from the cluster beam.

Supported samples are prepared by first evacuating the gas cell to a pressure of ~10^{-3} Torr and measuring the total metal flux passing through the cell by means of an Inficon film thickness monitor (FTM).   The cell pressure is then increased by bleeding He into the cell.   Initially, the FTM signal drops rapidly due to deceleration and rejection of the uncondensed metal atoms in the beam.   Next, there is a pressure region in which the signal remains constant.   In this region, while all uncondensed atoms have been stripped from the beam, essentially all of the clusters still reach the detector. Finally, the FTM signal again drops as the clusters themselves are stopped before they reach the surface of the FTM.

For the STM experiments described below, the pressure at which just half of the clusters reach the FTM was found and recorded. A previously unexposed substrate was then rotated into the position previously occupied by the FTM and exposed for a measured time to this cluster flux. This procedure eliminates codeposition of clusters and uncondensed atoms and slows the clusters to thermal speeds before depositing them on a substrate. After several samples are prepared, the MECS and the vacuum system are shut down and bled up to atmospheric pressure and the samples are removed for analysis. No attempt was made to prevent the supported clusters from adsorbing gas phase contaminants.

For the FEM experiments described below, the cluster beam was directed through a small collimation capillary into a separately pumped deposition chamber, which is kept at 1 x 10^{-8} Torr. A transfer cell equipped with a 2 1/s ion pump enabled a tungsten FEM tip to be: 1. inserted into the deposition chamber and positioned with its apex in the cluster beam, 2. withdrawn and transported at 2 x 10^{-7} Torr to a UHV field emission microscope, and 3. inserted into the field emission apparatus and positioned properly for field emission measurements (6).

Directly across the deposition chamber from the FEM tip was a fluorescent screen and viewport with which the field emission pattern from the tip could be observed during cluster deposition. The tip was first cleaned by joule heating. The two fold symmetry characteristic of a clean W(110) field emitter was observed and the voltage necessary to observe this pattern was recorded. The voltage was then reduced to about 1/2 of the original value and the tip was exposed to clusters that had been slowed to thermal speeds in the gas cell. Clusters landing near the apex of the tip appear as bright dots on the fluorescent screen. This procedure was repeated until an individual cluster positioned near the apex of the tip was obtained.

TEM Studies of Supported Gold Clusters

Vacuum evaporation of gold atoms onto graphite or amorphous carbon substrates results in the formation of individual gold clusters, which nucleate at specific sites on the surface. The shape of these clusters is nearly spherical. Buffat and Borel in a classic TEM experiment used this behavior to prepare samples of gold clusters having different average diameters and studied the melting transition of the clusters as a function of cluster diameter by observing the sharp electron diffraction rings blur when the supported clusters are heated (7).

Gold clusters grown in the MECS and deposited on thin carbon or silicon films are also essentially spherical. In a series of experiments Park (5) found that, if he deposited MECS clusters on a thin carbon film and then imaged the sample with TEM, he could accurately estimate the experimentally determined weight of the deposit by assuming: 1. that the clusters were spheres and 2. that the clusters had the same density as bulk gold.

High resolution TEM studies of supported clusters formed both by atomic deposition (8-9) and in a gas aggregation type source (10-11) indicate that gold clusters with diameters less than 5 nm exhibit a variety of multiply twinned structures and can transform rapidly between different configurations under the intense electron radiation within the microscope. This phenomenon has been termed quasimelting and is the subject of continuing investigation. Despite this complication, on carbon and on silica substrates, the shape of these small gold clusters remains essentially spherical.

STM Study of Supported Gold Clusters

The parallel developments of scanning tunneling microscopy (STM) (12) and atomic force microscopy (AFM) (13) greatly broaden our perspectives and abilities to probe the shape and electronic structure of individual supported clusters, as well as enhance our ability to manipulate these entities on the atomic scale.

An example of the power of STM for imaging the morphology of a supported cluster is shown in Figure 1. This is an STM image of a pair of Au clusters of ~12 nm diameter supported on a Pt substrate. The total region scanned is 88 x 88 nm. The steps on the Pt substrate are ~4 nm high.

The Pt and Au substrates used for the present research were prepared by inserting a clean wire, 1 mm in diameter, into a gas torch flame. A molten sphere with a diameter of 2 mm forms at the end of the wire and is allowed to air cool.

This technique gives atomically flat facets on the surface of the sphere. STM scans of the substrates before cluster deposition showed them to be free from any debris that resembled clusters. After deposition the Au clusters were found uniformly distributed on the flat facets of Au substrates but only at steps on the Pt surface.

The image in Figure 1 was obtained with a pocket-sized STM (14) operating in air and in constant current mode, using an electrochemically etched tungsten tip. Typical tunneling bias voltages used to image Au clusters ranged between 20 and 600 mV, while typical tunneling currents were 2 nA with the electrons tunneling from the tip into the sample. The STM images of Au clusters on Pt and on Au substrates are very stable and there are no apparent differences between images taken with different bias voltages. While the STM used in these studies is able to image a graphite surface with atomic resolution, atomic resolution was not achieved either on the metal substrates or on the clusters (15).

An interesting observation encountered in our STM investigation of Au clusters on Au surfaces is the flattened profile of the clusters, which makes it difficult to resolve clusters with diameters less than 2 nm. We find that, even though gold clusters are spherical in free space and on carbon and silica substrates, they flatten when deposited on Au substrates. As knowledge of the shape of supported clusters is important both for interpreting catalytic behavior and for understanding thin film growth, we decided to study this phenomenon.

A series of samples were prepared in which MECS clusters of different controlled diameters in the nanometer size range with "soft landed" at room temperature on atomically flat substrates (15). Figure 2 is a STM image of a sample of some of the larger clusters studied. The shape of an individual cluster is characterized by taking a line scan across the center of the cluster as indicated by the arrow in Figure 2a and illustrated in Figure 2b. This line scan is used to measure the diameter (D) of the boundary between the cluster image and the substrate and the maximum height of the cluster above the substrate (H) as determined by STM. The volume of the cluster (Ω) is calculated by assuming it to be a perfect spherical cap ($\Omega = (\pi H/6)(3D^2/4 + H^2)$)

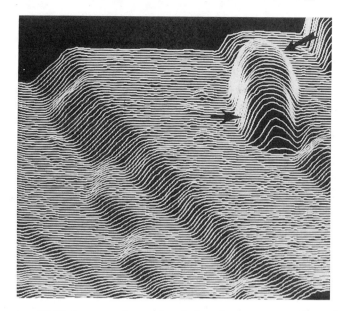

Figure 1. STM image of a pair of Au clusters on a Pt substrate.

(a)

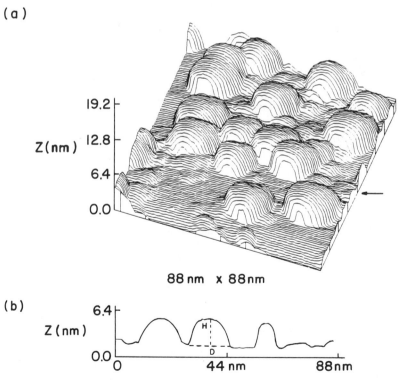

(b)

Figure 2. STM scan of Au clusters supported on a Au substrate.

The volume of the STM image of a sphere is larger than its free space volume ($V = (\pi D^3)/6$) and the D/H ratio obtained from the image is of order one. However, the volumes calculated from the STM images of gold clusters having free space diameters ranging from 2 nm to 12 nm agree to within experimental error with the free space volumes determined from TEM micrographs of the same MECS clusters deposited on thin carbon films (15). The D/H ratios of these clusters increase from 5 to 40 as the clusters become smaller. Thus, small gold clusters supported on gold assume a spherical cap shape and the smaller the cluster the larger its D/H.

We have been able to interpret this experimental data by means of a very simple continuum model. Assuming the cluster to be plastic with a uniform internal stress, the relationship between this internal stress and the radius of curvature of the cluster's free surface is given by the Young equation

$$\sigma = \frac{2\gamma}{R} \tag{1}$$

where σ is the internal stress, γ is the surface free energy, and R is the radius of curvature. Assuming the cluster to be plastic only above a critical yield stress, σ^*, and to be perfectly elastic below this threshold, Equation 1 with $\sigma = \sigma^*$ defines a critical radius of curvature, R^*. Clusters having free surface radii less than this critical value are stressed beyond their yield limit. Now, if the interfacial free energy between the cluster and the substrate is less than γ, these clusters can relax by spreading on the substrate. Once $\sigma = \sigma^*$, however, the cluster can no longer relax and its shape is frozen. Thus, small Au clusters on Au assume a spherical cap shape with free surface radius $R = R^*$.

The ability of this one parameter model to represent our experimental data is shown in Figure 3. This is a plot of the contact radii ($\rho = D/2$) of the gold clusters as measured by STM as a function of their free space volumes. The solid line in this figure is the prediction of the theoretical model, which is

$$\rho = (\frac{4R^*}{\pi} \Omega)^{0.25} \tag{2}$$

The critical radius used to fit the data is $R^* = 20$ nm.

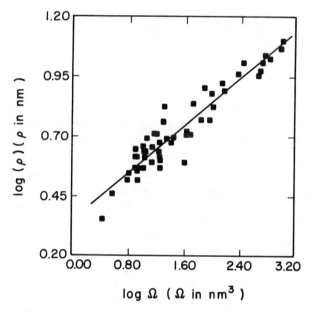

Figure 3. Plot of the contact radii of Au clusters supported on a Au substrate as a function of the volumes of the clusters.

FEM Study of Supported Gold Clusters

The onset of particle mobility often coincides with particle melting and results in rapid sintering of supported metal catalysts. Thus, prediction of the melting temperature of supported metal clusters is of more than academic interest.

It is well known that small particles melt at a lower temperature than the corresponding bulk solid. Buffat and Borel measured a monotonic lowering of melting temperature with decreasing diameter for gold clusters supported on carbon substrates (7). They accounted for their data by means of a thermodynamic model due to Pawlow (16), which states that the melting temperature of small particles is inversely proportional to the particle radius.

Baker pioneered direct measurements of the onset of particle mobility on substrates using controlled atmosphere electron microscopy (17). He has pointed out the close relationship between the onset of particle motion as determined in his studies and the Tammann temperature (18). It is important to establish whether melting temperature decreases monotonically with particle size as indicated by the data of Buffat and Borel (7) or is equal to the Tammann temperature as hypothesized by Baker (18).

We have developed a method by which field emission from a preformed metallic cluster, supported at the apex of a sharp field emission tip, can be used to study the melting temperature of supported clusters as a function of cluster size (6, 19). In this experiment, we measure the field emission current that is emitted by a single isolated cluster. The dependence of this current on the applied voltage is found to follow the well known Fowler-Nordheim law for field emission. Plotting the emission current as a function of applied voltage, the slope of the Fowler-Nordheim curve can be determined. This slope can be related to the shape of the cluster. When the field emission characteristics of a cluster are studied after each of a series of heating cycles in which the temperature of the W tip is raised from room temperature to successively higher temperatures and then cooled again to room temperature, it is found experimentally that the Fowler-Nordheim slope changes abruptly over a narrow temperature range, providing evidence that the cluster has melted and wet the surface.

Melting temperatures for gold clusters supported on tungsten determined in this way are plotted in Figure 4 as a function of cluster size. The melting temperatures are normalized to the bulk melting temperature of gold. These reduced melting temperatures are plotted vs inverse cluster radius. The circles with error estimates are the present measurements, while the x's are the data of Buffat and Borel. The theoretical curves in the Figure 4 represent variations of the continuum thermodynamic model of Pawlow.

The present measurements agree quite closely with the earlier measurements of Buffat and Borel up to a cluster radius of ~1 nm at which point the melting temperature becomes constant at ~0.4 of the melting temperature of bulk gold. Thus, these data serve to reconcile the apparent differences between the data of Buffat and Borel (7) and the observations of Baker (18).

Conclusions

The preliminary studies described above indicate the power of combining molecular-beam techniques for synthesizing metal clusters of known size and composition and techniques for studying individual supported clusters. It is to be expected that this fusion of experimental methods will lead to increased understanding of the complex world of supported metal catalysts.

STM study of the shape of small gold clusters supported on a flat gold substrate indicates the singular importance of surface free energy in considerations of cluster morphology. More experiments with a controlled gas environment, different cluster materials, different substrate materials, and AFM are needed. It is expected that these studies will have an important impact on both our understanding of supported metal catalysts and thin film technology.

FEM measurements of melting temperature establish for the first time the detailed dependence of this important quantity on particle size over the size range of interest to catalysis. Further studies of the electron emission from a single supported cluster have broad potential for determining the electronic structure of catalytic particles.

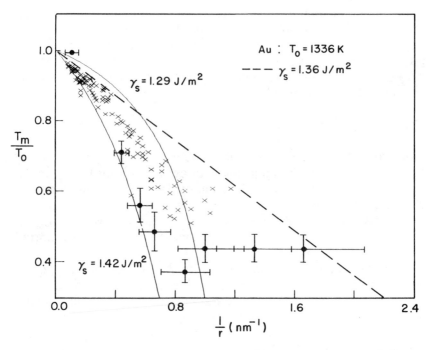

Figure 4. Plot of normalized melting temperatures of Au clusters as a function of the inverse of their radii.

Acknowledgments

The research described in this paper was made possible by the support of the National Science Foundation, the Department of Energy, and the Purdue Research Foundation.

Literature Cited

1. Andres, R.P.; Averback, R.S.; Brown, W.L.; Brus, L.E.; Goddard, W.A.,III; Kaldor, A.; Louie, S.G.; Moscovits, M.; Peercy, P.S.; Riley, S.J.; Siegel, R.W.; Spaepen, F.; and Wang, Y. J. Mater. Res. 1989, 4, 704-36.
2. Bowles, R.S.; Kolstad, J.J.; Calo, J.M.; Andres, R.P. Surf. Sci. 1981, 106 117-24.
3. Bowles, R.S.; Park, S.B.; Otsuka, N.; Andres, R.P. J. Molec. Catal. 1983, 20, 279-87.
4. Choi, E.; Andres, R.P. In Physics and Chemistry of Small Clusters, Jena, P.; Rao, B.K.; Khanna, S.N., Eds.; Plenum: New York, 1987; NATO ASI Series B, Vol. 158, 61-5.
5. Park, S.B.; Ph.D. Thesis, School of Chemical Engineering, Purdue University, Dec. 1988.
6. Castro, T.; Ph.D. Thesis, Department of Physics, Purdue University, Dec. 1989.
7. Buffat, Ph; Borel, J-P. Phys. Rev. A 1976, 13, 2287-98.
8. Bovin, J-O.; Wallenberg, R.; Smith, D.J. Nature 1985, 317, 47-9.
9. Lewis, J.; Smith, D.J. In Proceedings of the 47th Annual Meeting of the Electron Microscopy Society of America, Bailey, G.W., Ed.; San Francisco Press; San Francisco, 1989; 640-41.
10. Iijima, S.; Ichihashi, T. Phys. Rev. Letts.1986, 56, 616-19.
11. Iijima, S.; Ichihashi, T. Japanese J. of Appl. Phys. 1985, 24, L125-28.
12. Binnig, G.; Rohrer, H.; Gerber, Ch.; Weibel, E. Phys. Rev. Lett. 1983, 50, 120-23.
13. Binnig, G.; Quate, C.F.; Gerber, Ch. Phys. Rev. Lett. 1986, 56, 930-33.
14. Gerber, Ch.; Binnig, G.; Fuchs, H.; Marti, O.; Roher, H. Rev. Sci. Instrum. 1986, 57, 221-24.
15. Li, Y.Z.; Ph.D. Thesis, Department of Physics, Purdue University, Dec. 1989.
16. Pawlow, P. Z. Phys. Chem. 1909, 65, 1.
17. Baker, R.T.K.; Harris, P.S.; Thomas, R.R. Surf. Sci. 1974, 46, 311-16.
18. Baker, R.T.K. J. Catal. 1982, 78, 473.
19. Castro, T.; Reifenberger, R.; Choi, E.; Andres, R.P. Surf. Sci. (in press).

RECEIVED June 1, 1990

INDEXES

Author Index

345

Affiliation Index

Subject Index